AF794334

Proceedings of the International School on
Lasers and Quantum Optics

LASERS AND QUANTUM OPTICS

Proceedings of the International School on Lasers and Quantum Optics

LASERS AND QUANTUM OPTICS

CIF Series — Vol. 13

Mar del Plata, Argentina
22 – 31 Aug 1988

Editors
L M Narducci
E J Quel
J R Tredicce

World Scientific
Singapore • New Jersey • London • Hong Kong

Published by

World Scientific Publishing Co. Pte. Ltd.

5 Toh Tuck Link, Singapore 596224

USA office: 27 Warren Street, Suite 401-402, Hackensack, NJ 07601

UK office: 57 Shelton Street, Covent Garden, London WC2H 9HE

Library of Congress Cataloging-in-Publication Data
International School on Lasers and Quantum Optics (1988: Mar del Plata, Argentina)
Lasers and quantum optics: proceedings of the International School on
Lasers and Quantum Optics, Mar del Plata, Argentina, 22–31 Aug. 1988 /
editors, L. M. Narducci, E. J. Quel, J. R. Tredicce.
p. cm. -- (CIF series; vol. 13)
ISBN-13 978-9971-5-0952-1 -- ISBN-10 9971-5-0952-0
1. Lasers -- Congresses. 2. Quantum optics -- Congresses.
I. Narducci, L. M. II. Quel, E. J. III. Tredicce, J. R. IV. Title. V. Series.
QC685.159 1988
621.36'6'01539 -- dc20 90-31644
CIP

British Library Cataloguing-in-Publication Data
A catalogue record for this book is available from the British Library.

Copyright © 1990 by World Scientific Publishing Co. Pte. Ltd.

All rights reserved. This book, or parts thereof, may not be reproduced in any form or by any means, electronic or mechanical, including photocopying, recording or any information storage and retrieval system now known or to be invented, without written permission from the publisher.

For photocopying of material in this volume, please pay a copying fee through the Copyright Clearance Center, Inc., 222 Rosewood Drive, Danvers, MA 01923, USA. In this case permission to photocopy is not required from the publisher.

Preface

This volume is the result of the efforts of many committed scientists and educators who have donated freely of their time and talents for the success of the International School on Lasers and Quantum Optics which was held in Mar del Plata, Argentina in the period August 23-31, 1988.

The main purpose of the school was to provide students and young researchers from the Latin American countries with a modern overview several active areas of research at the fronteers of Quantum Optics. As the program of the school shows, theoretical and experimental topics were represented in almost equal proportions. The theoretical overviews focused both on the semiclassical theory of the laser and on the more fundamental quantum theory in order to develop the appropriate background for the discussion of the recent experiments of laser dynamics and for an appreciation of the statistical properties of laser radiation.

Many laser systems were analyzed in the lectures: these included solid state lasers, gas lasers, semiconductor and dye lasers. In addition, the very special physics of single atoms in a cavity was discussed together with some remarkable new results from the two-photon micromaser experiments.

The lecturers agreed to prepare sets of lecture notes which for the most part have been collected in this volume with very minor adjustments of the format. We asked the authors to maintain a close contact between the subject of their lectures and their written notes. For this reason, the style of the manuscripts retains much of the spontaneity of the corresponding verbal presentations.

We believe that this school has provided a first small step to ease the formidable problems faced by students and scientists in the South American countries as they attempt to share in the technical and scientific developments that are taking place everywhere. Our thanks go to the lecturers for their dedicated efforts to provide informative and stimulating discussions and for accepting the added task of preparing written lecture notes. We also owe our gratitude in equal measure to the many organizations that have contributed so generously for the success of the school. Our goal was to provide as much financial aid as possible to the participants in order to insure that travel and living expenses would not be a serious obstacle for qualified students. We believe we have succeeded with flying colors on this score thanks to the financial assistance of :

CIF	Centro Internacional de Fisica
OAS	Organization of the American States
CLAF	Centro Latino-Americano de Fisica
CONICET	Consejo Nacional de Investigaciones Cientificas y Tecnologicas
CIC	Comision de Investigaciones Cientificas de la Prov.de Buenos Aires
SH	Sheraton Hotels of Argentina
UNMP	Universidad Nacional de Mar del Plata
CITEFA	Centro de Investigaciones Cientificas y Tecnologicas de las Fuerzas Armadas
DU	Drexel University

We look forward to future opportunities of renewing our acquantaince with the many old and new friends that we met in Mar del Plata.

J.R. Tredicce,
L.M. Narducci

Physics Department
Drexel University
Philadelphia, Pa, USA

and

E.J. Quel

CEILAP-CITEFA
Zufriategui y Varela
1603 - Villa Martelli.
Prov. de Buenos Aires, Argentina

CONTENTS

Lecturer	Title of lecture
N.B. Abraham Bryn Mawr College Bryn Mawr, PA, USA	Inhomogeneous gas lasers; nonlinear dynamics
H.J. Carmichael University of Arkansas Fayetteville, Arkansas, USA Present address: University of Oregon Eugene, Oregon	The quantum theory of the laser
V. Degiorgio Universita' di Pavia Pavia, Italy	Quantum statistics of the electromagnetic field
S. Haroche Yale University New Haven, Conn., USA	Single-atom electrodynamics
H.Haug University of Frankfurt Frankfurt, FRG	Semiconductor laser physics
L.M. Narducci Drexel University Philadelphia, PA, USA	The Maxwell-Bloch theory of the homogeneously broadened laser
J.R. Tredicce Drexel University Philadelphia, PA, USA	Regular and chaotic dynamics of a laser

Seminar speakers	Title of lecture
D.K. Bandy Oklahoma State University Stillwater, OK, USA	Transverse effects in laser instabilities
L. Bobb Naval Air Development Center Warminster, PA, USA	Nonlinear optics in Xenon
P. Diodati CEILAP, Buenos Aires, Argentina	Solid state lasers

F. Manzano CEILAP, Buenos Aires, Argentina	The nitrogen laser
G.-L. Oppo Drexel University Philadelphia, PA, USA	Simple dynamics of single-mode lasers
E.J. Quel CEILAP, Buenos Aires, Argentina	Low-power CO_2 lasers
J.-M. Raimond Ecole Normale Superieure Paris, France	Realization of a two-photon maser oscillator
H. Ranea-Sandoval Universidad Nacional de Buenos Aires Argentina	Dye lasers
J.G. Rocca Colorado State University Fort Collins, Colorado, USA	Ion lasers
C. Rosito CEILAP, Buenos Aires, Argentina	High power CO_2 lasers
J. Tocho CIOP, Argentina	A survey of nonlinear optics
J. Van der Berg ETH, Lausanne, Switzerland	Infrared photochemistry

The school was followed by the "III Encuentro Latinoamericano de Laser y Aplicaciones". Some of the papers presented on the conference are included in this book.

THE MAXWELL-BLOCH THEORY OF THE HOMOGENEOUSLY BROADENED LASER

L.M. Narducci
Physics Department
Drexel University
Philadelphia, PA 19104 USA

L.A. Lugiato
Dipartimento di Fisica
Politecnico di Torino
10129 Torino, ITALY

and

D.K. Bandy
Physics Department
Oklahoma State University
Stillwater, OK 74078 USA

The purpose of these lecture notes is to provide an introduction to the field of laser instabilities and an overview of one of the most popular theoretical models: that of a homogeneously broadened, unidirectional ring laser. We discuss both multimode and single-mode operation, the possible steady states and their stability properties. In the process, we identify some valuable features of the plane-wave Maxwell-Bloch equations and single out some of their shortcomings. We conclude this survey with a review of current attempts at removing the remaining open problems with an extension of the plane-wave theory and the inclusion of transverse effects. We apply the extended model to the study of spatial symmetry breaking effects.

1. Introduction

After 30 years of active research[1] it is tempting to believe that very little might be left to be discovered in the field of lasers. In fact, to be sure, the main issues related to the principles of operation of this device have become standard classroom subjects. Yet, it would be inaccurate to say that all major theoretical problems have been completely solved. Let us consider, for example, a situation which will be the focus of most of these lectures. The laser is a source of light. It is designed to convert incoherent energy into coherent radiation. On the surface, and apart from the spectral purity and power of the respective output fields, there does not seem to be much difference between a laser and a flashlamp. Yet, when we apply a d.c. voltage to a flashlamp we expect and obtain a steady stream of light; when we apply a steady pump to a laser we expect, but we do not always get, a steady output. Often, in fact, c.w. pumped lasers may develop complicated intensity fluctuations even after removal of all the obvious sources of trouble, mechanical, electrical and thermal disturbances (at the quantum level there are other important and fundamental differences between laser and thermal light which are discussed in the lectures by Professors Carmichael and Degiorgio).

Perhaps the most obvious difference between a laser and a flashlamp is that the latter operates more or less in a state of thermal equilibrium, while the former does not. The atoms in a laser are maintained in a state of inversion by the pump. This allows the emergence of large temporal variations of the atomic population and opens the door to important nonlinear effects.

Ever since the early 1960's we have come to recognize that nonlinearities can be sources of varied and often extremely complex dynamical behaviors. The laser is an example of a system whose operation is predicated upon the nonlinear interaction of light and matter. In this respect, it is certainly not unique, as scores of interesting nonlinear systems can be found in just about any branch of science and technology. In some way, however, the laser is an ideal tool for the study of nonlinear dynamics for many practical reasons that will be emphasized by several lecturers at this school (see, for example, the lectures by Professors Tredicce and Abraham).

Over the last 10 years or so this has created the impetus for renewed investigations of the theoretical foundations of laser dynamics.[2] The highlights of this recent work will be reviewed in these lectures with emphasis on the interesting behaviors of the output intensity, or dynamical instabilities as these phenomena are commonly called. Instabilities have an illustrious history, one that even predates the invention of the laser. In fact, the spontaneous appearance of pulsations was already noted in masers.[3] With the advent of lasers and especially solid state devices these effects became almost commonplace, not only with flashlamp-pumped systems but also under c.w. excitation.[4]

The significance of laser instabilities as fundamental consequences of the inner workings of a laser was recognized since the early 1960's by a number of Russian scientists such as Uspensky,[5] Fain and Khanin,[6] Oraevsky[7] and others, and by Haken[8] and his school in Stuttgart. A major advance, an event that perhaps more than any other marked the rites of passage into maturity for this field of investigations took place in 1975 with the discovery by Haken[9] that the laser equations are isomorphic to the Lorenz equations[10] for convective hydrodynamical instabilities. If we just consider the important role played by the Lorenz equations in the modern development of the theory of chaos, we can then readily appreciate that Haken's discovery all at once enriched the study of laser dynamics by endowing it with a rich chaotic phenomenology of its own. Important experimental tests of these theoretical results have been carried out recently by Abraham and will be discussed in this school.

Why should we be interested in this type of "malfunctions"? We can cite at least three reasons:

1. Oscillations emerge spontaneously in steadily driven systems. They represent a break of temporal and, as we will see, also spatial symmetry.
2. Instabilities display a surprisingly varied range of behaviors as we change the control parameters.
3. These phenomena offer valuable clues on the internal mechanisms of laser action.

Some of the earliest laser models focused on the energy exchanges between collections of inverted two-level atoms and the cavity field (rate equations models).[11] In their simplest version the rate equations couple the population difference $D = N_2 - N_1$ between the two active levels to the photon number n according to the nonlinear system of differential equations

$$\frac{dD}{dt} = -2WnD - \frac{1}{T_1}(D - D_0) \tag{1.1a}$$

$$\frac{dn}{dt} = WnD - \frac{1}{T_c}n \tag{1.1b}$$

where W is proportional to the absorption and stimulated emission rate, T_1^{-1} is the rate of approach of the population difference to its equilibrium value, D_0, under the action of both spontaneous emission and the external pump; T_C is the escape time of the electromagnetic energy out of the cavity (See Fig. 1.1).

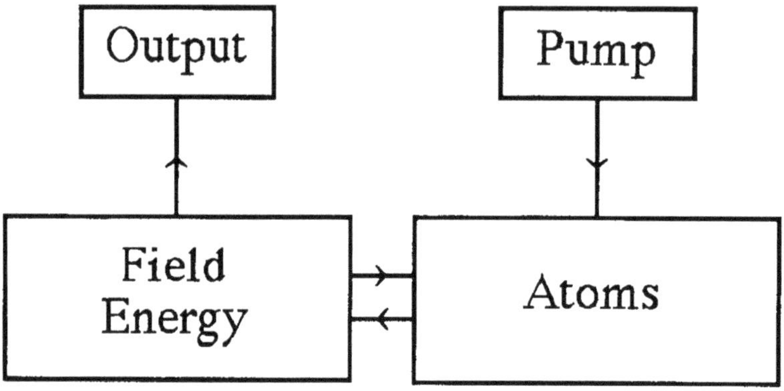

Fig. 1.1 Schematic representation of the energy flow between atoms and field according to the rate equations model.

The rate equations offer a good starting point for the description of some interesting features of laser action and for the inclusion of complicated details of the atomic structure. They will be discussed in greater details and their origin justified in Professor Oppo's lectures. Here we only mention their main theoretical predictions:

1. Above a certain pump level (the threshold inversion) the laser begins to operate, as evidenced by the growth of the photon number, n, from a small initial value.
2. The output intensity approaches a constant value after a transient that can involve either a monotonic or, at most, a damped oscillatory behavior. In either case no sustained oscillations are possible under cw pumping conditions.

It is important to mention that in the regime of damped oscillations (relaxation oscillations, as they are commonly referred to in the literature) the system can be forced to develop persistent pulsations. This, however, requires the application of some kind of time dependent forcing mechanism from the outside (e.g. a loss or gain modulation) with a frequency of the order of the relaxation oscillations frequency.

The conclusion at this point is that, at least in the simple form given by Eqs. (1.1), the rate equations cannot describe the existence of undamped oscillations and instabilities. In fact, within the context of the plane-wave model, to which this formulation implicitly refers, one is forced to conclude that this failure is related to the absence of coherence in the coupling between radiation and matter. This conclusion will have to be revised in the light of some recent studies which include the possibility of transverse variations of the field and of the atomic variables (see section 4 of these lectures and the references cited therein). In the plane-wave approximation, however, it seems inescapable that one must look at a deeper level than the rate equations to trace the origin of the unstable behaviors.

Actually this conclusion was reached in some of the earliest contributions to this problem. Grazyuk and Oraevsky[7] in 1964 discovered that the coherent single-mode laser equations, a simplified predecessor of the more modern Maxwell-Bloch theory, predict the appearance of instabilities and in fact also highly irregular pulsations under well defined mathematical conditions that involve the unsaturated gain of the active medium and the cavity and atomic relaxation rates (See section 3 of these notes). By coherent laser equations we mean a description built on the basic premise that the interaction between light and matter involves the coupling between an optical wave and the atomic dipoles, and that the evolution of the radiated field is determined by a macroscopic polarization, in agreement with Maxwell's electrodynamics.

This brings us to the central point of these lectures, the so-called Maxwell-Bloch model. For the purpose of producing a balanced presentation and for pedagogical reasons, we begin with the development of a theory of the homogeneously broadened laser in the plane-wave approximation; we discuss the main predictions concerning the steady state and the unstable behaviors for both single and multimode configurations, and identify some of the major unsolved issues of the Maxwell-Bloch theory. Further in our presentation we review the highlights of some recent work of ours that includes transverse effects, and compare the most significant differences between the conclusions of these two approaches.

2. The Maxwell-Bloch theory of a ring laser - Multimode operation

The conceptual foundations of the Maxwell-Bloch theory of the laser rest on the self-consistent approximation illustrated schematically in Fig. 2.1. An incident electromagnetic field creates a macroscopic polarization as a result of its interaction with a collection of microscopic dipoles. This polarization, in turn, acts as the source of a radiated field which interacts again with the microscopic dipoles, etc. Energy is fed into the active medium by an external source, the pump, and is extracted from the interaction volume in the form of electromagnetic energy radiated by the macroscopic polarization, usually through partially transparent output couplers.

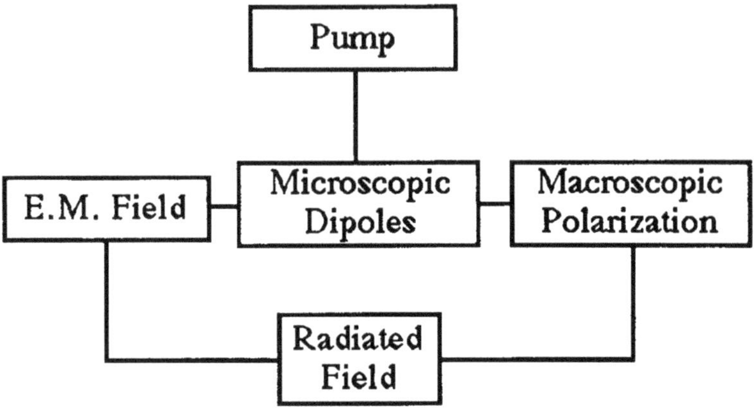

Fig. 2.1 Schematic representation of the self-consistent approach for the description of the interaction of light and matter.

The mathematical description of this model is based upon the classical wave equation for the slowly varying complex amplitude of the electromagnetic field and the Schroedinger equation for a collection of quantum mechanical two-level systems. This mix of classical and quantum theories is the essence of the semiclassical description of the laser[12]. Its self-consistency derives from the indentification of the driving Maxwell field with the field re-radiated by the macroscopic polarization which, in turn, is a suitable average of the microscopic atomic dipoles.

As it stands this model does not account for important physical effects: the atomic polarization in the absence of a driving field is known to decay as a result of random microscopic interactions (for example, collisions in a gas laser) while the population decays as a result of spontaneous emission and is restored by the action of the external pump. In addition, the radiated field escapes out of the active volume because of the partially transparent output mirrors of the resonator. It is customary to simulate the irreversible decay of the atomic variables by adding phenomenological relaxation terms to the so-called coherent equations (i.e. the equations describing the self-consistent loop of Fig. 2.1), while the presence of the lossy mirrors is accounted for by appropriate boundary conditions. Most traditional theories of the laser, at this point, add the assumption that the cavity field can be described as a plane wave, to a good approximation. For a long time this assumption was viewed as an acceptable way to tame the complexity of real resonators, but it now beginning to be reconsidered[13]. It appears, in fact, that some of the fine, and not so fine, points of laser dynamics are more sensitive to the transverse shape of the field and atomic variables than one might have expected. For the sake of simplicity, however, and in order to illustrate the advantages and drawbacks of this approach, we begin with a study of the plane-wave theory of the laser.

In these lectures we limit our considerations to a laser resonator shaped in the form of a ring as sketched in Fig. 2.2; the ring resonator has a total round-trip length Λ, and is bounded by two partially reflecting mirrors (mirrors 1 and 2) and by two additional ideally reflecting surfaces to complete the loop (mirrors 3 and 4). The field inside the cavity is assumed to circulate in only one direction with the help of a non-reciprocal element such as a Faraday isolator. If all four mirrors in Fig. 2.1 were ideal reflectors at all wavelengths, the resonator would be characterized by an infinite number of equispaced and ideally sharp resonances at the frequencies $\alpha_n = n2\pi c/\Lambda$ $(n = 0, 1, 2,...)$. The finite reflectivity of the mirrors causes a broadening of these resonances and forces them to acquire a characteristic width (cavity linewidth) which is a measure of the escape rate of the field out of the active volume. The active medium has a transition frequency ω_A between the two energy levels and is assumed to have a homogeneously broadened lineshape with a width $\gamma_\perp$ which is also the relaxation rate of the atomic polarization. For convenience of terminology, the empty cavity mode whose frequency ω_C lies closest to the center of the atomic gain line is called the resonant mode and is selected as the origin of the frequency axis; all other modes of the resonator are labelled collectively as off-resonant modes and their frequencies will be measured relative to ω_C.

The Maxwell-Bloch equations for the ring laser are (see, for example, reference 2)

$$\frac{\partial F}{\partial z} + \frac{1}{c}\frac{\partial F}{\partial t} = -\alpha P \tag{2.1a}$$

$$\frac{\partial P}{\partial t} = -\gamma_\perp \left\{ FD + (1+i\delta_{AC})P \right\} \tag{2.1b}$$

$$\frac{\partial D}{\partial t} = -\gamma_\parallel \left\{ -\frac{1}{2}(F^*P + FP^*) + D - 1 \right\} \tag{2.1c}$$

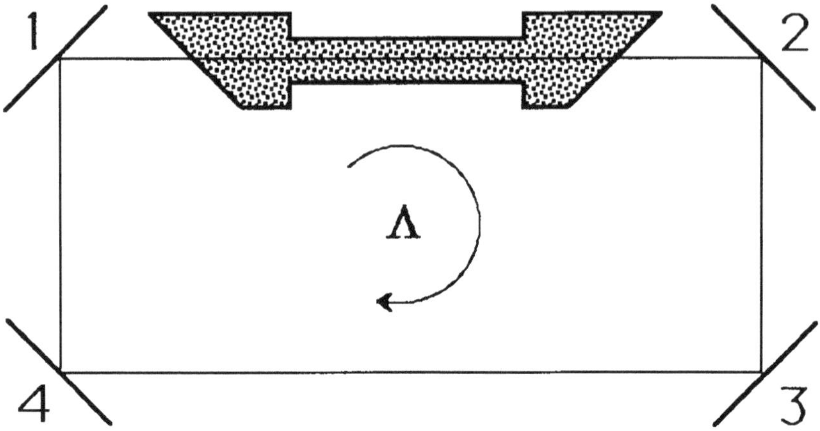

Fig. 2.2 Schematic representation of the ring cavity. The active medium has length L,while the ring's length is Λ.

where F is the slowly varying complex amplitude of the cavity field, scaled to the square root of the saturation intensity, P is the normalized amplitude of the macroscopic polarization, and D is the population difference between the upper and lower levels; α is the small signal gain constant per unit length, δ_{AC} denotes the frequency detuning between the atomic transition frequency and the resonant mode, in units of the atomic linewidth $\gamma_\perp$ [i.e. $\delta_{AC} = (\omega_A - \omega_C)/\gamma_\perp$], and $\gamma_\parallel$ is the rate of decay of the atomic population. Equations (2.1) describe the dynamics of any homogeneously broadened two-level laser medium and a plane-wave field; hence, for example, they are suitable for the description of a unidirectional single pass amplifier. We adapt them specifically to the ring laser of interest by the addition of the boundary conditions

$$F(0,t) = R\,F(L,t - \Delta t) \qquad (2.2)$$

where L is the length of the active medium, R is the power reflectivity coefficient of mirrors 1 and 2 and $\Delta t = (\Lambda - L)/c$ is the propagation time of light through the empty segment of the cavity. The role of the boundary conditions is "to force" Eqs. (2.1) to recognize the fact that the field that emerges out of the active volume must be re-injected into the medium after an appropriate delay and the attenuation caused by the partial reflectivity of mirrors 1 and 2. Note that, in this form, the Maxwell-Bloch equations and boundary condition account for propagation effects, through the active medium and the empty section of the resonator.

2.1 Steady State

Because the reference carrier frequency of the cavity field was selected, for convenience, to be ω_C the field envelope in steady state cannot be time independent except when ω_C turns out to be precisely the frequency of the laser radiation (as it might be expected, we shall see that this is the case under resonance conditions, i.e. when $\omega_C = \omega_A$). In general, the frequency of the laser field is expected to differ from ω_C by an unknown amount $\delta\Omega$ which must be calculated. Hence, the possible stationary states of Eqs (2.1) must be assumed to take the form

$$F(z,t) = F_{st}(z)\, e^{-i\delta\Omega t} \tag{2.3a}$$

$$P(z,t) = P_{st}(z)\, e^{-i\delta\Omega t} \tag{2.3b}$$

$$D(z,t) = D_{st}(z) \tag{2.3c}$$

where, according to Eqs. (2.3), F_{st}, P_{st}, D_{st} satisfy the equations

$$\frac{dF_{st}}{dz} = i\,\frac{\delta\Omega}{c}F_{st} - \alpha P \tag{2.4a}$$

$$0 = F_{st}D_{st} + (1+i\Delta)P_{st} \tag{2.4b}$$

$$0 = -\frac{1}{2}(F^*_{st}P_{st} + F_{st}P^*_{st}) + D_{st} - 1 \tag{2.4c}$$

and the parameter Δ is defined as

$$\Delta = \delta_{AC} - \delta\Omega/\gamma_\perp \tag{2.5}$$

Note that, in general, the field and atomic envelopes depend on the longitudinal coordinate; in fact, it is intuitively obvious that the field amplitude must grow as it propagates through the active medium and that it must decrease discontinuously upon reflection at mirrors 1 and 2.

The atomic variables can be calculated at once in terms of the stationary field profile from Eqs. (2.4b,c) with the result

$$P_{st}(z) = -F_{st}(z)\frac{1-i\Delta}{1+\Delta^2+|F_{st}(z)|^2} \tag{2.6a}$$

$$D_{st}(z) = \frac{1+\Delta^2}{1+\Delta^2+|F_{st}(z)|^2} \tag{2.6b}$$

where $F_{st}(z)$ is the solution of Eq. (2.4a) subject to the boundary condition

$$F_{st}(0) = R\,F_{st}(L)\exp[i\delta\Omega(\Lambda-L)/c] \tag{2.7}$$

Several conclusions are immediately obvious. The steady state polarization and the field envelope are generally out of phase with respect to each other by an amount that depends on the detuning parameter and the position of the operating laser line; they are in phase, instead, under resonant conditions ($\delta_{AC} = 0$) because, as it turns out, $\delta\Omega$ vanishes [see Eq. (2.18) and the following comment]. The steady state population saturates at high intensity levels in the sense that $D_{st}(z) \rightarrow 0$ as $|F_{st}(z)| \rightarrow \infty$. This is a consequence of the well known fact that at high field values the rates of stimulated emission and absorption tend to equalize. Note, also, that the level of saturation of the population difference depends on the detuning parameter, through Δ, with the largest degree of saturation corresponding, as expected, to the resonant case.

In order to calculate the operating frequency and the envelope of the steady state field everywhere along the active medium and at the output of the resonator, we substitute Eq. (2.6a) into Eq. (2.4a) and express the field amplitude in terms of its modulus and phase.[14-16] This appears to be the simplest way to proceed. Thus, we let

$$F_{st}(z) = \rho(z)\,e^{i\theta(z)} \tag{2.8}$$

and after a few simple steps we transform Eq. (2.4a) into the pair of coupled nonlinear equations

$$\frac{d\rho}{dz} = \alpha\frac{\rho}{1+\Delta^2+\rho^2} \tag{2.9a}$$

$$\frac{d\theta}{dz} = \frac{\delta\Omega}{c} - \alpha\frac{\Delta}{1+\Delta^2+\rho^2} \tag{2.9b}$$

Equation (2.9a) can be integrated at once to yield the implicit solution

$$(1+\Delta^2)\ln\left(\frac{\rho(z)}{\rho(0)}\right) + \frac{1}{2}\left(\rho^2(z) - \rho^2(0)\right) = \alpha z \tag{2.10}$$

Equation (2.9b), instead, appears to offer a formidable challenge because an explicit expression for $\rho(z)$ cannot be derived from Eq. (2.10). Fortunately, the two coupled equations (2.9) can be combined to yield the first integral of the problem

$$\ln\left(\frac{\rho(z)}{\rho(0)}\right) = -\frac{1}{\Delta}\left[\theta(z) - \theta(0) - \frac{\delta\Omega}{c}z\right] \tag{2.11}$$

which gives $\theta(z)$ in terms of the field modulus and the two input values $\rho(0)$ and $\theta(0)$. We still need to dispose of $\rho(0)$ and $\theta(0)$ and, of course, $\delta\Omega$ is still unknown. Actually, the input phase of the field, $\theta(0)$, is an irrelevant quantity because only the relative phase $\theta(z) - \theta(0)$ is physically significant, so that only two unknowns need to be calculated. These can be handled with the help of the boundary conditions (2.7) which, in terms of the field modulus and phase take the form

$$\rho(0) = R\,\rho(L) \tag{2.12a}$$

$$\theta(L) - \theta(0) = -\delta\Omega\frac{\Lambda\text{-}L}{c} + 2\pi j, \qquad (j = 0, \pm 1, \pm 2, ..) \tag{2.12b}$$

The additive term $2\pi j$ ($j = 0, \pm 1, \pm 2,..$) that appears on the right hand side of Eq. (2.12b) is a consequence of the boundary condition for the field phase which can be satisfied, in principle, by an infinite number of solutions. The physical meaning of this multiplicity and the constraints that lead to the finite (rather than infinite) set of realizable solutions will be discussed further on in this section. For now we observe that this result is important because it suggests the possibility of multiple steady states.

The output laser intensity can be calculated at once from Eq. (2.12) after setting $z = L$ and using the boundary condition (2.12a). The result is

$$\rho^2(L) = \frac{2}{1-R^2}\left(\alpha L - (1+\Delta^2)\,|\ln R|\right) \tag{2.13}$$

The unknown operating frequency or, equivalently, the value of Δ follows by setting $z = L$ in Eq. (2.11) and using the boundary condition (2.12b). The required result is

$$\ln R = \frac{1}{\Delta}\left(-\delta\Omega\frac{\Lambda\text{-}L}{c} + 2\pi j - \frac{\delta\Omega}{c}L\right) \tag{2.14}$$

or

$$\delta\Omega\left(1 + \frac{c\,|\ln R|}{\Lambda\gamma_\perp}\right) = \frac{c\,|\ln R|}{\Lambda}\delta_{AC} + \frac{2\pi c}{\Lambda}j \tag{2.15}$$

The quantity c |ln R|/Λ measures the escape rate of the cavity field out of the resonator, while $2\pi c/\Lambda$ is the frequency spacing between adjacent cavity resonances, also called the free spectral range. After introducing the symbols

$$\kappa = \frac{c\,|\ln R|}{\Lambda}, \qquad \alpha_1 = \frac{2\pi c}{\Lambda} \tag{2.16}$$

we obtain the required result

$$\delta\Omega_j = \omega_L^{(j)} - \omega_C = \frac{\kappa\delta_{AC} + \alpha_1\gamma_\perp j}{\gamma_\perp + \kappa}, \qquad (j = 0, \pm 1, \pm 2, \ldots) \tag{2.17}$$

where the index j is a label for the multiple solutions. Equation (2.17) is the well known mode-pulling formula, written in a slightly unconventional way. In most textbooks Eq. (2.17) is usually given in the form

$$\omega_L^{(j)} = \frac{(\omega_C + \alpha_1 j)\gamma_\perp + \omega_A\kappa}{\gamma_\perp + \kappa} \tag{2.18}$$

which shows that the operating laser frequency is the weighted average of the atomic resonant frequency and the frequency of one of the cavity modes. In resonance, the j=0 steady state solution has a frequency $\omega_L^{(0)} = \omega_C$, i.e. $\delta\Omega = 0$.

As already mentioned, the phase part of the boundary conditions can be satisfied by an infinite number of solutions, labelled by the indices j = 0, ±1, ±2,... etc. Actually, only a finite number of these solutions can also satisfy the threshold condition, i.e. the requirement that $\rho_j^2(L)$ be positive. The range of allowed values of j is such that

$$(\alpha L)_j \geq (1+\Delta_j^2)\,|\ln R| \equiv (\alpha L)_{thr,j} \tag{2.19}$$

Figure 2.3 illustrates schematically how various steady state solutions are organized in relation to the various empty cavity resonances, if their respective gain is sufficiently large to overcome the cavity losses.

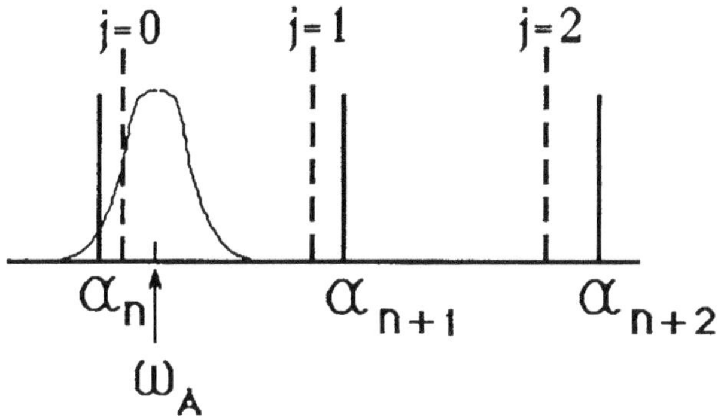

Fig. 2.3 Schematic representation of the multiple stationary solutions. The solid vertical lines represent three adjacent cavity modes. The dashed lines labelled j=0, 1, and 2 indicate the position of three steady state solutions; ω_A denotes the position of the center of the atomic line.

Each of the dashed lines in Fig. (2.3) denotes a possible steady state solution whose frequency, given by Eq. (2.18), is mode pulled toward the center of the gain line. It is important to stress that, while multiple steady state solutions coexist with one another in the sense that they all correspond to the same value of the gain, only one of them is excited in a true stationary state. However, the steady state analysis cannot identify the actual solution; for this we must resort to a study of the stability of the steady states and of the history of the excitation process, as we now explain in detail.

Depending on the value of the gain, αL, and of the free spectral range, we distinguish two quite different situations, as illustrated schematically in Figs. 2.4 a,b. In the first case, the intermode spacing is large enough that, at most, only one nontrivial steady state solution can exist for every value of the detuning parameter δ_{AC}. In the second case, more than one steady state can satisfy the threshold condition for selected values of the detuning.

The physical implication of Fig (2.4a) is unambiguous. Consider a detuning scan that begins with the resonant configuration, $\delta_{AC} = 0$. Initially the output intensity takes a maximum,value; as the magnitude of the detuning increases the intensity of the output decreases until, eventually, the laser is driven below threshold. A further increase of the detuning parameter brings the next steady state above threshold; the corresponding output intensity increases monotonically until δ_{AC} matches a full free spectral range of the ring cavity $(\alpha_1/\gamma_\perp)$. Note that the carrier frequency of the operating stationary state changes as the laser switches from the steady state j = 0 to j = 1. This is a consequence of the fact that, at some point during the detuning scan, the laser is turned off and when

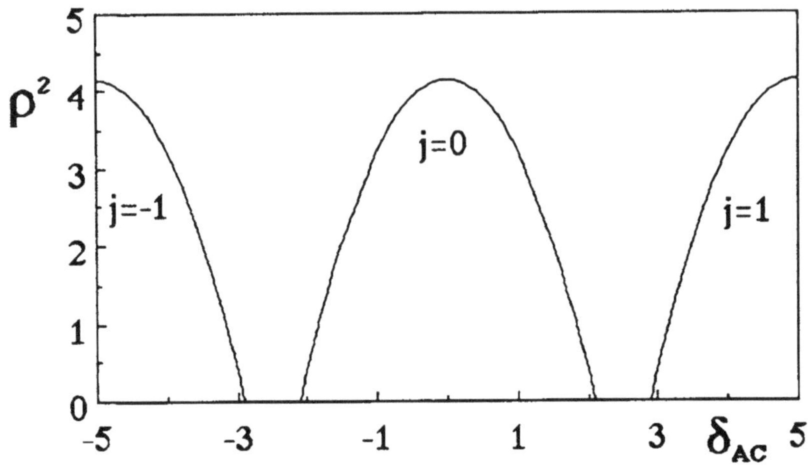

Fig. 2.4a Steady state output intensity ρ^2 for three stationary states labelled j=0,±1 as a function of the detuning parameter δ_{AC}. The value $\delta_{AC} = 0$ corresponds to the resonance between the atomic transition frequency and the reference cavity mode. The intermode spacing in units of $\gamma_\perp$ is equal to 5.

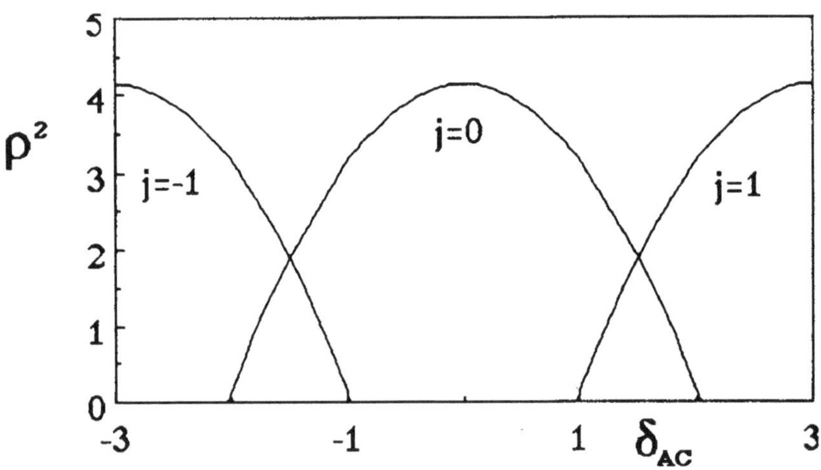

Fig. 2.4b Same as Figure 2.4a with an intermode spacing of 3 (in units of $\gamma_\perp$)

it begins to function again, it does so in correspondence to a different cavity resonance. Note that the sequence of steady states can be retraced in a fully reversible way if we reverse the direction of the detuning scan. The interpretation is less obvious in the case of Fig (2.4b). At the beginning of the detuning scan, with $\delta_{AC} = 0$, there is no ambiguity as to the steady state configuration of the laser (only one steady state, $j = 0$, satisfies the threshold condition). However, when δ_{AC} becomes sufficiently large, two coexisting steady states (or more, depending on the parameters of the problem) may develop. At this point we may anticipate two possible scenarios:

(i) the lasing mode retains control of the laser process well into the coexistence region or until, perhaps, its losses overcome the available gain;

(ii) the two coexisting steady states compete with one another.

In case (i) we can expect the appearance of discontinuous jumps of the steady state intensity and frequency, in addition to hysteresis upon reversing the direction of the scan; in case (ii) the competing steady states have different frequencies and cause undamped beats with a frequency approximately equal to the intermode spacing (we say "approximately" because the two coexisting states suffer from frequency pulling, i.e. their frequencies do not coincide with those of the empty cavity modes). The correctness of this conjecture can be verified only with the help of additional considerations on the stability of the steady state. We now turn our attention to this aspect of the problem.

2.2 The linear stability analysis

The stability of each steady state can be probed in the usual way, as with all nonlinear dynamical systems, by perturbing the dynamical variables around a stationary configuration, and by following the evolution of the perturbations according to their linearized equations of motion. The complex exponential rate constants, λ, of the infinitesimal perturbations provide information on the stability of the system in the sense that, if Re λ is negative for all the degrees of freedom, the steady state of interest is stable. If even one of the real parts of the rate constants is positive, the system is unstable and departs exponentially from the steady state in response to the applied perturbation.

The stability analysis of the full set of Maxwell-Bloch equations (2.1) is a difficult problem.[17] The main complications come from the longitudinal dependence of the steady state field and atomic variables on the longitudinal coordinate which makes even the linearized equations quite complicated to study in every case except for the resonant condition.

Fortunately, Bonifacio and Lugiato[18] identified a very useful mathematical limit which is responsible for great simplifications in the theoretical analysis of the Maxwell-Bloch equations. The main virtue of their procedure, which we have re-labelled the uniform field limit, is that it allows much progress to be made analytically. In addition, subsequent work[15,16] has shown that the predictions of the uniform field limit continue to be valid, qualitatively, even when strictly speaking the mathematical conditions are not ideally satisfied. It is this favorable aspect of the uniform field approximation that makes it an extremely useful tool for the study of the dynamical aspects of the Maxwell-Bloch equations. This limit is characterized by the requirements

$$\alpha L \to 0, \qquad T \to 0 \tag{2.20a}$$

with

$$2C \equiv \frac{\alpha L}{T} = \text{arbitrary} \tag{2.20b}$$

At first sight one may be tempted to regard this limit as an oversimplification of the problem because if one continues to lower the unsaturated gain per pass, eventually the laser will be driven below threshold. Instead, it is important to realize that this prescription requires the simultaneous reduction of both the unsaturated gain and the transmission losses, so that non trivial operating conditions with arbitrary output intensity can be generated, depending on the value of the gain parameter C.

The reason why we talk about "uniform field" is not difficult to understand: if the gain per pass is very small, a wave entering the active medium undergoes a very small amplification leading to negligible longitudinal variations for both the field amplitude and the atomic variables. If one reduces the transmission losses to the appropriate level one can insure that the laser will continue to function as much above threshold as desired. Naturally, under uniform field conditions, the transient evolution into steady state may take a very long time, as the field requires more and more round trips, as T approaches zero, in order to reach its asymptotic configuration. Of course, if the uniform field limit is to make sense in physical terms, the only losses of the resonator must arise from the finite transmittivity of the mirrors because any fixed residual loss of a different origin would unavoidably force the laser below threshold.

We carry out the linear stability analysis in two steps: first, we derive an appropriate set of modal equations in the uniform field limit, and then linearize these equations and derive the characteristic equation for the rate constants.

2.2a The modal equations

The natural setting for a plane-wave modal expansion involves isochronous boundary conditions of the periodicity type. Our boundary conditions (2.3) are not isochronous, because of the time delay, nor periodic because of the presence of the multiplicative factor R on the right hand side. Lugiato[19] showed how to get around these difficulties with a two-step process. First we define the new set of space-time coordinates

$$z' = z \tag{2.21a}$$

$$t' = t + \frac{\Lambda - L}{c}\frac{z}{L} \tag{2.21b}$$

and then the new field and atomic variables

$$F_1(z',t') = F(z',t').\exp(\frac{z'}{L}\ln R) \tag{2.22a}$$

$$P_1(z',t') = P(z',t')\exp(\frac{z'}{L}\ln R) \tag{2.22b}$$

$$D_1(z',t') = D(z',t') \tag{2.22c}$$

With a simple calculation it is easy to show that the transformed Maxwell-Bloch equations take the form

$$\frac{\partial F_1}{\partial t'} + \frac{cL}{\Lambda}\frac{\partial F_1}{\partial z'} = -\kappa\,(F_1 + 2CP_1) \tag{2.23a}$$

$$\frac{\partial P_1}{\partial t'} = -\gamma_\perp\{F_1 D_1 + (1+i\delta_{AC})\,P_1\} \tag{2.23b}$$

$$\frac{\partial D_1}{\partial t'} = -\gamma_\parallel\{-\tfrac{1}{2}(F_1^* P_1 + F_1 P_1^*)\,e^{2\frac{z'}{L}|\ln R|} + D_1 - 1\} \tag{2.23c}$$

where the symbol

$$2C = \frac{\alpha L}{|\ln R|} \tag{2.24}$$

is a measure of the gain parameter. The reason for the transformations (2.21) and (2.22) becomes clear if we observe that in terms of the new space-time coordinates the boundary condition becomes isochronous and in terms of the new field variable it acquires the standard periodicity form. In fact the transformed boundary condition is

$$F_1(0,t') = F_1(L,t') \tag{2.25}$$

Now the dynamical variables can be decomposed into a Fourier series, as commonly done in ordinary vibration problems. Thus, we let

$$\begin{pmatrix} F_1(z',t') \\ \\ P_1(z',t') \end{pmatrix} = e^{-i\delta\Omega t'} \sum_{n=-\infty}^{+\infty} e^{ik_n z'} \begin{pmatrix} f_n(t') \\ \\ p_n(t') \end{pmatrix} \tag{2.26a}$$

and

$$D_1(z',t') = \sum_{n=-\infty}^{+\infty} e^{ik_n z'} d_n(t') \tag{2.26b}$$

Of course, in order to satisfy the boundary condition (2.25), the wave numbers k_n must be chosen such that

$$k_n = \frac{2\pi c}{L} n\,, \qquad (n = 0, \pm 1, \pm 2, \ldots) \tag{2.27}$$

With this selection, the boundary condition (2.25) is satisfied automatically. Furthermore, the modal functions of the ring cavity in the new reference system are

$$u_n(z') = \frac{1}{\sqrt{L}} e^{ik_n z'} \tag{2.28}$$

and satisfy the orthogonality relation

$$\big(u_n(z'), u_m(z')\big) = \int_0^L dz'\, u_n^*(z')\, u_m(z') = \delta_{n,m} \tag{2.29}$$

The construction of the equations of motion for the modal amplitudes $f_n(t')$, $p_n(t')$, $d_n(t')$ is now a simple matter, which involves substituting Eqs. (2.26) into Eqs. (2.23) and taking into account the orthonormality of the cavity modal functions. The result is a complicated infinite set of equations which is exactly equivalent to the original Maxwell-Bloch equations (2.23) [or (2.1)].

In the uniform field limit these modal equations become

$$\frac{df_n}{dt'} = i\delta\Omega\, f_n - \kappa\,(f_n + 2Cp_n) \tag{2.30a}$$

$$\frac{df_n^*}{dt'} = -\, i\delta\Omega\, f_n^* - \kappa\,(f_n^* + 2Cp_n^*) \tag{2.30b}$$

$$\frac{dp_n}{dt'} = -\,\gamma_\perp\Big\{\sum_{n'=-\infty}^{+\infty} f_{n'}\, d_{n-n'} + \big[1+i\,(\delta_{AC} - \delta\Omega' - \alpha_n')\big]\, p_n\Big\} \tag{2.30c}$$

$$\frac{dp_n^*}{dt'} = -\,\gamma_\perp\Big\{\sum_{n'=-\infty}^{+\infty} f_{n'}^*\, d_{n-n'}^* + \big[1-i\,(\delta_{AC} - \delta\Omega' - \alpha_n')\big]\, p_n^*\Big\} \tag{2.30d}$$

$$\frac{dd_n}{dt'} = i\,\alpha_n d_n - \gamma_{||}\Big\{-\frac{1}{2}\sum_{n'}\big(f_{n'}^*\, p_{n+n'} + f_{n'}\, p_{n'-n}^*\big) + d_n - \delta_{n,0}\Big\} \tag{2.30e}$$

where the symbols $\delta\Omega'$ and α_n' represent the frequencies $\delta\Omega$ and α_n scaled to $\gamma_\perp$, and where the field and atomic amplitudes are defined according to relations of the type

$$x_n(t') = \mathbf{x}_n(t')\, e^{-ick_n t'} \equiv \mathbf{x}_n(t')\, e^{-i\alpha_n t'} \tag{2.31}$$

Finding the stationary states of Eqs. (2.30) looks like a very difficult task. Instead, close inspection shows that they can be represented by the simple formulae[16]

$$f_n^{(j)} = \big[2C - (1+\Delta_j^2)\big]^{1/2}\delta_{n,j} \tag{2.32a}$$

$$p_n^{(j)} = -\, f_n^{(j)}\,\frac{1-i\Delta_j}{1+\Delta_j^2 + |f_n^{(j)}|^2} \tag{2.32b}$$

$$d_n^{(j)} = \frac{1+\Delta_j^2}{1+\Delta_j^2 + |f_n^{(j)}|^2}\,\delta_{n,0} \tag{2.32c}$$

where

$$\Delta_j = \frac{\delta_{AC} - j\,\alpha_1/\gamma_\perp}{1+\kappa/\gamma_\perp} \tag{2.33}$$

and

$$\delta\Omega' \equiv \delta\Omega_j' = \frac{\kappa/\gamma_\perp}{1+\kappa/\gamma_\perp}(\delta_{AC} - j\alpha_1/\gamma_\perp) \tag{2.34}$$

Multiple steady state solutions exist if the threshold condition

$$2C > 1 + \Delta_j^2 \tag{2.35}$$

is satisfied for more than one value of j. This result agree with Eq. (2.19) after application of the uniform field condition.

Thus each steady state is characterized by five non-zero Fourier amplitudes leading to single- mode operation with a uniform field profile in steady state and to a purely harmonic structure both in space and time.

2.2b The eigenvalues of the linearized equations

We linearize Eq. (2.30) around the j-th stationary solution by setting

$$x_n(t') = x_n^{(j)}\delta_{n,j} + \delta x_n(t') \tag{2.36a}$$

$$d_n(t') = d_n^{(j)}\delta_{n,0} + \delta d_n(t') \tag{2.36b}$$

where x_n denotes either f_n, f_n^*, p_n, or p_n^*, and δx_n and δd_n are infinitesimal perturbations. Substitution of Eqs. (2.36) into Eqs. (2.30) leads to an infinite dimensional linear system of equations for the fluctuation variables δx_n and δd_n. The remarkable aspect of this problem is that the infinite system breaks up into separate blocks of five equations. In each block the fluctuation variables δf_{n+j}, δp_{n+j}, δf_{j-n}^*, δp_{j-n}^*, and δd_n for fixed values of the steady state index j and the modal index n are coupled to one another, but not to other variables, according to the equations[16]

$$\frac{df_{n+j}}{dt'} = i\,\delta\Omega_j\,\delta f_{n+j} - \kappa\,(\delta f_{n+j} + 2C\,\delta p_{n+j}) \tag{2.37a}$$

$$\frac{df_{j-n}^*}{dt'} = -i\,\delta\Omega_j\,\delta f_{j-n}^* - \kappa\,(\delta f_{j-n}^* + 2C\,\delta p_{j-n}^*) \tag{2.37b}$$

$$\frac{dp_{n+j}}{dt'} = -\gamma_\perp\left\{f_j^{(j)}\,\delta d_n + \delta f_{n+j} d_0^{(j)} + [1 + i\,(\Delta_j - \alpha_n')]\,\delta p_{n+j}\right\} \tag{2.37c}$$

$$\frac{dp^*_{j-n}}{dt'} = -\gamma_\perp \{ f_j^{(j)*} \delta d_n + \delta f^*_{j-n} d_0^{(j)} + [1 - i(\Delta_j + \alpha'_n)] \delta p^*_{j-n} \} \tag{2.37d}$$

$$\frac{dd_n}{dt'} = i\,\alpha_n\, \delta d_n - \gamma_\perp \{ -\tfrac{1}{2} (f_j^{(j)*} \delta p_{n+j} + p_j^{(j)} \delta f^*_{j-n} + f_j^{(j)} \delta p^{(j)*}_{j-n} +$$

$$+ p_j^{(j)} \delta f_{n+j}) + \delta d_n \} \tag{2.37e}$$

We solve Eqs. (2.37) with the usual ansatz

$$\begin{bmatrix} \delta f_{n+j}(t') \\ \delta f^*_{j-n}(t') \\ \delta p_{n+j}(t') \\ \delta p^*_{j-n}(t') \\ \delta d_n(t') \end{bmatrix} = e^{\lambda t'} \begin{bmatrix} \delta f^{(0)}_{n+j} \\ \delta f^{(0)*}_{j-n} \\ \delta p^{(0)}_{n+j} \\ \delta p^{(0)*}_{j-n} \\ \delta d^{(0)}_n \end{bmatrix} \tag{2.38}$$

and obtain a fifth-degree characteristic equation for the rate constants λ of the form

$$\sum_{i=0}^{5} A_i(\alpha'_n)\,(\lambda/\gamma_\perp)^i = 0 \tag{2.39}$$

where the coefficients A_i are complicated but known expressions that depend on the stationary state parameters and the frequency α_n of the n-th Fourier mode (or its scaled value α_n'). The steady state that is being probed, is stable if and only if the real parts of all five eigenvalues λ_{nj} are negative for all values of n. If the real part of one or more of the eigenvalues is positive the selected steady state is unstable against small perturbations. The instability reveals itself with the growth of sidebands at frequencies $\pm\alpha_n'$. Hence the field amplitude $F_1(z',t')$ departs from its uniform stationary configuration and develops a space-time structure.

A numerical study[16] of the fifth-degree equation (2.39) shows that for every value of α_n' three of the five eigenvalues have large and negative real parts. They correspond to the evolution of the linearized equations that are most closely related to the atomic degrees of freedom and can be ignored for the purpose of instabilities studies. The remaining two eigenvalues, for each value of the index n, can be associated with the linearized evolution of the field amplitude and phase. The telltale sign that motivates this identification is that the resonant phase eigenvalue has a zero real part because of the marginal stability of the phase variable in steady state for the resonant mode.

While the only physically meaningful eigenvalues are those for which α_n is an integer multiple of α_1, it is convenient , for the purpose of graphical display, to regard α_n as a continuous variable and to plot Re λ as a continuous function. Two typical results are shown in Figs. (2.5a,b) which illustrate the behavior of the two eigenvalues with the largest real parts as functions of the mode frequency α_n, viewed as a continuous variable, for $\delta_{AC}=0$ and for two values of the unsaturated gain. In Fig. (2.5a) all the complex rate constants have negative real parts and the steady state is stable. Figure (2.5b) corresponds to a larger value of the unsaturated gain; in this case, the cavity modes at frequency $\alpha_3 = \pm 3\ \alpha_1$ are unstable (the figure displays only the right half of the symmetric curves). This implies that an arbitrary perturbation at these frequencies grows exponentially during the linear regime, and the cavity field becomes a superposition of the steady state solution, oscillating at frequency ω_C, and of the growing sidebands whose frequencies are $\omega_C \pm \alpha_3$. Hence, the total output intensity shows the characteristic beat pattern due to the interference of these frequency components and the laser displays self-pulsing.

To learn about the nonlinear evolution, in this case, one has to solve the full Maxwell-Bloch equations as done for example in Ref. 15 and 16. The result of a calculation of this type, corresponding to the parameters chosen in Fig. 2.5b, shows periodic oscillations with a frequency that is approximately α_3 even in the nonlinear regime. For even larger gain values more than one sideband may become unstable simultaneously; the nonlinear dynamics can then become very complicated and even develop the characteristic signatures of deterministic chaos[15].

When the detuning δ_{AC} is either zero or sufficiently small with respect to the free spectral range of the cavity the instabilities are triggered by the positive real part of the amplitude eigenvalue. This situation usually requires that the unsaturated gain be about a factor of 10 larger than the threshold value for laser action. This is a common feature of the amplitude instabilities in the plane-wave Maxwell-Bloch model, one that appears to have no readily identifyable match in experimental situations. For this reason it has remained a major stumbling block in the attempts to interpret the observed pulsations in terms of these theoretical models.

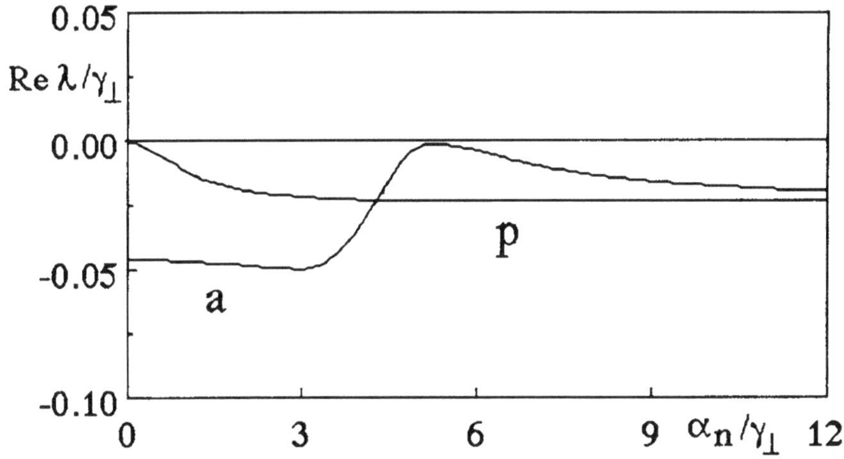

Fig. 2.5a The two largest real parts of the eigenvalues of the linearized equations are plotted as functions of α_n (in units of $\gamma_\perp$) viewed as a continuous variable for $\alpha L=0.8$, $R=0.95$, $\alpha_1=3\ \gamma_\perp$, $\gamma_\parallel/\gamma_\perp=1.5$ and $\delta_{AC}=0$. under the conditions of this simulation the steady state is stable. The line marked (a) denotes the amplitude eigenvalue; the line marked (p) denotes the phase eigenvalue.

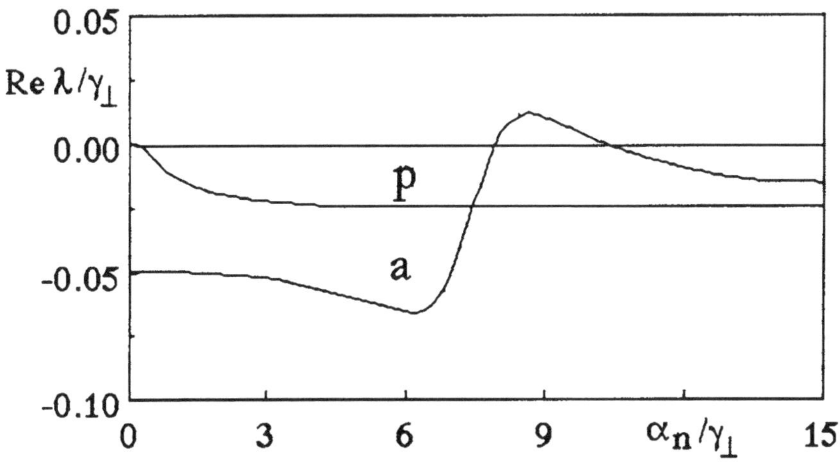

Fig. 2.5b Same as Figure 2.5a with $\alpha L=2.0$. For sufficiently high values of the gain, the real part of the amplitude eigenvalue becomes positive and the steady state becomes unstable by developing sidebands at $\alpha_{\pm 3}=\pm 3\alpha_1$.

We now turn our attention to a situation such as shown in Fig. 2.4b, where multiple steady states can exist for sufficiently large values of the detuning parameter. The linear stability analysis [i.e. the solution of the characteristic equation (2.39)] shows that the lasing mode retains control of the laser emission well into the region of coexistence of multiple steady states when the ratio $\gamma_{\|}/\gamma_{\perp}$ is sufficiently smaller than unity and that, instead, competition between coexisting steady states and self-pulsing develop when $\gamma_{\|}/\gamma_{\perp}$ is closer to unity. In both cases instabilities can emerge even at low gain values, but their origin is quite different from the mechanism described above.

In Figures 2.6 we show again the two largest real parts of the eigenvalues plotted as functions of the continuous frequency α_n. For small values of the detuning, the lasing mode is stable (Fig. 2.6a); for larger values of δ_{AC}, in the region where multiple steady states coexist, one of the real parts of the eigenvalues (the one corresponding to the cavity frequency α_1 in this figure) becomes positive, and the lasing mode loses stability. At this point the subsequent evolution of the system is determined by the stability properties of the coexisting stationary solution. If the coexisting solution is stable, the laser operation is transferred from the unstable to the stable state, with a discontinuous change of the asymptotic output intensity and operating frequency. If, on the other hand, the coexisting solution is itself unstable, undamped pulsations develop (the system cannot find stable fixed points).

A schematic illustration of the behavior of the intensity and operating frequency as a function of the detuning parameter is shown in Fig. 2.7 for the case in which the coexisting mode is stable. The self pulsing solutions corresponding to $\gamma_{\|}/\gamma_{\perp}$ of the order of unity tend to be simple periodic oscillations with some distortion due to the nonlinearity of the problem. An example is given in Ref. 16.

Direct inspection of the behavior of the linearized eigenvalues in Figs. 2.5b and 2.6b shows immediately that, while both kinds of unstable behaviors are of the multimode type, in the sense that they involve the running laser mode and at least a pair of sidebands, the type of eigenvalues whose real parts become positive is different in these two cases. In Fig. 2.5b the unstable eigenvalue is associated with the linear response of the field amplitude; in the case of Fig. 2.6b the unstable eigenvalue is connected with the field phase. For this reason, it makes sense to label the first type of behavior "amplitude instability" and the second "phase instability".

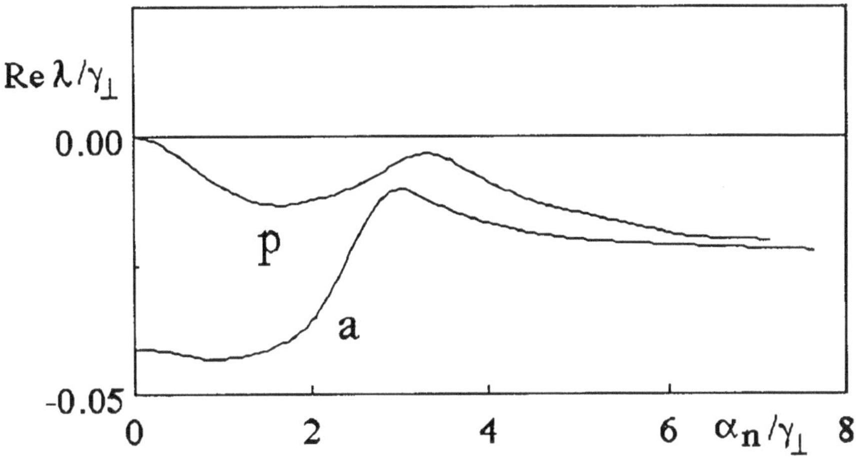

Fig. 2.6a For increasing values of the detuning parameter the phase eigenvalue eventually develops a positive real part. In this case the detuning is not sufficiently large to create an instability. The parameters are $\alpha L=0.5$, $R=0.95$, $\alpha_1=3\gamma_\perp$, $\gamma_\parallel/\gamma_\perp=0.8$, and $\delta_{AC}=0.7$.

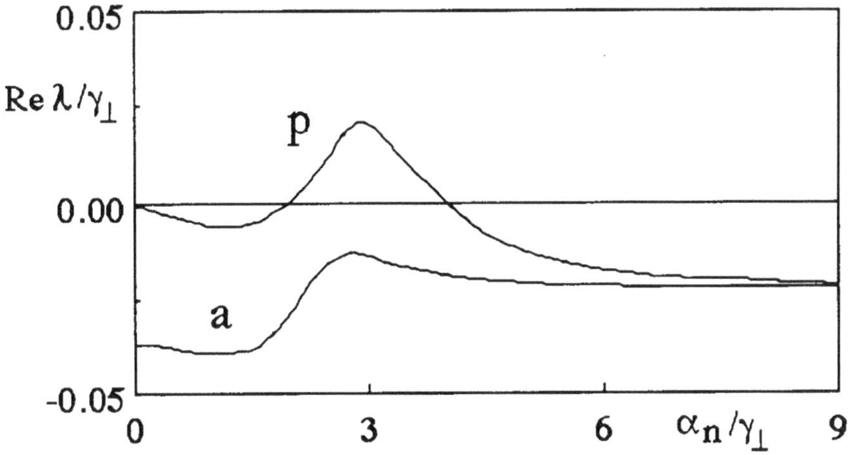

Fig. 2.6b Same as Fig. 2.6a with $\delta_{AC}=1.2$.

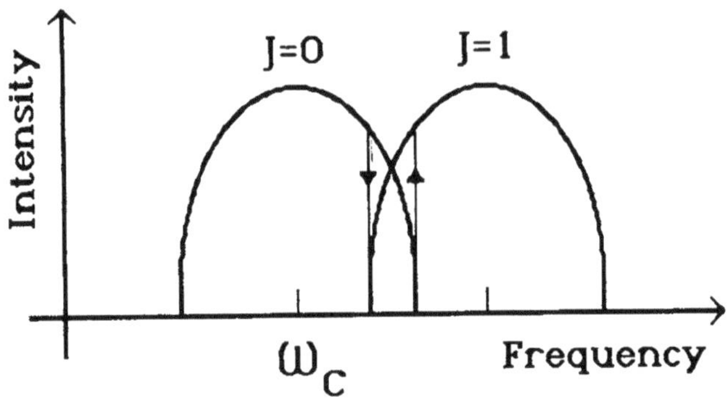

Fig. 2.7a Schematic behavior of the output intensity as a function of the detuning parameter.

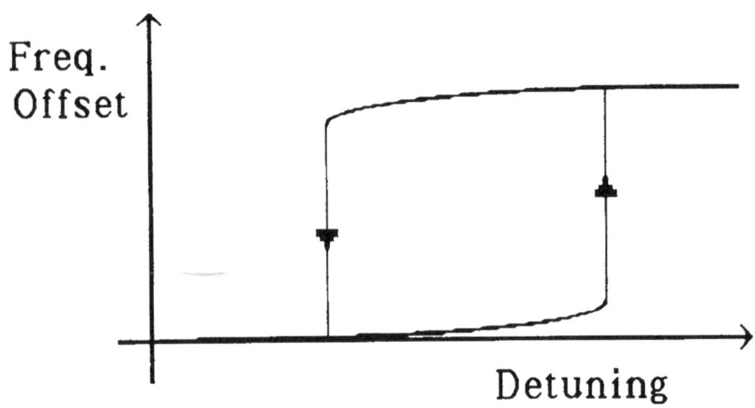

Fig. 2.7b Schematic behavior of the laser operating frequency as a function of the detuning parameter.

3. The single-mode laser equations

Historically, the single-mode laser model was the first to display unstable and very complex behaviors.[5-8] At that time the connection between the laser equations and the self-consistent Maxwell-Bloch description was not clear, but in view of our increased understanding of the foundations of laser theory we must raise the question of how the single-mode model follows from the exact Maxwell-Bloch equations.

Sufficient conditions for the validity of the single-mode model can be established with the help of the following steps. First, we impose the uniform field limit (2.20) and bring the modal equations for the field and atomic amplitudes into the form given by (2.30). Next, we require that all the modal amplitudes f_n, p_n, and d_n, with $n \neq 0$, vanish identically at $t' = 0$. Finally, we impose the limit

$$\frac{\Lambda\gamma_\perp}{c} \to 0 \tag{3.1}$$

whose effect is to move every cavity mode, except for the resonant one, so far away from the atomic gain line that any active role on their part is effectively excluded during the evolution.

Under these conditions the Maxwell-Bloch equations reduce to the single-mode model

$$\frac{df}{dt'} = -\kappa\,(f + 2Cp) \tag{3.2a}$$

$$\frac{dp}{dt'} = -\gamma_\perp (fd + p) \tag{3.2b}$$

$$\frac{dd}{dt'} = -\gamma_\parallel (-fp + d - 1) \tag{3.2c}$$

where, for simplicity, we have made the further assumption of resonance between the center of the atomic gain line and the only remaining cavity mode. The single-mode laser equations, also known as the Haken-Lorenz model, are unstable under the following two conditions[7,9]

$$\text{i)} \qquad \kappa > \gamma_\perp + \gamma_\parallel \qquad \text{(bad cavity condition)} \tag{3.3a}$$

$$\text{ii)} \qquad 2C > 1 + \frac{(\kappa + \gamma_\perp + \gamma_\parallel)(\kappa + \gamma_\perp)}{\kappa - \gamma_\perp - \gamma_\parallel} \qquad \text{(high gain condition)} \tag{3.3b}$$

The bad cavity condition is a constraint for the cavity linewidth relative to the sum of the atomic decay rates, and should not be interpreted literally as an indication that the cavity design is of poor optical quality (low finesse). It is true in fact, that for most optical lasers, the atomic linewidth $\gamma_\perp$ is so large, that Eq. (3.3a) can only be satisfied if the cavity has a very high damping rate (i.e. low reflectivity mirrors). However, for example, far infrared lasers can be characterized by very narrow atomic linewidths, so that the actual quality of the optical cavity does not have to be poor in absolute terms. A useful graphical display of Eqs (3.3b) is shown in Fig. 3.1 which displays the domain of instability of a single-mode laser in the space of the parameters 2C and $\kappa/\gamma_\perp$. For each value of the ratio $\gamma_\parallel/\gamma_\perp$, the solid curves represent the instability boundary of the laser. It is clear that, even under the most favorable conditions, the ratio $\kappa/\gamma_\perp$ must be sufficiently larger than unity, and that the gain must exceed the ordinary laser threshold value by about a factor of 10. Note that the limiting boundary corresponding to $\gamma_\parallel/\gamma_\perp \rightarrow 0$ is very close to curve (c) of Fig. 3.1, so that smaller values of $\gamma_\parallel/\gamma_\perp$ do not help significantly in lowering the instability threshold. Hence here also, as in the multimode case, instabilities are high gain phenomena.

It is interesting to summarize some of the dynamical aspects of the Haken-Lorenz model. For low values of the unsaturated gain, the laser operates stably and approaches its steady state monotonically; for higher values of the gain , the approach to steady state is characterized by damped relaxation oscillations. Eventually, above the instability threshold (often called the second laser threshold), undamped pulsations set in. Usually they take the form of erratic oscillations as shown in Fig. 3.2, but for selected values of the parameters, periodic behaviors can also develop[20]. These are common when $\gamma_\parallel/\gamma_\perp$ is smaller than about 0.2.

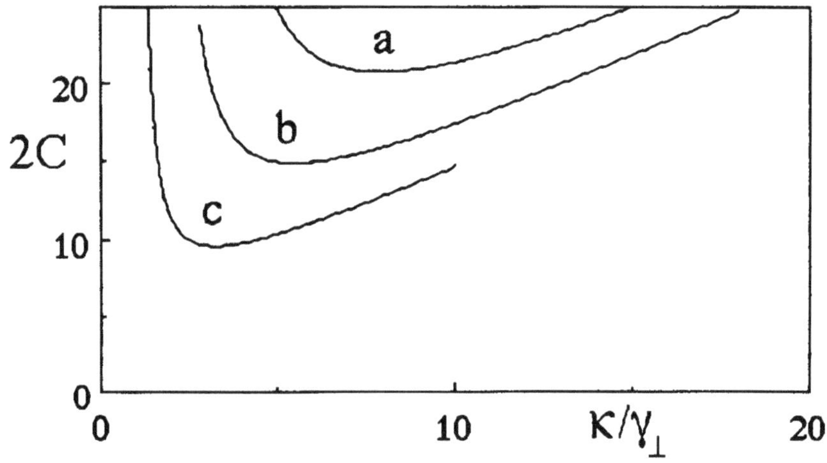

Fig. 3.1 Instability boundaries of the single-mode laser model are plotted in the plane of the parameters $\kappa/\gamma_\perp$ and 2C for (a) $\gamma_\parallel/\gamma_\perp$=2.0, (b) $\gamma_\parallel/\gamma_\perp$=1.0, (c) $\gamma_\parallel/\gamma_\perp$=0.1. The optimum (most unstable) configuration corresponds to $\kappa/\gamma_\perp$=3 and $2C_{thr} = 9$.

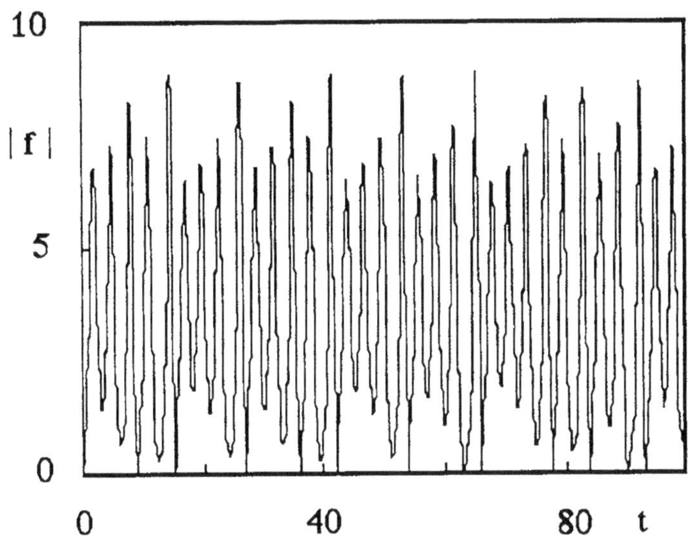

Fig. 3.2 Time evolution of the output intensity according to Eqs. (3.2). When 2C exceeds the second threshold value, the laser output often develops undamped chaotic pulsations. The parameters chosen in this simulation are 2C=15.0, $\gamma_{\parallel}/\gamma_{\perp}=0.5$, $\kappa/\gamma_{\perp}=4$.

In Fig. 3.3 we show a periodic solution of the Haken-Lorenz equations whose electric field amplitude executes symmetric oscillations in the positive and negative direction. This type of solution is called symmetric for reasons that are especially clear if one inspects the corresponding phase-space portrait (Fig. 3.3b). A small change of $\gamma_{\parallel}/\gamma_{\perp}$ brings about a symmetry breaking transition (Fig. 3.4), leading to unequal positive and negative excursions of the electric field, as shown more clearly by the phase portrait (Fig. 3.4b).

It is interesting to note that the intensity patterns corresponding to Figs. 3.3a and 3.4a would lead one to believe that a period doubling bifurcation has taken place. In fact this is not the case. Because the electric field amplitude is not directly observable, however, one must ask how a symmetry breaking transformation can be identified in a traditional measurement. The answer lies in the application of heterodyne mixing techniques with a stable reference source. Consider, in fact, the superposition of the signal of interest with a coherent reference wave. The total electric field at the detector takes the form

$$E_{tot}(t) = f(t)\, e^{-i\omega t} + A\, e^{-i\omega t} + c.c. \tag{3.4}$$

where A and ω_0 are the amplitude and the carrier frequency of the reference signal, respectively. The photocurrent produced by a standard square-law detector is given by

$$i(t) = \text{const} \times \left\{ |A|^2 + |f(t)|^2 + f(t)\, A^*\, e^{-i(\omega - \omega_0)t} + f(t)^* A\, e^{i(\omega - \omega_0)t} \right\} \tag{3.5}$$

The power spectrum of i(t) contains a delta-function contribution at zero frequency, due to the constant background intensity $|A|^2$, the homodyne power spectrum of $|f(t)|^2$ in the low frequency range, and finally the spectrum of the electric field f(t) centered at the frequency difference between the two optical carriers. The main difference between a symmetric and an asymmetric solution is that the former has a zero average value for the electric field while the latter does not. Hence, the heterodyne spectrum of a symmetric solution will display symmetric frequency components around $|\omega - \omega_0|$, but no spectral power at $|\omega - \omega_0|$, while the distinguishing signature of an asymmetric solution will be the appearance of a line at the frequency $|\omega - \omega_0|$, in addition to symmetric sidebands.

Further changes of $\gamma_\parallel/\gamma_\perp$ produce real bifurcations and eventually chaos. An example of a period doubled asymmetric solution is shown in Fig. 3.5. A period-5 window in a chaotic region is shown in Fig. 3.6.

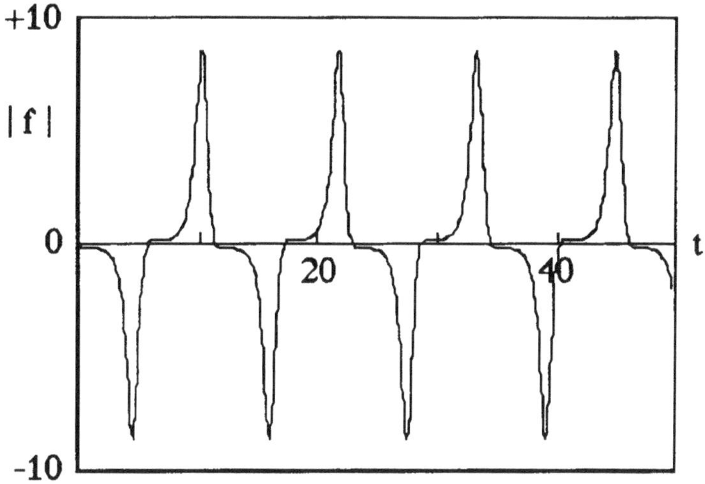

Fig. 3.3a Time dependence of the electric field amplitude for 2C=12.0, $\gamma_{||}/\gamma_{\perp}=0.14$, $\kappa/\gamma_{\perp}=4$. This is an example of a symmetric solution.

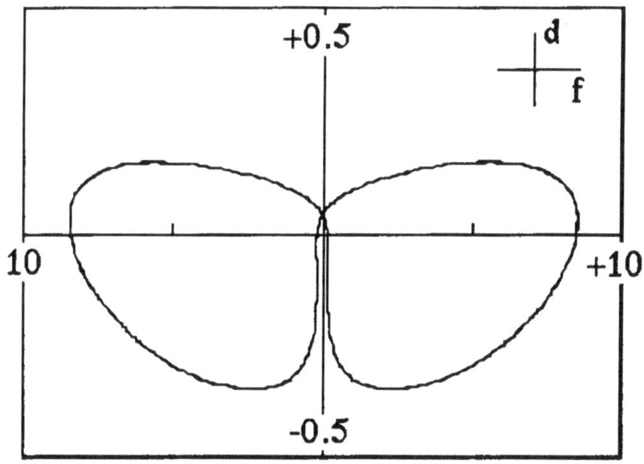

Fig. 3.3b Projection of the phase space trajectory shown in Fig. 3.3a in the (f,d) plane.

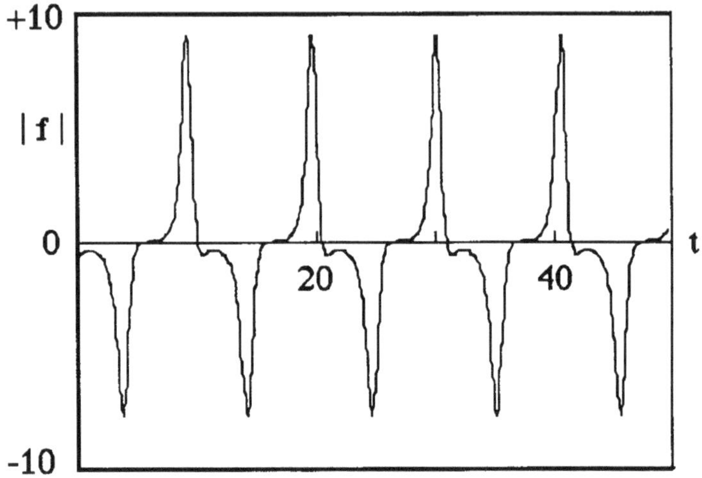

Fig. 3.4a Time dependence of the electric field amplitude for 2C=12.0, $\gamma_{\parallel}/\gamma_{\perp}=0.17$, $\kappa/\gamma_{\perp}=4$. This is an example of an asymmetric solution.

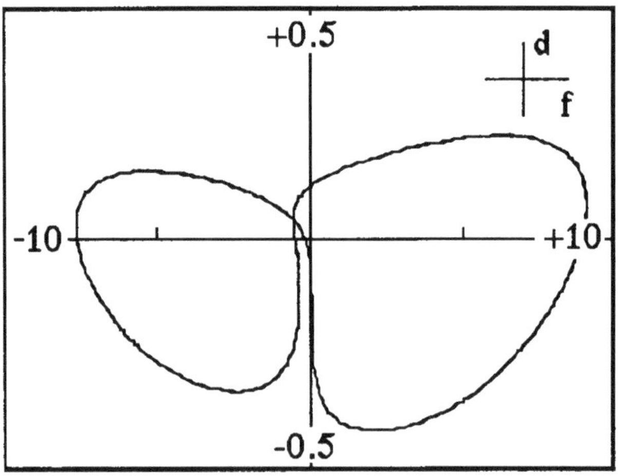

Fig. 3.4b Projection of the phase space trajectory shown in Fig. 3.4a in the (f,d) plane.

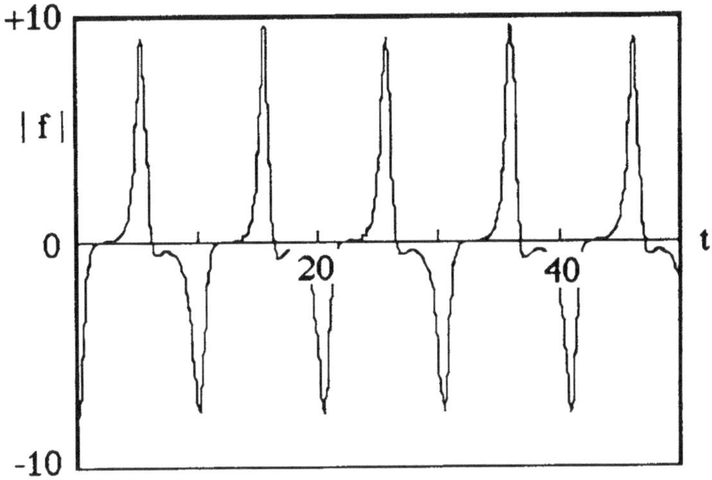

Fig. 3.5a Time dependence of the electric field amplitude for 2C=12.0, $\gamma_{||}/\gamma_{\perp}$=0.19, $\kappa/\gamma_{\perp}$=4. This is an example of a symmetric solution of period 2.

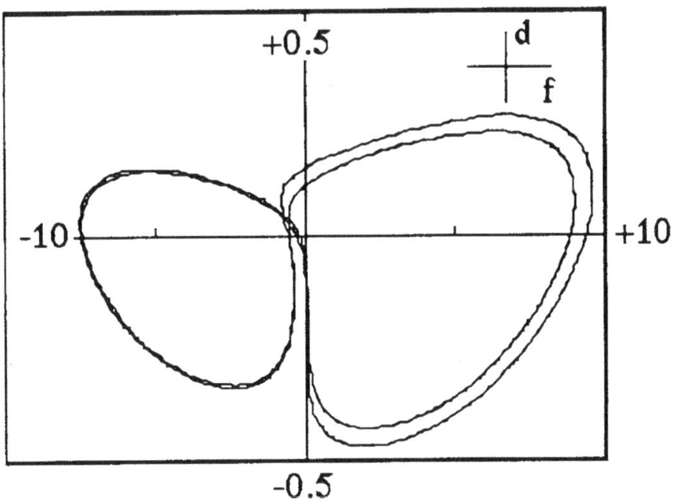

Fig. 3.5b Projection of the phase space trajectory shown in Fig. 3.5a in the (f,d) plane.

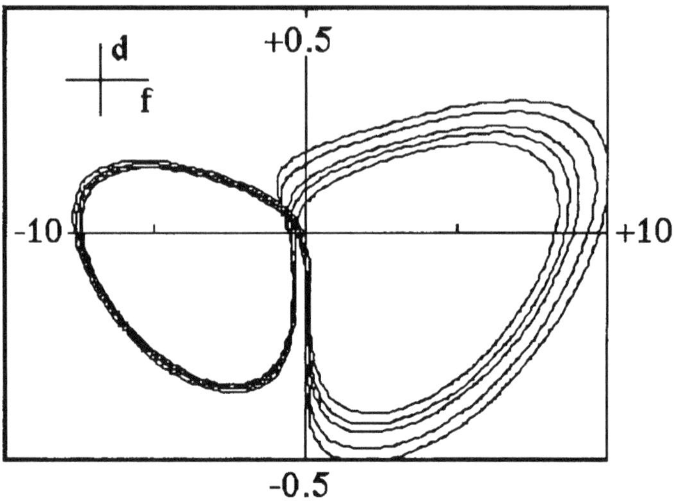

Fig. 3.6 Projection of the phase space trajectory of a period-5 solution in the (f,d) plane.

4. The inclusion of transverse effects

In the first three sections of these notes we have shown that the Maxwell-Bloch equations of the laser predict the existence of instabilities leading to both regular and chaotic pulsations, under single- or multimode operation, and in or out of resonance relative to the center of the atomic gain line. While the theoretical link between the observed laser pulsations and dynamical instabilities is very reasonable, the prediction that these phenomena can occur only for very large values of the pump parameter is a source of major difficulties as we try to interpret the experimental results.

An additional difficulty with the plane-wave Maxwell-Bloch equations is the unconditional stability of their rate equation limit. Accepting this result leads to the conclusion that all laser instabilities are manifestations of atomic coherence; this assertion is hardly in agreement with the behavior of ruby, Nd-YAG, among others, for which the validity of the rate equation description is well justified and which are quite prone to developing pulsed output.

Better agreement between theory and experiments is obtained when the laser undergoes a phase instability,[16] a phenomenon that requires operation under detuned conditions and which is responsible, for example, for the observed mode hopping in CO_2 laser systems.[21] However, the physical mechanism that drives a phase instability is quite different from that which is operative under resonant conditions (i.e. the amplitude instability[15]) so that the theory of the phase instabilities is not very helpful in clarifying the nature of the "high second laser threshold problem" under general conditions.

It is probably fair to say that the plane-wave approximation, long viewed as adequate in capturing the essential aspects of laser dynamics, may be a much stronger theoretical constraint than anticipated. In fact, important warning signals in this direction came from the lack of quantitative agreement between the predictions of the plane-wave studies of optical bistability[22] and the careful absolute measurements performed by Kimble and collaborators under steady state conditions.[23] If, however, these theoretical predictions were in disagreement with the experiments in a quantitative way, more serious qualitative discrepancies have emerged from the study of self-pulsing in bistable systems. Here the plane-wave theory predicts a rich variety of dynamical behaviors[24] which do not appear to have a counterpart among the observed patterns.[25] The combined message of a growing number of experimental[26] and theoretical[27] reports supports the belief that the inclusion of transverse effects is a necessary requirement for the successful description of a realistic laser system.

In this study we consider some physical and geometrical factors that are absent by definition from the plane-wave theory. To be specific, we focus on the effects of diffraction, on the wavefront curvature induced by the spherical mirrors and on the transverse and longitudinal gain variation caused by the pump mechanism. Of course, the mathematical description of this problem is considerably more involved that in the case of a plane-wave model and significant theoretical progress in general requires large-scale numerical calculations. An unavoidable drawback of numerical approaches is that a global understanding necessitates the analysis of wide regions of parameter space. For this reason it is very desirable to proceed in our investigations of transverse effects by analytical means as much as feasible. As expected, this cannot be done for arbitrary configurations of the laser system; fortunately, we can identify non trivial settings that allow significant theoretical progress without excessive numerical labor. Our approach will be based on the natural extension of the uniform field limit. Among the important consequences of this approach is the existence of steady state solutions of the single-mode type, a conclusion which also implies that the empty cavity modes are also exact modes of the filled cavity (in the uniform field limit).[13]

In these lectures we focus mainly on spatial symmetry breaking effects, phenomena that are quite relevant also to current studies of hydrodynamics, chemical reactions and developmental biology. For this purpose, we discuss in some details the empty cavity modal structure, the Maxwell-Bloch equations with the inclusion of transverse effects and a few selected solutions.

4.1 The empty cavity modes

This subsection contains a brief description of the calculations leading to the modal eigenfunctions and eigenfrequencies of the empty resonator in the paraxial approximation. For the sake of definitness we consider an empty ring resonator with the form sketched in Fig. 4.1. The length of the resonator is Λ, and the two spherical mirrors have the same radii of curvature R_0 and are spaced a distance L. In order to keep our calculations as simple as possible we consider first the case in which the cavity field has cylindrical symmetry, i.e. the cavity field depends only on the radial coordinate r, the longitudinal coordinate z and the time.

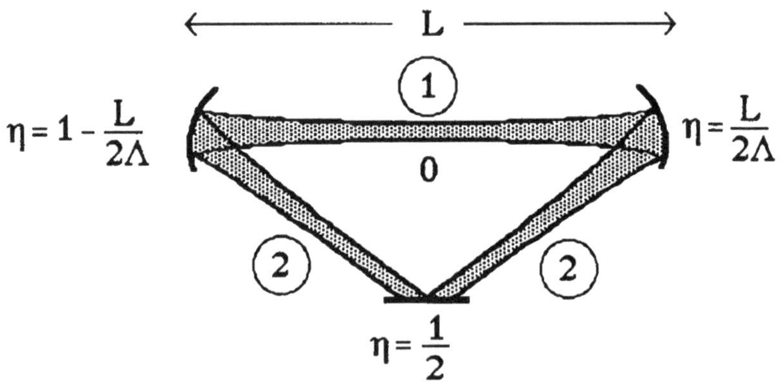

Fig. 4.1 Schematic representation of the empty cavity with two spherical mirrors of radius of curvature R_0. The length of the ring is Λ, while the distance between the two curved reflectors is L. The numeral 0 indicates the origin of the longitudinal coordinate system and η measures various positions of relevance to our discussion in units of Λ.

The cavity field obeys the wave equation

$$\nabla_2 E(r,z,t) - \frac{1}{c^2}\frac{\partial^2}{\partial t^2}E(r,z,t) = 0 \qquad (4.1)$$

where

$$\nabla_2 \equiv \frac{\partial^2}{\partial r^2} + \frac{1}{r}\frac{\partial}{\partial r} + \frac{\partial^2}{\partial z^2} \qquad (4.2)$$

and r is the radial coordinate along a direction transverse to the axis of the resonator. In the paraxial approximation we seek elementary solutions of the form

$$E(r,z,t) = A(r,z)\exp[i(kz - \omega t)] \qquad (4.3)$$

where A(r,z), a slowly varying function of z, is a solution of the equation

$$\left(\frac{\partial^2}{\partial r^2} + \frac{1}{r}\frac{\partial}{\partial r}\right)A(r,z) + 2ik\frac{\partial}{\partial z}A(r,z) = 0 \qquad (4.4)$$

and $k = \omega/c$. In terms of the scaled coordinates

$$\eta = \frac{z}{\Lambda}, \qquad \rho = \sqrt{\frac{\pi}{\Lambda\lambda}}\, r \qquad (4.5)$$

we have

$$\frac{\partial A}{\partial \eta} = \frac{i}{4}\left(\frac{\partial^2}{\partial \rho^2} + \frac{1}{\rho}\frac{\partial}{\partial \rho}\right)A \qquad (4.6)$$

We divide the ring resonator in successive longitudinal segments bounded by pairs of curved mirrors.[29] In the example of Fig. 4.1 there are two such segments labelled 1 and 2 in the figure (If the plane mirror were replaced by a curved reflector, the cavity would have to be divided into three segments). The solution of Eq. (4.6) over the range $-L/2\Lambda \le \eta \le +L/2\Lambda$ is given by

$$A^{(1)}(\rho,\eta) = \frac{2}{v(\eta)} L_p\left(\frac{2\rho^2}{v^2(\eta)}\right)\exp\left(-\frac{\rho^2}{v^2(\eta)} + i\frac{\rho^2}{u(\eta)}\right)\exp\left(-i(2p+1)\tan^{-1}\eta/\eta_1\right) \qquad (4.7)$$

where L_p (p = 0, 1, 2,...) is the Laguerre polynomial of order p and the indicated argument,

$$v(\eta) = \sqrt{\eta_1}\left[1 + (\eta/\eta_1)^2\right]^{1/2} \qquad (4.8a)$$

represents the beam waist at position η, and

$$u(\eta) = \frac{1}{\eta}(\eta_1^2 + \eta^2) \tag{4.8b}$$

measures the scaled radius of curvature of the wavefront. The parameter $\eta_1$is arbitrary at this point and must be specified with the help of the cavity boundary conditions; in the chosen system of units it plays the role of the Rayleigh length of the beam (i.e. the distance between the two transverse planes where the beam waist is larger than the minimum size by a factor $\sqrt{2}$), while

$$\sqrt{\eta_1} = v(0)$$

is the minimum waist size.

The solution of Eq. (4.6) over the second segment of the resonator has the same formal structure as Eq. (4.7) except for the value of the minimum waist and for an extra phase factor which is needed to join the solutions smoothly with one another. Thus in region 2 we have

$$A^{(2)}(\rho,\eta) = \frac{2}{\alpha(\eta)} L_p\left(\frac{2\rho^2}{\alpha^2(\eta)}\right) \exp\left(-\frac{\rho^2}{\alpha^2(\eta)} + i\frac{\rho^2}{\beta(\eta)}\right) \times$$

$$\exp\left[-i(2p+1)\tan^{-1}\frac{\eta - \Lambda/2L}{\eta_2}\right] e^{i\varphi} \tag{4.9}$$

where

$$\alpha(\eta) = \eta_2\left[1 + \left(\frac{\eta - 1/2}{\eta_2}\right)^2\right]^{1/2} \tag{4.10a}$$

$$\beta(\eta) = \frac{1}{\eta - 1/2}\left[(\eta - 1/2)^2 + \eta_2^2\right] \tag{4.10b}$$

and φ is a phase factor that must be adjusted to fit the boundary conditions. The shift of the longitudinal coordinate is made in conformity with the chosen origin of the η axis.

The Gauss-Laguerre solutions, as they are called, are characterized by the beam parameter $q(\eta)$ defined by[30]

$$\frac{1}{q(\eta)} = \frac{1}{u(\eta)} + i\frac{1}{v^2(\eta)} \tag{4.11}$$

The beam parameter satisfies elementary mapping rules under translation and reflection from a spherical mirror. If q_1 is the initial beam parameter, we have

i) $$q_2 = q_1 + d \qquad \text{(translation by a scaled length d)} \tag{4.12a}$$

and

ii) $$q_2 = \frac{q_1}{-\frac{2}{\rho_0} q_1 + 1} \qquad \text{(reflection at a curved surface with a scaled radius of curvature } \rho_0 = \frac{R_0}{\Lambda}\text{)} \tag{4.12b}$$

With the help of these rules we can follow the variations of the Gauss-Laguerre beam at every point along the resonator.

In order to complete the solution of our problem we must fix the parameters η_1 and η_2 in terms of the cavity geometry, we must calculate the phase factor φ and derive an expression for the empty cavity eigenfrequencies.

The calculation of η_1 can be carried out as follows: we start at η=0 with a certain beam parameter $q(\eta=0)$, we propagate this value and reflect it as many times as needed to complete a loop. At the end we require that the final beam parameter $q(\eta=1)$ be identical to $q(\eta=0)$. We let

$$f = \frac{L}{\Lambda}$$

and we have

$$q(\eta=0) = -iv^2(0) = -i\eta_1 \tag{4.13a}$$

$$q(\eta = \tfrac{1}{2} f - \varepsilon) = q(\eta = 0) + \tfrac{1}{2} f \tag{4.13b}$$

$$q(\eta = \tfrac{1}{2} f + \varepsilon) = \frac{q(\eta = \frac{1}{2} f - \varepsilon)}{-\frac{2}{\rho_0} q(\eta = \frac{1}{2} f - \varepsilon) + 1} \tag{4.13c}$$

etc..

$$q(\eta = 1) = q(\eta = 1 - \tfrac{1}{2} f + \varepsilon) + \tfrac{1}{2} f \equiv q(\eta = 0) \tag{4.13d}$$

where $-\varepsilon$ and $+\varepsilon$ indicate positions just before and just after reflection by the spherical mirrors, respectively.

The solution of this simple set of equations yields

$$\eta_1 = \frac{1}{2}\sqrt{\frac{(\rho_0 - f)(\rho_0 + f^2 - f)}{\rho_0 + f - 1}} \tag{4.14}$$

subject to the constraint $\rho_0 + f - 1 > 0$. In a similar way we can propagate the initial beam parameter (which is now known) up to point $\eta = 1/2$ of Fig. 4.1 and arrive at

$$\eta_2 = \frac{1}{\rho_0^2 \eta_1}\left\{\eta_0^2(1 - f - \rho_0)^2 + \frac{1}{4}(\rho_0 + f^2 - f)^2\right\} \tag{4.15}$$

Now we must impose the boundary conditions at each curved mirror (reflections at plane surfaces are assumed to leave the beam invariant). In the paraxial approximation of wave optics the reflection at a spherical mirror is approximately equivalent to a multiplication by the phase factor

$$\Re = \exp(-i\frac{2\rho^2}{\rho_0}) \tag{4.16}$$

Thus we impose that

$$\Re\, E^{(1)}(\eta = \frac{1}{2}f - \varepsilon, \rho, t) = E^{(2)}(\eta = \frac{1}{2}f + \varepsilon, \rho, t) \tag{4.17}$$

We observe that

$$v^2(\eta = \frac{1}{2}f - \varepsilon) = \alpha^2(\eta = \frac{1}{2}f + \varepsilon) \tag{4.18a}$$

$$-\frac{2}{\rho_0} + \frac{1}{u(\eta = \frac{1}{2}f - \varepsilon)} = \frac{1}{\beta(\eta = \frac{1}{2}f + \varepsilon)} \tag{4.18b}$$

With the help of these relations it is easy to verify that the first boundary condition (4.17) is satisfied if

$$\varphi = -(2p+1)\left[\tan^{-1}\frac{f}{2\eta_1} + \tan^{-1}(\frac{1-f}{2\eta_2})\right] \tag{4.19}$$

Now we impose the boundary condition at the second curved mirror, i.e.

$$\Re\, E^{(2)}(\eta = 1 - \frac{1}{2}f - \varepsilon, \rho, t) = E^{(1)}(\eta = -\frac{1}{2}f + \varepsilon, \rho, t) \tag{4.20}$$

and arrive at the result

$$\omega_{n,p} = \frac{2\pi c}{\Lambda}n + \frac{2c}{\Lambda}(2p+1)\left[\tan^{-1}(\frac{1-f}{2\eta_2}) + \tan^{-1}\frac{f}{2\eta_1}\right] \tag{4.21}$$

$$n, p = 0, 1, 2, \ldots$$

In the following development we are especially interested in situations where the longitudinal mode spacing, $2\pi c/\Lambda$, is considerably larger than the frequency separation between transverse modes. This is true if $\eta_1 >> 1$, which in turn implies large radii of curvature for the spherical mirrors. The modal structure in this case takes the form sketched in Fig. 4.2.

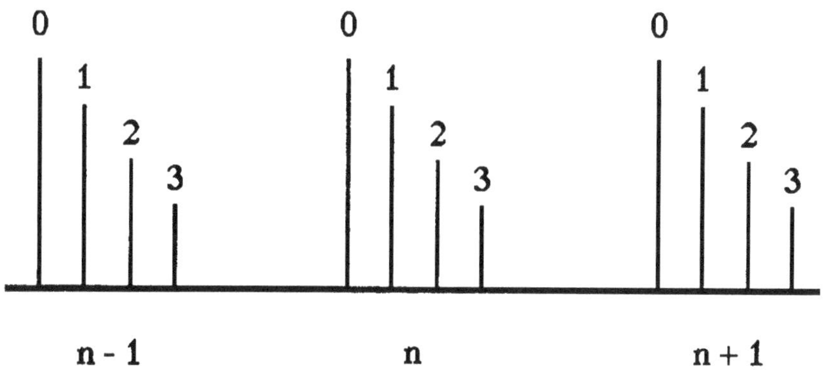

Fig. 4.2 Schematic representation of three consecutive longitudinal groups of modes; each set is labelled by the radial indices p=0,1,2,...

The inclusion of the angular dependence requires only a few modifications. The transverse Laplacian is now given by

$$\nabla_\perp^2 = \frac{\partial^2}{\partial\rho^2} + \frac{1}{\rho}\frac{\partial}{\partial\rho} + \frac{1}{\rho^2}\frac{\partial^2}{\partial\varphi^2} \tag{4.22}$$

and the empty cavity eigenfunctions have the form

$$A_{pm}(\rho,\varphi,\eta) = B_m(\varphi)\, C_{pm}(\rho,\eta)\, \exp[i\theta_{pm}(\rho,\eta)] \tag{4.23}$$

where

$$B_m(\varphi) = \begin{cases} \frac{1}{\sqrt{2\pi}} & m = 0 \\ \frac{1}{\sqrt{\pi}} \sin m\varphi & m > 0 \\ \frac{1}{\sqrt{\pi}} \cos m\varphi & m > 0 \end{cases} \tag{4.24a}$$

$$C_{pm}(\rho,\eta) = \frac{2}{v(\eta)} \left(\frac{2\rho^2}{v^2(\eta)}\right)^{m/2} \sqrt{\frac{p!}{(p+m)!}}\, L_p^m\left(\frac{2\rho^2}{v^2(\eta)}\right) \exp\left(-\frac{\rho^2}{v^2(\eta)}\right) \tag{4.24b}$$

and

$$\theta_{pm}(\rho,\eta) = \frac{\rho^2}{u(\eta)} - (2p+m+1) \tan^{-1}\eta/\eta_1 \tag{4.24c}$$

$$p = 0,1,2,\ldots; \qquad m = 0,1,2,\ldots.$$

The symbol $L_p{}^m$ indicates the associated Laguerre polynomial of order p and index m, and $v(\eta)$ and $u(\eta)$ are defined by Eqs. (4.8a) and (4.8b), respectively. The functions A_{pm} form a complete orthonormal set and obey the orthogonality relation

$$\int_0^{2\pi} d\varphi \int_0^{\infty} d\rho\, \rho\, A_{pm}^{(i)\,*}(\rho,\varphi,\eta)\, A_{p'm'}^{(i')}(\rho,\varphi,\eta) = \delta_{pp'}\delta_{mm'}\delta_{ii'} \tag{4.25}$$

The superscript i (i=1,2) distinguishes the solutions of type sin(mφ) from those of type cos(mφ), if $m > 0$ (for $m = 0$ the index i must be omitted). The eigenfrequencies of the ring resonator are given by

$$\omega_{n,p,m} = \frac{2\pi c}{\Lambda} n + \frac{2c}{\Lambda}(2p+m+1)\left[\tan^{-1}\frac{1-f}{2\eta_2} + \tan^{-1}\frac{f}{2\eta_1}\right] \tag{4.26}$$

where

$$n, p, m = 0, 1, 2, \ldots$$

For a resonator with a different mirror configuration from the one sketched in Fig. 4.1, Eq. (4.26) retains the same structure except for the geometry-dependent coefficient in the square brackets.

4.2 The modal equations

The generalized Maxwell-Bloch equations differ from their plane-wave counterpart for the inclusion of a transverse Laplacian which is responsible for diffractive effects. Here we consider a situation where both radial and angular variations may be important but the laser operates in resonance with the center of the atomic line.

The generalized Maxwell-Bloch equations are

$$-\frac{i}{4}\nabla_\perp^2 F + \frac{\partial F}{\partial \eta} + \frac{1}{v}\frac{\partial F}{\partial \tau} = -\alpha\Lambda P \tag{4.27a}$$

$$\frac{\partial P}{\partial \tau} = -(FD + P) \tag{4.27b}$$

$$\frac{\partial D}{\partial \tau} = -\gamma\left\{-\frac{1}{2}(F^*P + FP^*) + D - \chi\right\} \tag{4.27c}$$

where

$$\tau = \gamma_\perp t, \qquad v = \frac{c}{\Lambda\gamma_\perp} \tag{4.28}$$

and χ is the equilibrium population difference created by the pump. The laser cavity is shown schematically in Fig. 4.3.

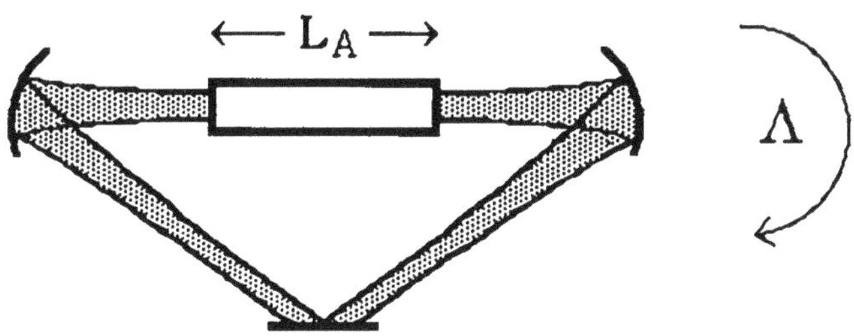

Fig. 4.3 Schematic representation of the laser acvity. The active medium is confined to the rectangular region of length L_A.

For the sake of simplicity we assume that the Rayleigh range η_1 is much larger than the length of the active medium so that the beam waist is essentially constant over the range occupied by the active medium. In this case the modal functions in the region of interest take the simpler form

$$\bar{A}_{p,0} = \frac{1}{\sqrt{2\pi}} 2\, L_p^0(2\rho'^2) \exp(-\rho'^2) \tag{4.29a}$$

$$\bar{A}_{pm}^{(1)} = \frac{1}{\sqrt{\pi}} \sin(m\varphi)\, 2\, (2\rho'^2)^{m/2} \sqrt{\frac{p!}{(p+m)!}}\, L_p^m(2\rho'^2) \exp(-\rho'^2) \tag{4.29b}$$

$$\bar{A}_{pm}^{(2)} = \frac{1}{\sqrt{\pi}} \cos(m\varphi)\, 2\, (2\rho'^2)^{m/2} \sqrt{\frac{p!}{(p+m)!}}\, L_p^m(2\rho'^2) \exp(-\rho'^2) \tag{4.29c}$$

where we have defined

$$\rho' = \frac{\rho}{\sqrt{\eta_1}} \tag{4.30a}$$

and

$$\bar{A}_{pm}^{(i)}(\rho',\varphi) = \sqrt{\eta_1}\; A_{pm}^{(i)}(\rho',\varphi) \qquad (i = 1,2 \text{ for } m>0;\text{ no index for } m=0) \tag{4.30b}$$

We expand the cavity field according to the Fourier sum

$$F(\rho',\varphi,\eta,\tau) = \sum_{pmi} \bar{A}_{pm}^{(i)}(\rho',\varphi,\eta)\, g_{pm}^{(i)}(\eta,\tau) \tag{4.31}$$

and with the help of Eqs. (4.27a) and the properties of the modal functions we obtain the following equations for the expansion coefficients

$$\frac{\partial g_{pm}^{(i)}}{\partial \eta} + \frac{1}{v}\frac{\partial g_{pm}^{(i)}}{\partial \tau} = -\alpha\Lambda \int_0^{2\pi} d\varphi \int_0^{\infty} d\rho\, \rho\, A_{pm}^{(i)*}\, P \tag{4.32}$$

These must be solved with the boundary conditions (See Ref. 13, Appendix A)

$$g_{pm}^{(i)}\left(-\frac{1}{2}\frac{L_A}{\Lambda}, \tau\right) = R\, e^{-i\delta_{pm}}\, g_{pm}^{(i)}\left(\frac{1}{2}\frac{L_A}{\Lambda}, \tau - \gamma_\perp \frac{\Lambda - L_A}{c}\right) \tag{4.33}$$

where

$$\delta_{pm} = (\omega_{pm} - \omega_{00})\,\frac{\Lambda}{c} \tag{4.34}$$

and R denotes the power reflectivity of each spherical mirror (the plane mirror is assumed to be a perfect reflector, for simplicity).

Our immediate goal is to derive a set of equations for the coupled field-atom system that is amenable to the uniform field approximation, as we have already done in dealing with the plane-wave limit of the Maxwell-Bloch equations. In addition we want to limit our considerations to the full adiabatic limit (i.e. $\partial P/\partial\tau = 0$, $\partial D/\partial\tau = 0$) This procedure is straightforward but rather lengthy. Here we consider the resulting modal equations for a set of frequency degenerate modes, i.e. modes for which

$$2p + m \equiv 2q = \text{const.} \tag{4.35}$$

We are interested in the evolution of this collection of modes assuming that thei common frequency coincides with the center of the atomic line and that all the others are sufficiently removed that their interaction with the gain curve can be neglected. We will assume that, at the beginning of the evolution, the laser operates in the cylindrically symmetric steady state configuration (p = q, m= 0). In the uniform field limit the complex field amplitudes obey the coupled equations

$$\frac{df_{pm}^{(i)}}{d\tau'} = -\kappa\left\{f_{pm}^{(i)} - 2C\int_0^{2\pi} d\varphi \int_0^{\infty} d\rho\, \rho\, \chi(\rho)\, \bar{A}_{pm}^{(i)*} \frac{F}{1+|F|^2}\right\} \tag{4.35}$$

where

$$F = \sqrt{\frac{\pi}{2}} \sum_{pmi} \bar{A}_{pm}^{(i)} f_{pm}^{(i)} \tag{4.36a}$$

$$\chi(\rho) = \exp\left[-\frac{2\rho^2}{\psi^2}\right] \tag{4.36b}$$

and ψ is a measure of the transverse extension of the inverted medium.

Initially the laser is in a steady state characterized by the modal amplitude $f_{q,0}$ which is the solution of the state equation

$$1 = 2C\int_0^{2\pi} d\varphi \int_0^{\infty} d\rho\, \rho\, \chi(\rho) \frac{\bar{A}_{q,0}^2}{1 + \frac{\pi}{2} \bar{A}_{q,0}^2 |f_{q,0}|^2} \tag{4.37}$$

The behavior of the system for $\tau' > 0$ can be analyzed by a standard linear stability analysis. For this purpose we seek solutions of Eq. (4.35) of the form

$$f_{pm}^{(i)} = f_{q0}^{(i)} \delta_{p,q} \delta_{m,0} + \delta f_{pm}^{(i)} \tag{4.38}$$

where $\delta f_{pm}^{(i)}$ is a small perturbation, and after some simple steps we obtain the linearized equations

$$\frac{\partial}{\partial\xi}\delta f_{pm}^{(i)} = -\delta f_{pm}^{(i)} + \Gamma_1\,\delta f_{pm}^{(i)} - \Gamma_2\,\delta f_{pm}^{(i)*} \tag{4.39a}$$

$$\frac{\partial}{\partial\xi}\delta f_{pm}^{(i)*} = -\delta f_{pm}^{(i)*} + \Gamma_1\,\delta f_{pm}^{(i)*} - \Gamma_2\,\delta f_{pm}^{(i)} \tag{4.39b}$$

where $\xi = \kappa\tau'$ and

$$\Gamma_1 = 2C\int_0^{2\pi} d\varphi \int_0^{\infty} d\rho\;\rho\,\chi(\rho)\,\frac{\bar{A}_{pm}^{(i)\,2}}{\left(1+\frac{\pi}{2}\bar{A}_{q0}^{2}f_{q0}^{2}\right)^2} \tag{4.40a}$$

$$\Gamma_2 = 2C\int_0^{2\pi} d\varphi \int_0^{\infty} d\rho\;\rho\,\chi(\rho)\,\frac{\pi}{2}\frac{\bar{A}_{pm}^{(i)\,2}\,\bar{A}_{q0}^{2}\,f_{q0}^{2}}{\left(1+\frac{\pi}{2}\bar{A}_{q0}^{2}f_{q0}^{2}\right)^2} \tag{4.40b}$$

(Note that the integrals (4.40) take the same value for i = 1 and i = 2; note also that we have assumed f_{q0} to be real without loss of generality).

The linear response of the system to an initial perturbation is described by the eigenvalues

$$\lambda_{pm}^{(1)} = -1 + \Gamma_1 + \Gamma_2 \tag{4.41a}$$

$$\lambda_{pm}^{(2)} = -1 + \Gamma_1 - \Gamma_2 \tag{4.41b}$$

Because

$$\lambda_{pm}^{(1)} - \lambda_{pm}^{(2)} = 2\Gamma_2 > 0 \tag{4.42}$$

it follows that the eigenvalue $\lambda_{pm}^{(1)}$ is the first to become positive (these eigenvalues are real) and therefore is responsible for the emergence of possible instabilities. It is clear by inspection of the quantity $\Gamma_1+\Gamma_2$ that, for sufficiently large values of the gain parameter 2C, the eigenvalue $\lambda_{pm}^{(1)}$ can become positive. In fact, numerical simulations show that the threshold for this instability can be only a few percent higher than the ordinary laser threshold.[28] Above this instability threshold the initial state (p=q, m=0) becomes unstable and other modal amplitudes with p≠q, m≠0 begin to grow. Eventually, after a certain transient, the initial field (see Fig. 4.4a) approaches a new multimode stable state whose transverse intensity profile is shown in Fig. 4.4b. This is an interesting example of a Turing instability[31], i.e. an instability that leads to a broken spatial symmetry.

It is interesting to observe that the stable asymmetric stationary state of Fig. 4.4b can be destabilized in turn by the injection into the laser of a coherent external signal of sufficient strength and a transverse shape which is matched to the symmetric (q,0) configuration. When this is done the new stationary field output reacquires the symmetric form shown by Fig. 4.4a. We note that the restoration of the original cylindrical symmetry is a threshold effect that requires a finite value of the injected signal strength.

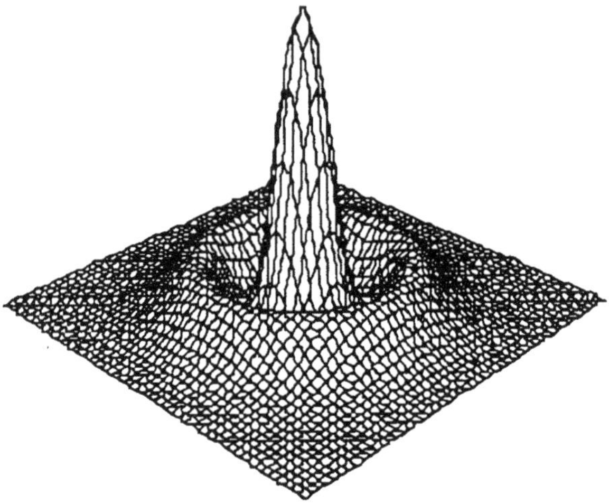

Fig. 4.4a The transverse profile of the output intensity for $\psi = 1.5$ shows the unstable single-mode stationary solution for $2C/(2C)_{thr} = 1.5$

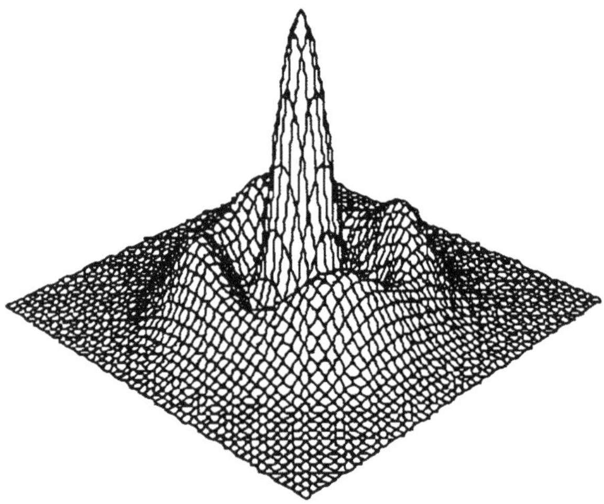

Fig. 4.4b Same as Fig. 4.4a at the end of the time evolution. The cylindrical symmetry has been broken

References

1. O. Svelto, Principles of Lasers, Plenum Press, New York, 1982 (Preface).
2. For an extensive review on the subject of laser dynamics see N.B. Abraham, P. Mandel and L.M. Narducci, in Progress in Optics, Vol XXV, edited by E. Wolf, North Holland, 1988, p. 3
3. See, for example, G. Makhov, C. Kikuchi, J. Lambe and R.W. Terhune, Phys. Rev. 109, 1399 (1958).
4. For some early studies see: (a) R.J. Collins, D.F. Nelson, A.L. Schawlow, W. Bond, C.G.B. Garrett and W. Kaiser, Phys. Rev. Lett 5, 303 (1960); (b) D.F. Nelson and W.S. Boyle, Appl. Opt. 1, 181 (1962); (c) T.N. Zubarev and A.K. Sokolov, Sov. Phys. Doklady 9, 1006 (1965).
5. (a) A.V. Uspensky, Radio Eng. Electron. Phys. (USSR) 8, 1145 (1963); (b) A.V. Uspensky, Radio Eng. Electron. Phys. (USSR) 9, 605 (1964); (c) V.V. Korobkin and A.V. Uspensky, Spv. Phys. JETP 18, 693 (1964).
6. V.M. Fain and Ya J. Khanin, Sov. Phys. JETP 14, 1069 (1962).
7. (a) A.Z. Grazyuk and A.N. Oraevsky, in Quantum Electronics and Coherent Light, P.A. Miles, ed., Academic Press, New York, 1964; (b) A.Z. Grazyuk and A.N. Oraevsky, Radio Eng. Electron. Phys. (USSR) 9, 424 (1964).
8. (a) H. Haken, Z. Phys. 190, 327 (1966); (b) H. Risken, C. Schmid and W. Weidlich, Z. Phys. 194, 337 (1966).
9. H. Haken, Phys. Lett. 53A, 77 (1975).
10. E. Lorenz, J. Atm. Sci. 20, 130 (1963).
11. For a modern review of the laser rate equations, see Ref. 1, chapter 5.
12. This philosphy was developed during the very early days of the Laser. See, for example, (a) H. Haken and H. Sauermann, Z. Phys. 173, 261 (1963); (b) A.N. Oraevsky, Molecular Oscillators, Nauka, Moskow, 1964; (c) W.E. Lamb, Jr., Phys. Rev. 134, 1429 (1964); (d) F.T. Arecchi and R. Bonifacio, IEEE J. Quantum Electron. QE-1, 169 (1965); (e) Ya.I. Khanin, Dynamics of Quantum Oscillators, Soviet Radio, Moskow, 1975.
13. L.A. Lugiato, F. Prati, L.M. Narducci, P. Ru, J.R. Tredicce and D.K. Bandy, Phys. Rev. A37, 3847 (1988).
14. L.A. Lugiato, in Progress in Optics, Vol. XXI, E. Wolf, ed., North Holland, Amsterdam, 1984, p. 69.

15. L.A. Lugiato, L.M. Narducci, E.V. Eschenazi, D.K. Bandy and N.B. Abraham, Phys. Rev. A32, 1563 (1985).
16. L.M. Narducci, J.R. Tredicce, L.A. Lugiato, N.B. Abraham and D.K. Bandy, Phys. Rev. A33, 1842 (1986).
17. (a) L.A. Lugiato, L.M. Narducci, and M.F. Squicciarini, Phys. Rev. A34, 3101 (1986); (b) R.R. Snapp, Ph.D. dissertation, University of Texas at Austin, 1986 (unpublished).
18. (a) R. Bonifacio and L.A. Lugiato, Lett. Nuovo Cimento, 21, 505 (1978); (b) R. Bonifacio and L.A. Lugiato, Lett. Nuovo Cimento, 21, 517 (1978).
19. (a) L.A. Lugiato, Opt. Comm. 33, 108 (1980); (b) L.A. Lugiato, Z. Phys. B41, 85 (1980).
20. L.M. Narducci, H. Sadiky, L.A. Lugiato, and N.B. Abraham, Opt. Comm. 55, 370 (1985).
21. J.R. Tredicce, L.M. Narducci, D.K. Bandy, L.A. Lugiato and N.B. Abraham, Opt. Comm. 56, 435 (1986).
22. (a) R. Bonifacio and L.A. Lugiato, Opt. Comm. 19, 172 (1976); (b) L.A. Lugiato, in Progress in Optics, Vol XXI, edited by E. Wolf, North Holland, Amsterdam, 1984, p. 69.
23. (a) D.E. Grant and H.J. Kimble, Opt. Lett. 7, 353 (1982); (b) D.E. Grant and H.J. Kimble, Opt. Comm. 44, 485 (1983).
24. L.A. Lugiato, L.M. Narducci, D.K. Bandy and C.A. Pennise, Opt. Comm. 43, 281 (1982).
25. (a) A.T. Rosenberger, L.A. Orozco and H.J. Kimble, Phys. Rev. A28, 2569 (1983); (b) L.A. Orozco, A.T. Rosenbeger and H.J. Kimble, Phys. Rev. Lett. 53, 2547 (1984); (c) L.A. Orozco, H.J. Kimble, A.T. Rosenberger, L.A. Lugiato, M.L. Asquini, M. Brambilla and L.M. Narducci, Phys. Rev. A39, 1235 (1989).
26. M. LeBerre, E. Ressayre, A. Tallet, K. Tai, H.M. Gibbs, M.C. Rushford and N. Peyghambarian, J. Opt. Soc. Am. B1, 591 (1984); G. Grynberg, E. Le Bihan, P. Verkerk, P. Simoneau, J.R.R. Leite, D. Bloch, S. Le Boiteux, and M. Ducloy, Opt. Commun. 67, 363 (1988); G. Giusfredi, J.F. Valley, R. Pon, G. Khitrova and H.M. Gibbs, J. Opt. Soc. Am. B5, 1181 (1988).
27. J.V. Moloney, M.R. Belic and H.M. Gibbs, Opt. Comm. 41, 379 (1982); J.V. Moloney and H.M. Gibbs, Phys. Rev. Lett. 48, 1607 (1982); J.V. Moloney, Phys. Rev. Lett. 53, 556 (1984); L.A. Lugiato and R. Lefever, Phys. Rev. Lett. 58, 220 (1987); L.A. Lugiato and C. Oldano, Phys. Rev. A37, 3896 (1988); J.V. Moloney,

H. Adachihara, D.W. McLaughlin and A.C. Newell, in Chaos, Noise and Fractals, edited by E.R. Pike and L.A. Lugiato, Hilger, London, 1988, p. 137.

28. L.A. Lugiato, F. Prati, L.M. Narducci and G.L. Oppo, Opt. Commun 69, 387 (1989).

29. P. Ru, L.M. Narducci, J.R. Tredicce, D.K. Bandy and L.A. Lugiato, Opt. Commun 63, 310 (1987)

30. A.E. Siegman, Lasers, University Science Books, Mill Valey, Calif., 1986.

31. A.M. Turing, Phil. Trans. R. Soc. London, Ser. B237, 37 (1952).

QUANTUM THEORY OF THE LASER

H. J. Carmichael†
Norman Bridge Laboratory of Physics 12-33,
California Institute of Technology, Pasadena, CA 91125.

The quantum theory of the laser is now over twenty years old. It has been discussed many times before, both in original research papers and review articles. Given this fact, it is a little difficult to know what sort of lectures I should give. I could organize my lectures as a review. However, this approach is perhaps not best suited to the tutorial intent of this school. Instead, I have decided to work through my own version of laser theory, adding yet another treatment of the subject to the many that already exist.

We are going to look into a small, and some might say dark corner of laser physics. We want to understand the fundamental statistical character of laser light resulting from the probabilistic nature of quantum mechanics. In the real world the noise in lasers generally has more to do with practical engineering concerns, such as mechanical stability, hydrodynamic stability in a dye flow, and so on. The noise we are interested in — the quantum noise — is what remains after all of this is gone.

Treatments of quantum noise in the laser fall into three catagories. First there are phenomenological treatments. These make no attempt at a first principles solution of a problem in quantum mechanics. They begin with Einstein's idea of the energy quantum and the processes of spontaneous emission, stimulated emission, and absorption. From this starting point, using a little intuition and some elementary classical statistics, results such as the Schalow-Townes linewidth[1] can be derived. The second and third categories include treatments that begin from a microscopic formulation in the language of quantum electrodynamics. The first principles formulation is reduced to a manageable mathematical description for the laser field using a suitable set of simplifying approximations. There are two distinct treatments of this type: Scully-Lamb theory,[2] built around density matrix equations, and various phase space theories[3–5] which use a quasidistribution — a function of a complex field amplitude — to represent the quantum mechanical state of the laser field.

Treatments from first principles involve a heavy overhead in mathematical formalism which tends to obscure what is really very simple physics. My first two lectures will therefore be devoted to developing some intuition on the basis of an Einstein rate equation for the number of photons in the laser mode. We will see how to arrive at results for photon statistics and the laser linewidth from such an approach. I will then sketch, without too much mathematical detail, one approach to the full quantum statistical theory. We will arrive at the same results for photon statistics and the laser linewidth by a much more cumbersome route. Although more cumbersome, the route from first principles does teach us some useful things. It uncovers the approximations implicit in the phenomenological treatment and provides a more general theory that holds when the approximations are not valid. It will also introduce us to techniques from quantum statistical mechanics that are widely used to solve other problems in quantum optics.

Although I will not explicitly follow any earlier work the theory presented in the second half of these lectures is essentially the phase space theory covered in sections IV and V of Haken's volume on laser theory.[6] The phase space approach draws on a body of background material that I will not have time to cover in detail in just four lectures. This material, together with an expanded version of these lectures will appear shortly in a volume of lecture notes that I hope to complete shortly.[7]

1. Laser Theory from Einstein Rate Equations

A conventional gas discharge generates light by spontaneous emission from an excited medium. The central idea behind laser action is to extract energy by stimulated emission. Two essential conditions must be met: The excitation (pumping) of the medium must produce population inversion so that stimulated emission will dominate absorption, and the energy density in the laser mode must be raised sufficiently for the stimulated emission rate to exceed the total loss rate, including the removal of energy in the output beam. The means for achieving population inversion are as diverse as the available types of lasers. There is often much physics involved in the details of an inversion mechanism. For our purpose, however, this is a practical concern; certainly central to the design of a real laser, but not important for understanding generic quantum statistical properties of the laser field. We will model the pumping process by a simple idealized scheme involving two or three levels.

Control over the energy density of the laser mode is provided by an optical cavity. The output beam represents loss from this cavity, and therefore the laser mode is modeled as a damped oscillator, driven by the inverted medium. Since energy is injected by stimulated emission, the rate at which energy is deposited in the field depends on the amount of energy already present; this introduces a nonlinearity. The dynamical paradigm is that of a *driven damped nonlinear oscillator*. Our goal is to find the oscillator equation in quantized form.

1.1 Rate Equations and Laser Threshold. Lamb's semiclassical laser theory derives a classical oscillator equation from Maxwell's equations driven by a nonlinear polarization.[8,9] Our first attempt at a quantum theory will follow an equally simple, but different approach. This approach has the advantage that it is formulated from the outset in quantum mechanical language and naturally includes the source of quantum fluctuations. We focus on the energy exchange between the laser mode and the lasing medium, and accept quantum ideas at the level of Einstein theory: the energy quantum (photon) is the basic unit of excitation for the laser oscillator, and the exchange of quanta by spontaneous emission, stimulated emission, and absorption[10] describes the interaction of the laser light with the lasing atoms.

A collection of N three-level atoms provides the simplest model for the laser medium that can produce population inversion by incoherent pumping. We adopt the level scheme illustrated in Fig. 1. The laser transition $|2\rangle \rightarrow |1\rangle$ interacts on resonance with a single cavity mode with frequency ω_C; γ_{31}, γ_{32}, and γ_{21} are decay rates for the various atomic levels, Γ_p is the incoherent pump rate, and $\overline{W}B_{21}$ is the stimulated emission and absorption rate for the laser transition, where $\overline{W}$ is the energy density per unit atomic linewidth in the laser mode, and $B_{21} = B_{12}$ is the Einstein B coefficient. If N_1, N_2, and N_3 denote the atomic state populations

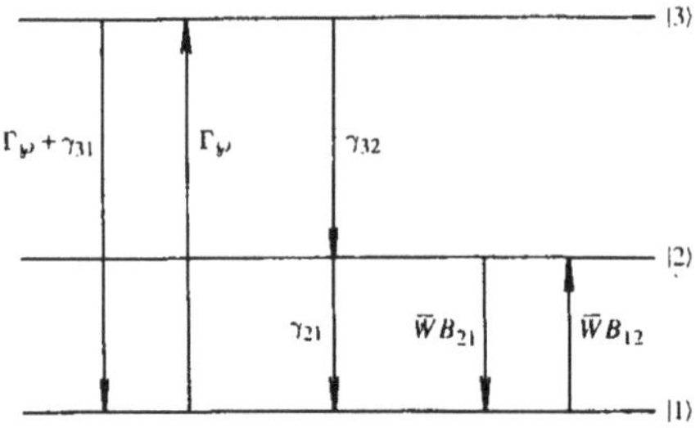

Fig. 1. Three-level model for the laser medium. The transition $|2\rangle \to |1\rangle$ is the laser transition.

$(N_1 + N_2 + N_3 = N)$, the atomic rate equations take the form

$$\dot{N}_1 = -\Gamma_p N_1 + \gamma_{21} N_2 + (\Gamma_p + \gamma_{31}) N_3 + \overline{W} B_{21} (N_2 - N_1), \tag{1a}$$

$$\dot{N}_2 = -\gamma_{21} N_2 + \gamma_{32} N_3 - \overline{W} B_{21} (N_2 - N_1), \tag{1b}$$

$$\dot{N}_3 = -(\Gamma_p + \gamma_{31} + \gamma_{32}) N_3 + \Gamma_p N_1. \tag{1c}$$

The number of photons in the laser mode is determined by the balance between stimulated emission and loss to the output beam:

$$\dot{n} = -2\kappa n + \overline{W} B_{21} (N_2 - N_1), \tag{2}$$

where 2κ is the photon loss rate from the cavity. For the present we neglect spontaneous emission into the laser mode.

The energy density $\overline{W}$ is given in terms of the photon number by:

$$\overline{W} = n \frac{\hbar \omega_C}{V_Q \pi (\gamma_h / 2)}, \tag{3}$$

where V_Q is the quantization volume for the laser mode, and

$$\gamma_h = \Gamma_p + \gamma_{21} + \gamma_p \tag{4}$$

is the homogeneous width (full width at half maximum) for the laser transition; γ_p is the contribution to the homogeneous linewidth due to phase destroying collisions. Note that it is the homogeneous width $\gamma_h/2$ and not the laser linewidth that appears in the denominator of Eq. (3). We are used to seeing Einstein theory applied in situations involving broadband excitation, where it is the energy density per unit bandwidth of the exciting field, $\overline{W}(\omega)$, that enters the stimulated emission rate. But the interaction between the laser mode and the laser transition is not broadband. Rate equations are also valid for narrowband excitation, so long as the homogeneous width is much broader that the natural width.[11] Under these conditions it is the average of $\overline{W}(\omega)$ over the atomic absorption line that enters the stimulated emission rate:

$$\begin{aligned}\overline{W} &= \int_0^\infty d\omega\,\overline{W}(\omega)\frac{\gamma_h/2\pi}{(\gamma_h/2)^2+(\omega-\omega_A)^2}\\ &= \int_0^\infty d\omega\left[n\frac{\hbar\omega}{V_Q}\frac{\gamma_L/2\pi}{(\gamma_L/2)^2+(\omega-\omega_L)^2}\right]\frac{\gamma_h/2\pi}{(\gamma_h/2)^2+(\omega-\omega_A)^2}\\ &= n\frac{\hbar\omega_L}{V_Q\pi(\gamma_h/2)}\frac{1}{1+\Delta^2},\end{aligned}\tag{5}$$

where $\Delta = 2(\omega_A - \omega_L)/\gamma_h$, and ω_L and $\gamma_L/2$ are the laser frequency and linewidth. To evaluate the integral we have assumed that $\gamma_L \ll \gamma_h$. For $\omega_L = \omega_A = \omega_C$ Eq. (5) reduces to Eq. (3).

To calculate B_{21} we use the relationship between Einstein A and B coefficients[12] to write ($\omega_A = \omega_C$)

$$\begin{aligned}B_{21} &= \left(\frac{3\mu^2}{\mu_{12}^2}\right)\frac{\pi^2c^3}{\hbar\omega_C^3}A_{21}\\ &= \left(\frac{3\mu^2}{\mu_{12}^2}\right)\frac{\pi^2c^3}{\hbar\omega_C^3}\left(\frac{1}{4\pi\epsilon_0}\frac{4\omega_C^3\mu_{12}^2}{3\hbar c^3}\right)\\ &= \frac{2V_Q\pi}{\hbar\omega_C}g^2,\end{aligned}\tag{6}$$

where

$$g = \sqrt{\frac{\omega_C\mu^2}{2\hbar\epsilon_0 V_Q}}\tag{7}$$

is the coupling constant (assumed real) between the laser field and the laser transition. μ_{12} is the atomic dipole matrix element, and since we are considering a single laser mode with a particular polarization $\hat{e}$, the factor $3\mu^2/\mu_{12}^2$, $\mu = \hat{e}\cdot\vec{\mu}_{12}$, is introduced on the right-hand side of Eq. (6) to convert the dipole orientation average $\langle\mu^2\rangle = \mu_{12}^2/3$ to μ^2. Now Eqs. (3) and (6) give the *stimulated emission rate*:

$$\overline{W}B_{21}(N_2-N_1) = n\frac{4g^2}{\gamma_h}(N_2-N_1).\tag{8}$$

The stimulated emission term in Eqs. (1) and (2) introduces the nonlinearity that causes the laser threshold behavior. This term also couples the rate equations for atomic populations to the equation for the photon number. When there are no photons in the laser mode the atomic populations settle into an equilibrium state balancing decay and incoherent pumping. Solving Eqs. (1a) – (1c) with $n = 0$ and $\dot{N}_1 = \dot{N}_2 = \dot{N}_3 = 0$ gives the *unsaturated steady-state inversion*

$$N_2^0 - N_1^0 = N\frac{\gamma_{32}(\Gamma_p-\gamma_{21})-\gamma_{21}(\Gamma_p+\gamma_{31})}{\gamma_{32}(\Gamma_p+\gamma_{21})+\gamma_{21}(2\Gamma_p+\gamma_{31})}.\tag{9}$$

Once some photons appear in the laser mode the atomic populations change, as the laser transition begins to saturate. If $n(t)$ changes slowly compared with atomic decay and pumping rates, the atomic populations will follow adiabatically, maintaining an equilibrium with the instantaneous photon number. Assuming that these conditions hold (they do in the vicinity of threshold) we adiabatically eliminate the atomic populations. Solving (1a) – (1c) with $\dot{N}_1 = \dot{N}_2 = \dot{N}_3 = 0$ and $n \neq 0$ gives the *saturated steady-state inversion*

$$N_2 - N_1 = (N_2^0 - N_1^0)\frac{1}{1 + n/n_{sat}}, \tag{10}$$

where the *saturation photon number* n_{sat} is given by

$$n_{sat} = \frac{\gamma_h\big[\gamma_{32}(\Gamma_\wp + \gamma_{21}) + \gamma_{21}(2\Gamma_\wp + \gamma_{31})\big]}{8g^2(\gamma_{32} + \gamma_{31} + 3\Gamma_\wp/2)}. \tag{11}$$

Substituting from Eqs. (8) and (10) into Eq. (2) we obtain the photon number rate equation

$$\dot{n} = -2\kappa n\left(1 - \frac{\wp}{1 + n/n_{sat}}\right), \tag{12}$$

where the *pump parameter* $\wp$ is defined by

$$\wp = \frac{2g^2}{\gamma_h \kappa}(N_2^0 - N_1^0). \tag{13}$$

The steady-state photon number n_{ss} satisfies the quadratic equation

$$n_{ss}(1 - \wp + n_{ss}/n_{sat}) = 0. \tag{14}$$

Solutions to this equation show the threshold behavior illustrated in Fig. 2.

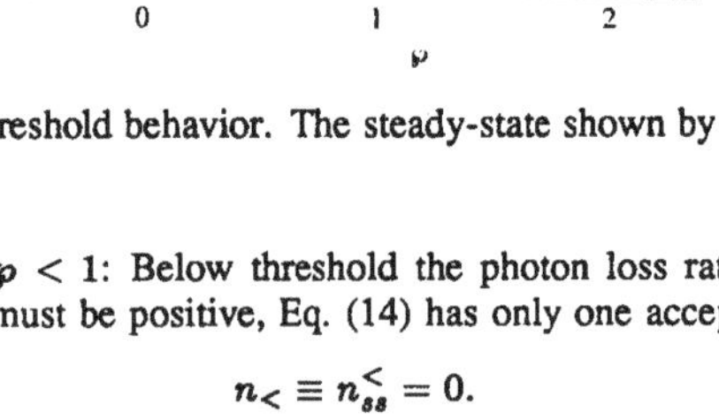

Fig. 2. Laser threshold behavior. The steady-state shown by the broken line is unstable.

Below Threshold — $\wp < 1$: Below threshold the photon loss rate exceeds the stimulated emission rate. Since n_{ss} must be positive, Eq. (14) has only one acceptable solution

$$n_< \equiv n_{ss}^< = 0. \tag{15}$$

From Eqs. (10) and (13) the below threshold inversion is

$$\big(N_2 - N_1\big)_< = N_2^0 - N_1^0 = \frac{\gamma_h \kappa}{2g^2}\wp. \tag{16}$$

AT THRESHOLD — $\wp = 1$: At threshold the photon loss rate is equal to the stimulated emission rate. Equation (14) has the doubly degenerate solution

$$n_{thr} \equiv n_{ss}^{thr} = 0, \tag{17}$$

and the threshold inversion is

$$\left(N_2 - N_1\right)_{thr} = N_2^0 - N_1^0 = \frac{\gamma_h \kappa}{2g^2}. \tag{18}$$

ABOVE THRESHOLD — $\wp > 1$: Above threshold the stimulated emission rate determined by the *unsaturated* inversion exceeds the photon loss rate. There are two acceptable solutions for n_{ss}:

$$\begin{aligned} {}_{ss} &= 0, && \text{(unstable)} \\ n_{ss} &= n_{sat}(\wp - 1). && \text{(stable)} \end{aligned}$$

The $n_{ss} = 0$ state is unstable; when linearized about this state, Eq. (12) gives $\dot{n} = 2\kappa(\wp - 1)n > 0$, and therefore any small change of the photon number from zero is amplified. Amplification stops when the saturation term $(1 + n/n_{sat})^{-1}$ brings the gain minus loss back to zero, with

$$n_{>} \equiv n_{ss}^{>} = n_{sat}(\wp - 1). \tag{19}$$

The saturated inversion above threshold is

$$\begin{aligned} \left(N_2 - N_1\right)_{>} &= (N_2^0 - N_1^0)\frac{1}{1 + n/n_{sat}} \\ &= (N_2^0 - N_1^0)\frac{1}{\wp} \\ &= \frac{\gamma_h \kappa}{2g^2} \\ &= \left(N_2 - N_1\right)_{thr}. \end{aligned} \tag{20}$$

The saturated inversion is "clamped" to the threshold value (a phenomenon known as *inversion clamping*).

1.2 Spontaneous Emission and Thermal Photons. Equation (12) describes the amplification that leads to the lasing state above threshold. But what is to be amplified? Where does the first photon come from? According to Eq. (12) the laser will not turn on above threshold if the laser cavity does not initially contain at least one photon.

We have neglected two photon sources. We have omitted the thermal source that brings the laser mode to thermal equilibrium when the pumping is turned off. This is corrected by adding the term $2\kappa\bar{n}$ to Eq. (12), where

$$\bar{n} = \left(e^{\hbar\omega_C/k_BT} - 1\right)^{-1} \tag{21}$$

is the mean number of photons in the laser cavity when it is in thermal equilibrium at the ambient temperature T. We have also omitted spontaneous emission. Spontaneously emitted photons are included by adding the term $\gamma_{spon}N_2$ to Eq. (12); γ_{spon} is the spontaneous emission rate into the laser mode. With the addition of these two terms the photon number rate equation reads

$$\dot{n} = -2\kappa n\left(1 - \frac{\wp}{1 + n/n_{sat}}\right) + 2\kappa\bar{n} + \gamma_{spon}N_2. \tag{22}$$

At optical frequencies $\bar{n}$ is negligible, therefore the first photon must be supplied by spontaneous emission. At microwave frequencies $\bar{n}$ is not negligible; even at low temperatures ($T \sim 4\,\text{K}$) a few thermal photons are present.

The rate γ_{spon} is not equal to the Einstein A coefficient for the $|2\rangle \to |1\rangle$ transition. γ_{spon} only accounts for spontaneous emission into the laser mode, while A is the emission rate to all modes in free space. We can calculate γ_{spon}, however, using the standard method for relating Einstein A and B coefficients. We write the stimulated emission term in Eq. (22) in the form given in Eq. (8), and then

$$\dot{n} = -2\kappa n\left[1 - \frac{2g^2}{\gamma_h\kappa}(N_2 - N_1)\right] + 2\kappa\bar{n} + \gamma_{spon}N_2. \tag{23}$$

Now if the atomic populations are maintained in thermal equilibrium, with

$$N_1/N_2 = e^{(E_2-E_1)/k_BT} = e^{\hbar\omega_C/k_BT}, \tag{24}$$

Eq. (23) must bring the cavity photons into equilibrium with the atoms. Thus, this equation must have the steady-state solution $n_{ss} = \bar{n}$. This requires

$$\gamma_{spon} = \frac{4g^2}{\gamma_h}\bar{n}\left(\frac{N_1}{N_2} - 1\right) = \frac{4g^2}{\gamma_h}.$$

Using Eqs. (21) and (24), the *spontaneous emission rate into the laser mode* is therefore

$$\gamma_{spon}N_2 = \frac{4g^2}{\gamma_h}N_2. \tag{25}$$

With the help of Eqs. (10) and (13) we write this in the form

$$\begin{aligned}\gamma_{spon}N_2 &= \frac{2g^2}{\gamma_h}(N_2 + N_1) + \frac{2g^2}{\gamma_h}(N_2 - N_1)\\ &= 2\kappa\left(C\frac{N_{21}}{N_{21}^0} + \frac{1}{2}\frac{\wp}{1 + n/n_{sat}}\right),\end{aligned} \tag{26}$$

where

$$C = \frac{N_{21}^0 g^2}{\gamma_h\kappa}, \tag{27}$$

and

$$N_{21} = N_2 + N_1 \tag{28}$$

is the number of atoms distributed between the levels of the laser transition. For the three-level model illustrated in Fig. 1,

$$N_{21} = N_{21}^0 \frac{1 + n/n'_{sat}}{1 + n/n_{sat}}, \tag{29}$$

where

$$N_{21}^0 = N \frac{\gamma_{32}(\Gamma_\wp + \gamma_{21}) + \gamma_{21}(\Gamma_\wp + \gamma_{31})}{\gamma_{32}(\Gamma_\wp + \gamma_{21}) + \gamma_{21}(2\Gamma_\wp + \gamma_{31})}, \tag{30}$$

and

$$n'_{sat} = \frac{\gamma_h\left[\gamma_{32}(\Gamma_\wp + \gamma_{21}) + \gamma_{21}(\Gamma_\wp + \gamma_{31})\right]}{8g^2(\gamma_{32} + \gamma_{31} + \Gamma_\wp)}. \tag{31}$$

Substituting Eq. (26) into Eq. (22), and using Eq. (29), we have

$$(2\kappa)^{-1}\dot{n} = -n\left(1 - \frac{\wp}{1 + n/n_{sat}}\right) + \left(\bar{n} + C\frac{1 + n/n'_{sat}}{1 + n/n_{sat}} + \frac{1}{2}\frac{\wp}{1 + n/n_{sat}}\right). \tag{32}$$

This equation can be simplified considerably if the laser is not operated too far above threshold. Typically n_{sat} and n'_{sat} are large numbers, and under normal operating conditions $n/n_{sat} \ll 1$, and $n/n'_{sat} \ll 1$. Then we can neglect the saturation terms in the second bracket on the right-hand side of Eq. (32). [We will discuss how to estimate n_{sat} shortly (see section 1.4); for a He-Ne laser $n_{sat} \sim 10^8$.] Of course, we cannot neglect the saturation of stimulated emission that appears in the first bracket. It is this gain saturation that prevents the photon number from growing without bound above threshold. However, for $n/n_{sat} \ll 1$ we only need to include the nonlinearity giving gain saturation to first order. Then the final form of the *laser rate equation for photon number*, including thermal and spontaneous emission sources, is

$$(2\kappa)^{-1}\dot{n} = -n(1 - \wp + \wp n/n_{sat}) + \bar{n} + n_{spon}. \tag{33}$$

We have written Eq. (26) as

$$\gamma_{spon}N_2 = 2\kappa(C + \tfrac{1}{2}\wp) = 2\kappa n_{spon}; \tag{34}$$

$n_{spon} = C + \frac{1}{2}\wp$ is the *spontaneous emission photon number* — the number of photons in the laser mode due to spontaneous emission well below threshold.

The laser can now turn on above threshold using a thermal or spontaneous photon to start the amplification process. The steady-state photon number satisfies the quadratic equation

$$n_{ss}(1 - \wp + \wp n_{ss}/n_{sat}) - (\bar{n} + n_{spon}) = 0, \tag{35}$$

with the single (positive) solution parameter:

$$n_{ss} = -\frac{1}{2\wp}n_{sat}(1 - \wp) + \frac{1}{2\wp}n_{sat}\sqrt{(1 - \wp)^2 + 4\wp(\bar{n} + n_{spon})/n_{sat}}. \tag{36}$$

In contrast to Fig. 2, the plot of Eq. (36) in Fig. 3 shows a smooth transition through the threshold region. The sharpness of the transition is determined by the ratio $(\bar{n} + n_{spon})/n_{sat}$. This ratio determines the range of the pump parameter over which the second term in the square root on the right-hand side of Eq. (36) is important. We define the *laser threshold region* by

$$|1-\wp| < |1-\wp|_{thr} = 2\sqrt{\frac{\bar{n}+n_{spon}}{n_{sat}}}. \tag{37}$$

This should be a small number. If it is not, even spontaneous emission into the laser mode is sufficient to saturate the laser transition, which would completely miss the objective of building up a field by stimulated emission. The range of the threshold region in a He-Ne laser operated on the 0.63 μm line is $|1-\wp|_{thr} \sim 10^{-4}$.[13] We can estimate the change in photon number over the threshold region using Eq. (36) to write

$$\begin{aligned} n^{\pm}_{thr} &= \pm\tfrac{1}{2}n_{sat}|1-\wp|_{thr} + \tfrac{1}{2}n_{sat}|1-\wp|_{thr}\sqrt{2} \\ &= \tfrac{1}{2}n_{sat}|1-\wp|_{thr}(\sqrt{2}\pm 1), \end{aligned} \tag{38}$$

where n^{+}_{thr} and n^{-}_{thr} are the steady-state photon numbers at the upper and lower boundaries of the threshold region, respectively. Then

$$\frac{n^{+}_{thr}}{n^{-}_{thr}} = \frac{\sqrt{2}+1}{\sqrt{2}-1} \approx 6. \tag{39}$$

The photon number changes by approximately an order of magnitude through the threshold region. (Since $\bar{n}$ is negligible at optical frequencies, if $n_{sat} \sim 10^8$ and $|1-\wp|_{thr} \sim 10^{-4}$, then $n_{spon} \sim 1$.)

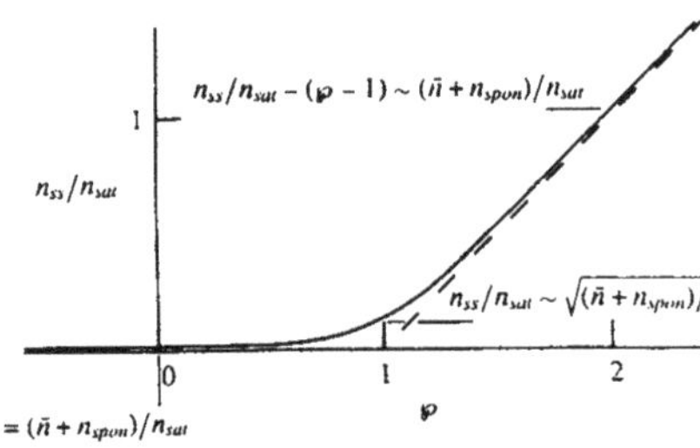

Fig. 3. Passage from the nonlasing to the lasing state with thermal and spontaneous emission sources included.

Equation (36) gives a number of results that are in rather remarkable agreement with those obtained from the full quantum statistical theory.

BELOW THRESHOLD — $1-\wp \gg |1-\wp|_{thr}$: With the inclusion of thermal and spontaneous photons the steady-state photon number is no longer zero below threshold. Expanding the square root in Eq. (36) to first order, we find

$$n_{<} \equiv n^{<}_{ss} = \frac{\bar{n}+n_{spon}}{1-\wp} = \frac{\bar{n}+C+\frac{1}{2}\wp}{1-\wp}. \tag{40}$$

This agrees exactly with the result obtained from the full quantum statistical theory (see section 3.1).

AT THRESHOLD — $\wp = 1$: The threshold photon number is given by

$$n_{thr} \equiv n_{ss}^{thr} = \sqrt{n_{sat}(\bar{n} + n_{spon})} = \sqrt{n_{sat}(\bar{n} + C + \tfrac{1}{2}\wp)}. \tag{41}$$

This result is larger than the result obtained from the full quantum statistical theory by the factor $\sqrt{\pi/2} \approx 1.25$ (see section 4.2). (For $n_{sat} \sim 10^8$, $n_{spon} \sim 1$, and $\bar{n}$ negligible, $n_{thr} \sim 10^4$.)

ABOVE THRESHOLD — $1 \gg \wp - 1 \gg |1 - \wp|_{thr}$: We restrict our attention to the region not too far above threshold. The requirement $\wp - 1 \ll 1$ ensures that $n_{ss}/n_{sat} \ll 1$. Expanding the square root in Eq. (36) to first order gives the correction to the solution $n_> = n_{sat}(\wp - 1)$ [Eq. (19)] due to thermal and spontaneous emission sources:

$$n_> - n_{sat}(\wp - 1) \equiv n_{ss}^> - n_{sat}(\wp - 1) = \frac{\bar{n} + n_{spon}}{\wp - 1} = \frac{\bar{n} + C + \frac{1}{2}\wp}{\wp - 1}. \tag{42}$$

This result is a factor of four larger than that derived from the full quantum statistical theory (see section 5.3).

1.3 Quantum Fluctuations: A Stochastic Model. The rate equation treatment including thermal and spontaneous emission sources provides a surprisingly accurate picture of laser operation. Nevertheless, it is, of course, limited in two major respects. First, the rate equation approach deals only with energy. It has nothing to say about the field. In particular, it provides no direct information on the laser linewidth. Second, Eq. (33) is a deterministic equation for a definite photon number n. In quantum mechanical language this assumes that the laser mode is always in a Fock state $|n(t)\rangle$. Surely this is not correct. Actually, so far we have made no attempt to incorporate the probabilistic character of quantum mechanics into our theory, and are therefore in no position to speculate on quantum states. Equation (33) should be interpreted as an equation for the mean photon number.

One way to build a probabilistic theory around Eq. (33) is to "invent" an underlying birth-death equation for the probabilities P_n that there are n photons in the laser mode. Such an equation should produce Eq. (33), at least in some approximation, as the equation of motion for $\langle n\rangle = \sum_{n=0}^{\infty} nP_n$. This approach leads to a connection with Scully-Lamb theory. We invent the underlying birth-death equation by first writing Eq. (33) as an equation for the mean photon number:

$$(2\kappa)^{-1}\langle\dot{n}\rangle = -\langle n\rangle\big(1 - \wp + \wp\langle n\rangle/n_{sat}\big) + \bar{n} + n_{spon}. \tag{43}$$

Something must be done about the nonlinear term $\langle n\rangle^2$ appearing on the right-hand side, since such a term cannot appear in the exact equation for $\langle n\rangle$. In a statistical theory, we expect the equation for the mean photon number to couple to equations for higher moments of n; $\langle n\rangle^2$ must therefore be a factorized approximation for $\langle n^2\rangle$. The correct equation for the mean photon number must read

$$(2\kappa)^{-1}\langle\dot{n}\rangle = -\langle n\rangle(1 - \wp) + \wp\langle n^2\rangle/n_{sat} + \bar{n} + n_{spon}. \tag{44}$$

Each term in Eq. (44) now suggests a corresponding term in the underlying birth-death equation. We write $F(n) \longrightarrow \langle f(n)\rangle$ to mean $\sum_{n=0}^{\infty} nF(n) = \langle f(n)\rangle$. Then

$$\langle \dot{n}\rangle \longleftarrow \dot{P}_n, \tag{45a}$$

$$\langle n\rangle \longleftarrow nP_n - (n+1)P_{n+1}, \tag{45b}$$

$$\langle n^2\rangle \longleftarrow (n-1)^2 P_{n-1} - n^2 P_n, \tag{45c}$$

$$1 \longleftarrow nP_{n-1} - (2n+1)P_n + (n+1)P_{n+1}. \tag{45d}$$

The choice of terms on the right-hand sides of this correspondence is not entirely "black magic." Familiarity with the density matrix element equation for the damped harmonic oscillator leads us to Eqs. (45b) and (45d).[14] The nonlinear term in Eq. (45c) can be found with a little trial and error. Note that each of the right-hand sides sums to zero, guaranteeing the conservation of total probability ($\sum_{n=0}^{\infty} \dot{P}_n = 0$). [The requirement that this be so removes trivial ambiguities like the possibility of choosing P_n and nP_n on the right-hand sides of Eqs. (45b) and (45c)]. We now have a probabilistic model for the laser photon number in the form of the birth-death equation

$$\begin{aligned}(2\kappa)^{-1}\dot{P}_n = &-(1-\wp)[nP_n - (n+1)P_{n+1}] - (\wp/n_{sat})[(n-1)^2 P_{n-1} - n^2 P_n] \\ &+ (\bar{n} + n_{spon})[nP_{n-1} - (2n+1)P_n + (n+1)P_{n+1}].\end{aligned} \tag{46}$$

To see the connection with Scully-Lamb theory we set $N_1 = N_1^0 = 0$ in Eqs. (13) and (28). Then if $\bar{n}$ is negligible, Eq. (46) can be written as

$$\dot{P}_n = A'[nP_{n-1} - (n+1)P_n] - B'[(n-1)^2 P_{n-1} - n^2 P_n] - C'[nP_n - (n+1)P_{n+1}], \tag{47}$$

with $A' = 2\kappa\wp = 4g^2 N_2^0/\gamma_h$, $B' = 2\kappa\wp/n_{sat} = 4g^2 N_2^0/\gamma_h n_{sat}$, and $C' = 2\kappa$. This equation agrees with the first-order expansion, for $nB'/A' = n/n_{sat} \ll 1$, of the Scully-Lamb laser equation [Eq. (4) of Ref. 2a]. Actually, the agreement is not exact. The Scully-Lamb equation has n^2 and $(n+1)^2$ in place of $(n-1)^2$ and n^2, respectively, but this difference is negligible for operation not too far below threshold, where $(n \pm 1)^2 \approx n^2$. Most of the parameters in the Scully-Lamb definitions of A', B', and C' [Eq. (5) of Ref. 2a] may be identified in a one-to-one correspondence with parameters in our own theory. We have $N_2^0 \leftrightarrow r_a/\gamma_a$, $\gamma_h/2 \leftrightarrow \gamma_{ab}$, $n_{sat} \leftrightarrow 4g^2/\gamma_a\gamma_b$, and $2\kappa \leftrightarrow \nu/Q$. Because of the different pumping model used by Scully and Lamb there is no direct correspondence between the atomic decay rates appearing in their expression for the saturation photon number ($n_{sat} = 4g^2/\gamma_a\gamma_b$) and ours [Eq. (11)].

The steady-state probabilities P_n^{ss} defined by setting $\dot{P}_n = 0$ in Eq. (46) are found with reference to Fig. 4. The steady state is maintained by the balancing of transitions between neighboring photon states:

$$P_{n+1}^{ss} = \frac{\bar{n} + n_{spon} - \wp n^2/n_{sat}(n+1)}{\bar{n} + n_{spon} + 1 - \wp} P_n^{ss}. \tag{48}$$

For operation not too far below threshold we can set $n^2/(n+1) \approx (n+1)$, and then Eq. (48) has the solution

$$P_n^{ss} = P_0^{ss} \prod_{k=1}^{n} \frac{\bar{n} + n_{spon} - \wp k/n_{sat}}{\bar{n} + n_{spon} + 1 - \wp}; \tag{49}$$

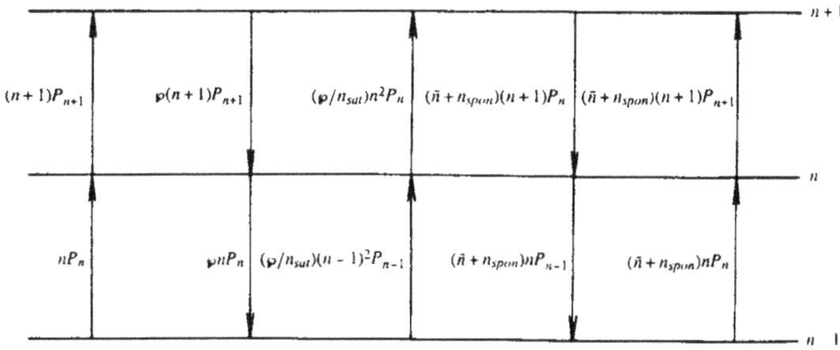

Fig. 4. Birth-death process described by Eq. (46). All transition rates are in units of 2κ.

P_0^{ss} is determined by normalization. The changing form of the photon number distribution through the threshold region is illustrated in Fig. 5. Various approximate expressions for P_n^{ss} capture this evolution well. The derivation of these results from Eq. (49) can be tried as an exercise:

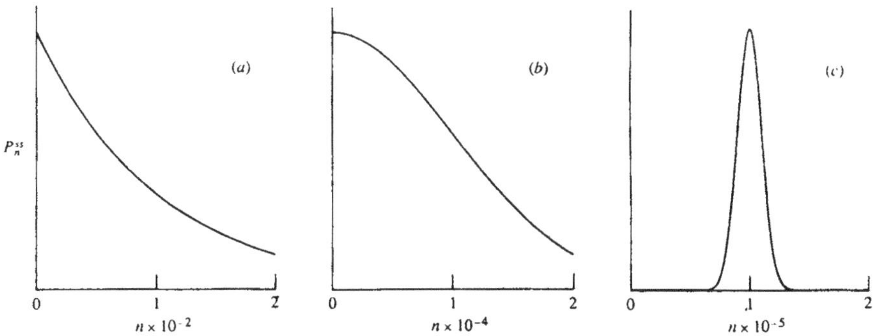

Fig. 5. Laser photon number distribution for $\bar{n} + n_{spon} = 1$, $n_{sat} = 10^8$ ($|1 - \wp|_{thr} \sim 10^{-4}$): (a) below threshold, $1 - \wp = 10^{-2}$; (b) at threshold, $\wp = 1$; (c) above threshold, $\wp - 1 = 10^{-3}$.

BELOW THRESHOLD — $1 \gg 1 - \wp \gg |1 - \wp|_{thr}$: The photon number is distributed according to the "thermal" distribution

$$P_n^{ss} = \frac{\left(\langle n \rangle_<\right)^n}{\left(1 + \langle n \rangle_<\right)^{n+1}}, \tag{50a}$$

where

$$\langle n \rangle_< \equiv \langle n \rangle_{ss}^< = \frac{\bar{n} + n_{spon}}{1 - \wp} = \frac{\bar{n} + C + \frac{1}{2}\wp}{1 - \wp}. \tag{50b}$$

Note the agreement between Eq. (50a) and Eq. (40) (also see section 3.1)

AT THRESHOLD — $\wp = 1$: The photon number is essentially distributed continuously, with

$$P_n^{ss} = \sqrt{\frac{2}{\pi}} \frac{1}{\sqrt{n_{sat}(\bar{n} + n_{spon})}} \exp\left[-\frac{1}{2}\frac{n^2}{n_{sat}(\bar{n} + n_{spon})}\right]. \tag{51a}$$

The mean threshold photon number is given by

$$\langle n \rangle_{thr} \equiv \langle n \rangle_{ss}^{thr} = \sqrt{\frac{2}{\pi}}\sqrt{n_{sat}(\bar{n} + n_{spon})} = \sqrt{\frac{2}{\pi}}\sqrt{n_{sat}(\bar{n} + C + \tfrac{1}{2})}. \tag{51b}$$

Note the agreement within a factor of $\sqrt{2/\pi}$ between Eq. (51b) and Eq. (41) (also see section 4.2).

ABOVE THRESHOLD — $1 \gg \wp - 1 > 0$: The distribution given by Eq. (51a) generalizes as

$$P_n^{ss} = \sqrt{\frac{2}{\pi}} \frac{1}{\sqrt{n_{sat}(\bar{n} + n_{spon})}} \left[1 + \Phi\left(\sqrt{2}\frac{\wp - 1}{|1 - \wp|_{thr}}\right)\right]^{-1} \exp\left[-\frac{1}{2}\frac{\left(n - n_{sat}(\wp - 1)\right)^2}{n_{sat}(\bar{n} + n_{spon})}\right], \tag{52}$$

where Φ denotes the Gaussian probability integral.[15] For $\wp - 1 \gg |1 - \wp|_{thr}$ the photon number has mean and variance

$$\langle n \rangle_> \equiv \langle n \rangle_{ss}^> = n_{sat}(\wp - 1), \tag{53a}$$

$$\langle n^2 \rangle_> - \left(\langle n \rangle_>\right)^2 \equiv \langle n^2 \rangle_{ss}^> - \left(\langle n \rangle_{ss}^>\right)^2 = n_{sat}(\bar{n} + n_{spon}) = n_{sat}(\bar{n} + C + \tfrac{1}{2}\wp). \tag{53b}$$

Note the agreement (to dominant order) between Eq. (53a) and Eq. (42). Also, Eqs. (53a) and (53b) indicate that the photon number distribution above threshold is significantly broader than a Poisson (also see section 5.3).

Equation (46) provides a simple probabilistic theory. But it is still limited to a description of energy states. A second way to build a probabilistic theory from Eq. (43) is to "invent" an underlying stochastic model for the laser field amplitude $\tilde{\alpha} = e^{i\omega_C t}\alpha$. This model can answer questions about the field, such as: what is the laser linewidth? It provides a connection with the phase space version of the full quantum statistical theory that we will discuss shortly. The stochastic model is formulated mathematically in terms of a Fokker-Planck equation, whose solution is the probability density $\tilde{P}(\tilde{\alpha}, \tilde{\alpha}^*)$ for the laser field amplitude $\tilde{\alpha}$. We must construct a Fokker-Planck equation that reproduces Eq. (43) with

$$\langle n \rangle \equiv \langle |\tilde{\alpha}|^2 \rangle = \int d^2\tilde{\alpha}\, |\tilde{\alpha}|^2 \tilde{P}(\tilde{\alpha}, \tilde{\alpha}^*); \tag{54}$$

more precisely, our Fokker-Planck equation must give the mean value equation

$$(2\kappa)^{-1}\langle |\dot{\tilde{\alpha}}|^2 \rangle = -\langle |\tilde{\alpha}|^2 \rangle (1 - \wp) - \wp \langle |\tilde{\alpha}|^4 \rangle / n_{sat} + \bar{n} + n_{spon}, \tag{55}$$

which reduces to Eq. (43) after the factorization $\langle |\tilde{\alpha}|^4 \rangle \approx \left(\langle |\tilde{\alpha}|^2 \rangle\right)^2$.

Some of you are no doubt unfamiliar with the theory of stochastic processes and the properties of Fokker-Planck equations. Unfortunately we do not have time to make a diversion into this subject here. Most books on classical statistical physics will have something to say on this topic. The books by Gardiner [15] and Risken [16] are useful references. Briefly, a Fokker-Planck equation is a partial differential equation for a probability density $P(\vec{x},t)$ in the form

$$\frac{\partial P(\vec{x},t)}{\partial t} = \left[-\sum_{i=1}^{n} \frac{\partial}{\partial x_i} A_i(\vec{x}) + \frac{1}{2} \sum_{i,j=1}^{n} \frac{\partial^2}{\partial x_i \partial x_j} D_{ij}(\vec{x}) \right] P(\vec{x},t), \tag{56}$$

where $\vec{x}$ is a vector of n random variables, and the $A_i(\vec{x})$ and $D_{ij}(\vec{x})$ are functions of these variables that control the rate of "drift" of the probability density, and the diffusion (or spreading) of the probability density, respectively. The simplest Fokker-Planck equation in one dimension has $A_1(x) = 0$ and $D_{11}(x) = D$ constant. It gives Fick's law for the diffusion of a Brownian particle.

It is not difficult to come up with the Fokker-Planck equation we seek. The first two-terms on the right-hand side of Eq. (55) describe cavity loss and saturable stimulated emission gain. These are *coherent* processes. They remove and add energy by coherently decreasing and increasing the field amplitude, and should be contributed by deterministic (drift) terms in the Fokker-Planck equation. In contrast, the third term on the right-hand side of Eq. (55) describes the *incoherent* thermal and spontaneous emission energy sources. This term should be contributed by a noise (diffusion) term in the Fokker-Planck equation. Writing $F(\tilde{\alpha},\tilde{\alpha}^*) \longrightarrow \langle f(\tilde{\alpha},\tilde{\alpha}^*)\rangle$ to mean $\int d^2\tilde{\alpha}|\tilde{\alpha}|^2 F(\tilde{\alpha},\tilde{\alpha}^*) = \langle f(\tilde{\alpha},\tilde{\alpha}^*)\rangle$, the term by term correspondence is

$$\langle |\dot{\tilde{\alpha}}|^2 \rangle \longleftarrow \dot{\tilde{P}}(\tilde{\alpha},\tilde{\alpha}^*), \tag{57a}$$

$$\langle |\tilde{\alpha}|^2 \rangle \longleftarrow -\frac{1}{2}\left(\frac{\partial}{\partial\tilde{\alpha}}\tilde{\alpha} + \frac{\partial}{\partial\tilde{\alpha}^*}\tilde{\alpha}^*\right)\tilde{P}(\tilde{\alpha},\tilde{\alpha}^*), \tag{57b}$$

$$\langle |\tilde{\alpha}|^4 \rangle \longleftarrow -\frac{1}{2}\left(\frac{\partial}{\partial\tilde{\alpha}}\tilde{\alpha} + \frac{\partial}{\partial\tilde{\alpha}^*}\tilde{\alpha}^*\right)|\tilde{\alpha}|^2\tilde{P}(\tilde{\alpha},\tilde{\alpha}^*), \tag{57c}$$

$$1 \longleftarrow \frac{\partial^2}{\partial\tilde{\alpha}\partial\tilde{\alpha}^*}\tilde{P}(\tilde{\alpha},\tilde{\alpha}^*). \tag{57d}$$

The Fokker-Planck equation suggested by Eq. (55) is then

$$\kappa^{-1}\frac{\partial\tilde{P}}{\partial t} = \left[\left(\frac{\partial}{\partial\tilde{\alpha}}\tilde{\alpha} + \frac{\partial}{\partial\tilde{\alpha}^*}\tilde{\alpha}^*\right)\left(1 - \wp + \wp|\tilde{\alpha}|^2/n_{sat}\right) + 2(\bar{n} + n_{spon})\frac{\partial^2}{\partial\tilde{\alpha}\partial\tilde{\alpha}^*}\right]\tilde{P}. \tag{58}$$

This is the *laser Fokker-Planck equation*. From this equation the results given by Eqs. (50) – (53) can be recovered, together with results, such as the linewidth, for the laser field. [The laser linewidth is calculated in sections (3.2) and (5.3).] It is important to realize that what we have here is essentially a classical stochastic description for the laser field. Although, one might choose to couch the theory in quantum mechanical language, the fluctuations of the laser field are the fluctuations of a classical stochastic complex field amplitude. Generally, quantum probability cannot be accommodated within a classical stochastic description and so some "reduction" of the quantum mechanics must take place as we pass from a fully quantum mechanical microscopic formulation to the macroscopic laser Fokker-Planck equation. We will see how this reduction is made in the next two lectures. This is not to say that we cannot label the noise on the laser field, or part of it, as quantum noise. Quantum mechanics leaves its mark in the diffusion coefficient n_{spon}. The size of this coefficient, specifically its nonzero value, is a statement from quantum mechanics. This statement is found in the relationship between Eqs. (8) and (25): *stimulated emission gain is necessarily accompanied by spontaneous emission noise*. This is something that Einstein theory tells us. But Einstein theory is *ad hoc* — it is not integrated with the mathematical formulations of mechanics and electromagnetism. Our job now is to derive the stochastic model defined by Eq. (58) from the theory of quantum electrodynamics.

1.4 Two-Level Model and Laser Parameters. First let us look again at the model for the laser medium. It is cumbersome to carry a detailed description of the pumping process through into the full quantum statistical theory. Moreover, the model shown in Fig. 7.1 is, first, an idealization of any real pumping process, and second, not the only reasonable idealized model that might be chosen (Ref. 17 describes a commonly used four-level model). We therefore build our microscopic theory around a two-level model, excluding all peripheral states needed to achieve population inversion. The model is illustrated in Fig. 6. It may be derived from our three-level model by taking $\gamma_{32} \gg \Gamma_p + \gamma_{31}, \gamma_{21}$. In this limit negligible population resides in state $|3\rangle$, and the pumping rate to the upper level of the laser transition is determined by Γ_p. We set

$$\gamma_\uparrow = \Gamma_p, \qquad \gamma_\downarrow = \gamma_{21}, \tag{59}$$

and Eqs. (4), (9), (11), (29) and (30), and (31) now read

$$\gamma_h = \gamma_\uparrow + \gamma_\downarrow + \gamma_p, \tag{60a}$$

$$N_2^0 - N_1^0 = N\frac{\gamma_\uparrow - \gamma_\downarrow}{\gamma_\uparrow + \gamma_\downarrow}, \tag{60b}$$

$$n_{sat} = \frac{\gamma_h(\gamma_\uparrow + \gamma_\downarrow)}{8g^2}, \tag{60c}$$

$$N_{21} = N_{21}^0 = N, \tag{60d}$$

$$n'_{sat} = n_{sat} = \frac{\gamma_h(\gamma_\uparrow + \gamma_\downarrow)}{8g^2}. \tag{60e}$$

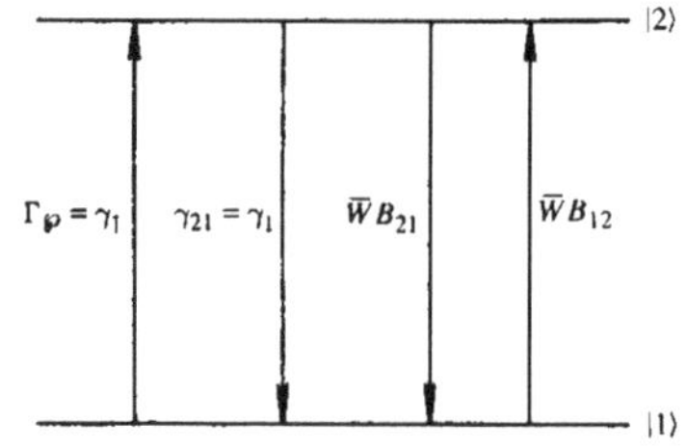

Fig. 6. Two-level model for the laser medium.

It is useful to summarize the parameters controlling the behavior of the stochastic model given by Eq. (58). The full quantum statistical theory will use a small noise expansion to describe fluctuations around a deterministic macroscopic state — a rigorous version of the addition of thermal and spontaneous emission noise to the noiseless laser behavior illustrated in Fig. 2. As we look at each parameter, we might take this opportunity to identify the role it plays in the context of this expansion. To this end we introduce a scaled field amplitude $\tilde{\tilde{\alpha}}$ with

$$n_{sat}^{\frac{1}{2}}\tilde{\tilde{\alpha}} = \tilde{\alpha}. \tag{61}$$

Then Eq. (58) reads

$$\kappa^{-1}\frac{\partial \tilde{\tilde{P}}}{\partial t} = \left[\left(\frac{\partial}{\partial \tilde{\tilde{\alpha}}}\tilde{\tilde{\alpha}} + \frac{\partial}{\partial \tilde{\tilde{\alpha}}^*}\tilde{\tilde{\alpha}}^*\right)\left(1 - \wp + \wp|\tilde{\tilde{\alpha}}|^2\right) + 2n_{sat}^{-1}(\bar{n} + n_{spon})\frac{\partial^2}{\partial \tilde{\tilde{\alpha}}\partial \tilde{\tilde{\alpha}}^*}\right]\tilde{\tilde{P}}. \tag{62}$$

Equation (62) depends on four parameters: $\wp$, $\bar{n}$, n_{spon}, and n_{sat}. The saturation photon number determines the importance of the diffusion term — the size of the noise. The macroscopic limit (with no fluctuations) is reached for $n_{sat} \to \infty$. In this limit the behavior of the laser is controlled by a single intensive parameter, the pump parameter

$$\wp = \frac{2g^2}{\gamma_h \kappa}(N_2^0 - N_1^0) = 2C\frac{\gamma_\uparrow - \gamma_\downarrow}{\gamma_\uparrow + \gamma_\downarrow}, \tag{63}$$

with

$$C = \frac{Ng^2}{\gamma_h \kappa}. \tag{64}$$

When n_{sat} is finite there are fluctuations about the macroscopic state. We will see shortly that the rigorous expansion used to describe these fluctuations is different for operation below threshold, at threshold, and above threshold. However, in each case the fluctuations scale as an inverse power of n_{sat}. After this scaling has been removed, the strength of the fluctuations is determined by two intensive parameters: the thermal photon number [Eq. (21)], and the spontaneous emission photon number

$$\begin{aligned} n_{spon} &= C + \tfrac{1}{2}\wp \\ &= \frac{2Ng^2}{\gamma_h \kappa}\frac{\gamma_\uparrow}{\gamma_\uparrow + \gamma_\downarrow}. \end{aligned} \tag{65}$$

Our two-level model for the laser medium suppresses the details of the pumping mechanism. It does, however, allow the pump parameter to be changed in two different ways. In the first $\gamma_\uparrow$ provides the control. This applies when the lower level of the lasing transition is populated in thermal equilibrium, with $N_1^0 = N$ when $\gamma_\uparrow = 0$. When $\gamma_\uparrow = \gamma_\downarrow$ the populations N_1^0 and N_2^0 are equalized, and gain is available once $\gamma_\uparrow$ exceeds $\gamma_\downarrow$. Generally the laser operates with $\gamma_\uparrow - \gamma_\downarrow \ll \gamma_\uparrow + \gamma_\downarrow$. Then, from Eq. (63), $2C$ must be large so that the small difference between $\gamma_\uparrow$ and $\gamma_\downarrow$ translates into sufficient gain to reach threshold. Equation (65) gives $n_{spon} = C + \frac{1}{2}\wp \gg 1$.

The second mode of operation applies in situations where the lower laser level is an excited state that is not normally populated in thermal equilibrium. If decay out of this state is fast enough, it remains unpopulated while the laser is in operation. This requires that we take $\gamma_\uparrow \gg \gamma_\downarrow$ in our two-level model. The pumping excites atoms from a state of lower energy than the lower level of the laser transition into the upper laser level. Thus, control over the pump parameter $\wp$ is provided by changing $N_2^0 = N$ (consequently changing C). A four-level scheme is needed to give the simplest complete description of this process.[18] Since $N_1^0 = 0$ ($\gamma_\uparrow \gg \gamma_\downarrow$), we find $n_{spon} = \wp$; at threshold, $n_{spon} = 1$. This is the situation in Scully-Lamb theory.

Since the saturation photon number determines the importance of fluctuations let us spend a little time to give it special consideration. If we express μ_{12}^2 in terms of the radiative decay rate $\gamma_\downarrow$, Eqs. (60c) and (7) give

$$\begin{aligned} n_{sat} &= \frac{\gamma_h(\gamma_\uparrow + \gamma_\downarrow)}{8}\frac{2\hbar\epsilon_0 V_Q}{\omega_C}\left(\frac{\mu_{12}}{\hat{e}\cdot\vec{\mu}_{12}}\right)^2\left(\frac{1}{4\pi\epsilon_0}\frac{4\omega_C^3}{3\hbar c^3}\frac{1}{\gamma_\downarrow}\right) \\ &= \frac{1}{2}\left(\frac{\pi w_0^2}{\sigma_{12}}\right)\left(\frac{L/c}{1/\gamma_\downarrow}\right)\frac{\gamma_h(\gamma_\uparrow + \gamma_\downarrow)}{\gamma_\downarrow^2}, \end{aligned} \tag{66}$$

where we have written $V_Q = \pi w_0^2 L$, where πw_0^2 is the cross-sectional area of the cavity mode and L is the cavity length;

$$\sigma_{12} = \frac{(\hat{e}\cdot\vec{\mu}_{12})^2}{\mu_{12}^2/3}(\lambda^2/2\pi) \tag{67}$$

is the weak-field absorption cross-section for the laser transition. The physical content of Eq. (66) can be appreciated by setting $\gamma_\uparrow = 0$. Then the laser medium acts as a two-level absorber. The ratio $(1/\gamma_\downarrow) \div (L/c)$ is the number of times a cavity photon revisits each atom during an atomic lifetime. Equation Eq. (66) can be written as

$$\begin{aligned} n/n_{sat} &= 2n\left(\frac{\sigma_{12}}{\pi w_0^2}\right)\frac{1/\gamma_\downarrow}{L/c} \\ &= 2\times\begin{pmatrix}\text{number of photons presented to each atom within} \\ \text{an absorption cross-section per atomic lifetime}\end{pmatrix}. \end{aligned} \tag{68}$$

Thus, the condition $n = n_{sat}$ means that half a photon is presented to the atom within an absorption cross-section per atomic lifetime. [Note that the greatest rate at which a two-level atom can scatter photons is *half* a photon per lifetime, corresponding to full saturation ($n/n_{sat} \gg 1$) where the probability for the atom to be in its excited state is one half.]

The saturation photon number must be large if the fluctuations are to be small. In this regard there is an important difference between the contributions from $\bar{n}$ and n_{spon} in Eq. (62). The thermal photon number may be extremely small. At optical frequencies $\bar{n} \sim 10^{-33}$ at $T = 300$K. Thus, thermal fluctuations are not intrinsic to the quantum dynamics of the laser and may be negligible quite independent of the size of n_{sat}. On the other hand, the fluctuations represented by n_{spon} are intrinsic to the quantized dynamics; they cannot be reduced (beyond a certain limit) other than by increasing n_{sat}. To see this note that at threshold $n_{spon} = C + \frac{1}{2}$, and, from Eq. (63), C has a minimum value $C = \frac{1}{2}$. Therefore, if the laser is to lase, it is necessary that $n_{spon} \geq 1$. The minimum $n_{spon} = 1$ occurs when the laser operates with $N_1^0 = 0$.

The laser is therefore intrinsically aware of its "size," as measured by n_{sat}. This is because the laser is a *nonlinear* device whose energy is *quantized* in "lumps" of finite size. There is an interplay between nonlinearity and the size of the lumps. The nonlinearity derives from the saturation of the laser transition. The degree of saturation depends on the energy density per unit atomic linewidth at the site of each laser atom. A characteristic energy density $\overline{W}_{sat}$ is needed to "turn on" the nonlinearity. The number of photons (lumps) that must be present in the cavity to provide this energy density scales proportional to the mode volume. The same energy density may be achieved in a cavity with a large mode volume or a cavity with a small mode volume. The smaller mode volume provides the same $\overline{W}_{sat}$ with fewer photons. In a small volume the fluctuations associated with the coming and going of the lumps are large relative to the mean energy needed to turn on the nonlinearity.

Equation (66) provides an estimate of n_{sat}. Choosing $w_0 = 1\,\text{mm}$, $L = 15\,\text{cm}$, $\lambda = 0.6\mu\text{m}$, $\gamma_{\downarrow} \sim 10^7\,\text{s}^{-1}$, and $\gamma_{\uparrow} \sim \gamma_h \approx 3 \times 10^8\,\text{s}^{-1}$, numbers appropriate for a He-Ne laser, we obtain $n_{sat} \sim 10^8$.

2. Quantum Statistical Theory of the Laser

Every physics student meets the standard closed system problems in quantum mechanics. Square well potentials, the hydrogen atom, the harmonic oscillator — these are all standard fare in an introductory course in quantum mechanics. Building a quantum theory of the laser from first principles is slightly different. The laser is an open system. We can write down a Hamiltonian describing the interaction of the laser atoms and the quantized laser field. But this is only part of the story. Energy flows into and out of this system. It comes in through the pumping of the atoms and leaves either as atomic fluorescence, or in the beam of light emitted from the laser cavity. We deal with an open system like this using methods from quantum statistical mechanics.

There are three main pieces to the quantum statistical theory we will discuss in the remaining lectures. The first is the derivation of an operator master equation. This is an equation satisfied by the density operator for the system of laser atoms and laser field, including the effects of the environmental interactions that couple energy into and out of the system. The master equation is derived from the Schrödinger equation for the laser system plus its environment by tracing over the environmental operators. The second step is to transform the operator master equation into a solvable form. We use a phase space approach to represent the density operator by a (quasi)distribution function over two complex variables for the field and polarization amplitudes, and one real variable for the inversion. The operator master equation is then replaced by a partial differential equation satisfied by this distribution function. Finally, we obtain approximate solutions to the phase space equation in the three regions of laser operation — below threshold, at threshold, and above threshold.

My coverage of this material in the time remaining will have to be sketchy. As I mentioned earlier, this phase space theory is covered, with extensive references, in Haken's book on laser theory.[6] An expanded version of my own calculations will be published shortly together with background material.[7]

2.1 Model Hamiltonian and the Laser Master Equation. We define our laser model by a Hamiltonian for the system of laser atoms and laser field (H_S), the environment or reservoir

(H_R), and the interaction between the laser system and its environment (H_{SR}):

$$H = H_S + H_R + H_{SR}. \tag{69}$$

The laser system comprises a single ring-cavity mode for the field, coupled to N identical two-level atoms. The choice of a ring cavity avoids complications associated with the different coupling strengths seen by atoms in a standing wave field. In the dipole and rotating-wave approximations, the Hamiltonian for the coupled atom-field system is given by

$$H_S = \tfrac{1}{2}\hbar\omega_C J_z + \hbar\omega_C a^\dagger a + i\hbar g(a^\dagger J_- - aJ_+); \tag{70}$$

a and $a^\dagger$ are annihilation and creation operators for the field, and

$$J_\pm = \sum_{j=1}^{N} e^{\pm i\phi_j}\sigma_{j\pm}, \qquad J_z = \sum_{j=1}^{N} \sigma_{jz}, \tag{71}$$

where σ_{j-}, σ_{j+}, and σ_{jz} are pseudo-spin operators for the jth atom, positioned at $\vec{r}_j$, and $\phi_j = \vec{k}_C \cdot \vec{r}_j$. For simplicity we are assuming exact resonance between the atoms and the field.

There are three important environmental interactions. First, the laser mode emits light through the output coupler. We model this loss by a weak interaction (H_{SR}^F) between the laser mode and the continuum of electromagnetic field modes (H_R^F) outside the cavity. Second, the atoms loose energy by spontaneous emission (or fluorescence) out the sides of the cavity. We model the atomic loss in a similar way by an interaction (H_{SR}^A) between the atoms and the continuum of electromagnetic field modes (H_R^A) traversing the cavity. Third, we model the pumping by an interaction H_{SR}^p with an energy source H_R^p. If atomic dephasing collisions occur a fourth interaction $(H_R^{dephase}$ and $H_{SR}^{dephase})$ must be added. Without specifying the explicit form of the environmental interactions we have

$$H_R = H_R^F + H_R^A + H_R^p + H_R^{dephase}, \tag{72}$$

and

$$H_{SR} = H_{SR}^F + H_{SR}^A + H_{SR}^p + H_{SR}^{dephase}. \tag{73}$$

The master equation is derived from the Schrödinger equation $\dot{\chi} = (1/i\hbar)[H, \chi]$ for the density operator χ describing the state of the laser system plus its environment. The reduced density operator

$$\rho(t) = \mathrm{tr}_R[\chi(t)], \tag{74}$$

satisfies the equation of motion [19]

$$\dot{\tilde{\rho}} = -\frac{1}{\hbar^2}\int_0^t dt' \mathrm{tr}_R\{[\tilde{H}_{SR}(t), [\tilde{H}_{SR}(t'), \tilde{\chi}(t')]]\}, \tag{75}$$

where the tilde denotes operators in the interaction picture. This equation cannot be solved exactly. We would like to have an equation for ρ with no explicit dependence on the state of the environment. Since the environment is largely unaffected by its interaction with the laser system, it seems reasonable to make the replacement $\tilde{\chi}(t') \rightarrow \tilde{\rho}(t')R_0$, where R_0 is the initial state of the environment. This is the *Born approximation*. Equation (75) is still difficult to solve because it involves an integration over the past history of $\tilde{\rho}$. This dependence on the past enters because fluctuations in the environment generally have a finite correlation time. However, if their correlation time is short relative to the fastest time scale for change in $\tilde{\rho}(t)$, we may make the replacement $\tilde{\rho}(t') \rightarrow \tilde{\rho}(t)$. This is the *Markoff approximation*. In the Born-Markoff approximation the trace over R_0 and the integral over t' can be carried out explicitly,[19] and leads to the *laser master equation*:

$$\begin{aligned}\dot{\rho} = &-i\tfrac{1}{2}\omega_C[J_z,\rho] - i\omega_C[a^\dagger a,\rho] + g[a^\dagger J_- - aJ_+,\rho]\\ &+\kappa(2a\rho a^\dagger - a^\dagger a\rho - \rho a^\dagger a) + 2\kappa\bar{n}(a\rho a^\dagger + a^\dagger\rho a - a^\dagger a\rho - \rho a a^\dagger)\\ &+\frac{\gamma_\downarrow}{2}\left(\sum_{j=1}^{N} 2\sigma_{j-}\rho\sigma_{j+} - \frac{1}{2}J_z\rho - \frac{1}{2}\rho J_z - N\rho\right)\\ &+\frac{\gamma_\uparrow}{2}\left(\sum_{j=1}^{N} 2\sigma_{j+}\rho\sigma_{j-} + \frac{1}{2}J_z\rho + \frac{1}{2}\rho J_z - N\rho\right) + \frac{\gamma_p}{2}\left(\sum_{j=1}^{N}\sigma_{jz}\rho\sigma_{jz} - N\rho\right).\end{aligned} \tag{76}$$

2.2 Phase Space Equation of Motion for the Laser. Phase-space representations are widely used in quantum optics.[20] They allow an operator equation such as Eq. (76) to be transformed into a partial differential equation for a distribution function. Averages of quantum mechanical operators are calculated from the distribution function by evaluating phase space integrals in the manner of classical statistical mechanics. Of course, there must be some difference from classical statistics. The difference is that the distribution function is not strictly a probability density; in some representations it can take negative values or be a highly singular function involving derivatives of δ-functions.

A variety of phase space representations are in use. Each has its own ordering rule for associating phase space averages and averages of quantum mechanical operators. We will use a representation that gives operator averages in what is called normal order. Haken[6] introduced this representation for the laser problem by generalizing the Glauber-Sudarshan P representation[21,22] for quantized fields to include atomic variables. The distribution $P(\alpha,\alpha^*,v,v^*,m)$ is defined by

$$P(\alpha,\alpha^*,v,v^*,m) = \frac{1}{2\pi^5}\int d^2z\int d^2\xi\int d\eta\,\chi_N(z,z^*,\xi,\xi^*,\eta)e^{-iz^*\alpha^*}e^{-iz\alpha}e^{-i\xi^*v^*}e^{-i\xi v}e^{-i\eta m}, \tag{77}$$

where

$$\chi_N(z,z^*,\xi,\xi^*,\eta) = \mathrm{tr}\left(\rho e^{iz^*a^\dagger}e^{iza}e^{i\xi^*J_+}e^{i\eta J_z}e^{i\xi J_-}\right) \tag{78}$$

is the normally-ordered characteristic function. We evaluate normally-ordered operator averages by taking derivatives of the characteristic function:

$$\begin{aligned}\langle a^{\dagger p'} a^{q'} J_+^p J_z^r J_-^q \rangle &= \operatorname{tr}\left(\rho a^{\dagger p'} a^{q'} J_+^p J_z^r J_-^p\right)\\ &= \frac{\partial^{p'+q'+p+r+q}}{\partial (iz^*)^{p'} \partial (iz)^{q'} \partial (i\xi^*)^p \partial (i\eta)^r \partial (i\xi)^q} \chi_N(z,z^*,\xi,\xi^*,\eta)\Bigg|_{\substack{z=z^*=0\\ \xi=\xi^*=\eta=0}} \end{aligned} \tag{79}$$

Since $P(\alpha,\alpha^*,v,v^*,m)$ is the five-dimensional Fourier transform of $\chi_N(z,z^*,\xi,\xi^*,\eta)$, we can substitute the inverse transform for $\chi_N(z,z^*,\xi,\xi^*,\eta)$ on the right-hand side of Eq. (79). This leads to the relationship between phase space averages and normally-ordered operator averages:

$$\begin{aligned}\langle a^{\dagger p'} a^{q'} J_+^p J_z^r J_-^q \rangle &= \int d^2\alpha \int d^2 v \int dm\, \alpha^{*p'} \alpha^{q'} z^{*p} m^r v^q P(\alpha,\alpha^*,v,v^*,m)\\ &\equiv \left(\overline{\alpha^{*p'} \alpha^{q'} v^{*p} m^r v^q}\right)_P . \end{aligned} \tag{80}$$

The conversion of Eq. (76) into a phase space equation of motion involves a lot of algebra but no new physics. At first sight the resulting equation does not seem to move us any closer to a solvable laser model; the full *phase space equation of motion for the laser* reads

$$\begin{aligned}\frac{\partial P}{\partial t} = \Bigg[& L_A\left(v,v^*,m,\frac{\partial}{\partial v},\frac{\partial}{\partial v^*},\frac{\partial}{\partial m}\right)\\ & + L_F\left(\alpha,\alpha^*,\frac{\partial}{\partial \alpha},\frac{\partial}{\partial \alpha^*}\right) + L_{AF}\left(\alpha,\alpha^*,v,v^*,m,\frac{\partial}{\partial \alpha},\frac{\partial}{\partial \alpha^*},\frac{\partial}{\partial v},\frac{\partial}{\partial v^*},\frac{\partial}{\partial m}\right)\Bigg] P, \end{aligned} \tag{81}$$

where,

$$\begin{aligned} & L_A\left(v,v^*,m,\frac{\partial}{\partial v},\frac{\partial}{\partial v^*},\frac{\partial}{\partial m}\right)\\ &\quad = i\omega_C\left(\frac{\partial}{\partial v}v - \frac{\partial}{\partial v^*}v^*\right) + \frac{\gamma_\downarrow}{2}\left[\left(e^{2\frac{\partial}{\partial m}} - 1\right)(N+m) + \frac{\partial}{\partial v}v + \frac{\partial}{\partial v^*}v^*\right]\\ &\quad\quad + \frac{\gamma_\uparrow}{2}\Bigg[\left(e^{-2\frac{\partial}{\partial m}} - 1\right)(N-m) + \frac{\partial^4}{\partial v^2 \partial v^{*2}} e^{2\frac{\partial}{\partial m}}(N+m)\\ &\quad\quad + 2\left(e^{-2\frac{\partial}{\partial m}} + \frac{\partial^2}{\partial v \partial v^*} - \tfrac{1}{2}\right)\left(\frac{\partial}{\partial v}v + \frac{\partial}{\partial v^*}v^*\right) + 2N\frac{\partial^2}{\partial v \partial v^*}\Bigg]\\ &\quad\quad + \frac{\gamma_p}{2}\left[\frac{\partial}{\partial v}v + \frac{\partial}{\partial v^*}v^* + \frac{\partial^2}{\partial v \partial v^*} e^{-2\frac{\partial}{\partial m}}(N+m)\right], \end{aligned} \tag{82a}$$

$$L_F\left(\alpha,\alpha^*,\frac{\partial}{\partial \alpha},\frac{\partial}{\partial \alpha^*}\right) = (\kappa + i\omega_C)\frac{\partial}{\partial \alpha}\alpha + (\kappa - i\omega_C)\frac{\partial}{\partial \alpha^*}\alpha^* + 2\kappa\bar{n}\frac{\partial^2}{\partial \alpha \partial \alpha^*}, \tag{82b}$$

and

$$L_{AF}\left(\alpha,\alpha^*,v,v^*,m,\frac{\partial}{\partial\alpha},\frac{\partial}{\partial\alpha^*},\frac{\partial}{\partial v},\frac{\partial}{\partial v^*},\frac{\partial}{\partial m}\right)$$
$$= -g\left\{\left[\left(e^{-2\frac{\partial}{\partial m}}-1\right)v^* + \frac{\partial}{\partial v}m - \frac{\partial^2}{\partial v^2}v\right]\alpha + \frac{\partial}{\partial\alpha}v\right.$$
$$\left. + \left[\left(e^{-2\frac{\partial}{\partial m}}-1\right)v + \frac{\partial}{\partial v^*}m - \frac{\partial^2}{\partial v^{*2}}v^*\right]\alpha^* + \frac{\partial}{\partial\alpha^*}v^*\right\}. \qquad (82c)$$

Equation (81) is considerably more complicated than the Fokker-Planck equation (58). First, it describes not only the laser field, but also the atoms. Second, it is not actually a Fokker-Planck equation at all since it includes derivatives to all orders. We cannot hope to solve this equation exactly. Our task now is to introduce approximations that enable us to extract useful information from this equation.

3. Laser Fokker-Planck Equation Below Threshold

To reduce Eq. (81) to the Fokker-Planck equation (58) we must justify the truncation of derivatives beyond second order and then adiabatically eliminate the variables describing the laser medium. Actually, we are not going to pass directly from Eq. (81) to Eq. (58). We eliminate derivatives beyond second order using van Kampen's system size expansion.[23] This is a systematic small noise expansion that generally leads directly to a *linear* Fokker-Planck equation. The laser Fokker-Planck equation (58) is nonlinear. It is possible to arrive at this nonlinear equation from Eq. (81) by dropping derivatives and performing the adiabatic elimination of atomic variables. But this approach does not treat the fluctuations in a systematic way. Equation (58) retains terms of the same order as some of the terms that are dropped in its derivation. A systematic expansion leads directly to the linearized version of Eq. (58). This expansion will tell us if, and when, the linearization breaks down. We will find that it holds below threshold.

3.1 System size expansion below threshold. Van Kampen's system size expansion makes an expansion in powers of a parameter Ω that characterizes the system's size. Each microscopic (extensive) state variable x determines a macroscopic (intensive) variable $\bar{x}$ via a scaling relationship

$$x = \Omega^p \bar{x}. \qquad (83)$$

$\bar{x}$ is decomposed into a deterministic motion $\bar{x}_0(t)$ plus fluctuations by writing

$$\bar{x} = \bar{x}_0(t) + \Omega^{-q}\eta. \qquad (84)$$

After making the change of variables $x = \Omega^p\bar{x}_0(t) + \Omega^{p-q}\eta$ in the microscopic theory, it is possible to determine p, q, and the macroscopic law governing the motion $\bar{x}_0(t)$ in a self-consistent manner.

We noted in section 1.4 that n_{sat} provides a measure of the system size for the laser; it is a natural choice for the parameter Ω. It can be shown that the scaling for the field (p_1, q_1), polarization (p_2, q_2), and inversion variables (p_3, q_3) must be made with $p_1 = q_1 = q_2 = q_3 = \frac{1}{2}$ and $p_2 = p_3 = 1$. We write

$$\alpha = n_{sat}^{\frac{1}{2}}\bar{\alpha}, \qquad \sqrt{\frac{\gamma_h}{\gamma_\uparrow + \gamma_\downarrow}}2\sqrt{2}Cv = N\bar{v}, \qquad 2Cm = N\bar{m}, \tag{85}$$

with

$$\bar{\alpha} = \langle \bar{a}(t)\rangle + n_{sat}^{-\frac{1}{2}}z, \qquad \bar{v} = \langle \bar{J}_-(t)\rangle + n_{sat}^{-\frac{1}{2}}\nu, \qquad \bar{m} = \langle \bar{J}_z(t)\rangle + n_{sat}^{-\frac{1}{2}}\mu, \tag{86}$$

where the operators $\bar{a}$, $\bar{J}_-$, and $\bar{J}_z$ are scaled in the same way as the corresponding phase space variables. Equations (60c) and (64) determine the relationship between N and n_{sat}:

$$N = n_{sat}4C\xi, \qquad \xi = \frac{2\kappa}{\gamma_\uparrow + \gamma_\downarrow}. \tag{87}$$

When we substitute for α, α^*, v, v^*, and m in Eq. (81), the equation of motion for the distribution

$$\bar{P}(z, z^*, \nu, \nu^*, \mu, t) = N^3 n_{sat}^{-\frac{3}{2}}\frac{1}{16C^3}P\Big(\alpha(z,t), \alpha^*(z^*,t), v(\nu,t), v^*(\nu^*,t), m(\mu,t), t\Big) \tag{88}$$

contains terms in $n_{sat}^{1/2}$, n_{sat}^0, and powers (greater than 0) of $n_{sat}^{-1/2}$. The terms of order $n_{sat}^{1/2}$ must vanish identically; otherwise they diverge in the limit $n_{sat} \to \infty$. This requirement determines the macroscopic equations — the *laser equations without fluctuations*. In a frame rotating at the frequency ω_C, these read

$$\kappa^{-1}\frac{d\langle\tilde{\bar{a}}\rangle}{dt} = -\langle\tilde{\bar{a}}\rangle + \langle\tilde{\bar{J}}_-\rangle, \tag{89a}$$

$$(\gamma_h/2)^{-1}\frac{d\langle\tilde{\bar{J}}_-\rangle}{dt} = -\langle\tilde{\bar{J}}_-\rangle + \langle\bar{J}_z\rangle\langle\tilde{\bar{a}}\rangle, \tag{89b}$$

$$(\gamma_\uparrow + \gamma_\downarrow)^{-1}\frac{d\langle\bar{J}_z\rangle}{dt} = -\big(\langle\bar{J}_z\rangle - \wp\big) - \tfrac{1}{2}\big(\langle\tilde{\bar{J}}_+\rangle\langle\tilde{\bar{a}}\rangle + \langle\tilde{\bar{J}}_-\rangle\langle\tilde{\bar{a}}^\dagger\rangle\big), \tag{89c}$$

where

$$\tilde{\bar{a}} = e^{i\omega_C t}\bar{a}, \qquad \tilde{\bar{J}}_- = e^{i\omega_C t}\bar{J}_-. \tag{90}$$

Equations (89) replace Eqs. (1) and (2) in the rate equation theory. They are more complete than the rate equations since they describe the field and polarization rather than just the photon number and atomic populations. We noted below Eq. (4) that rate equations are valid when the homogeneous width is much broader than the natural width. When this condition is satisfied the polarization may be adiabatically eliminated from Eqs. (89), giving

$$(\gamma_\uparrow + \gamma_\downarrow)^{-1}\frac{d\langle\bar{J}_z\rangle}{dt} = -\langle\bar{J}_z\rangle\big(1 + |\langle\bar{a}\rangle|^2\big) + \wp, \tag{91a}$$

$$(2\kappa)^{-1}\frac{d|\langle\bar{a}\rangle|^2}{dt} = -|\langle\bar{a}\rangle|^2\big(1 - \langle\bar{J}_z\rangle\big). \tag{91b}$$

These are equivalent to Eqs. (1) and (2). [The three-level model must be reduced to the two-level model we are now using.]

The terms of order n_{sat}^0 in the phase space equation determine the Fokker-Planck equation describing fluctuations about the macroscopic state. The Fokker-Planck equation is linear, with drift and diffusion coefficients that depend on the solution to Eqs. (89). The steady state solutions to Eqs. (89) are

$$\langle\tilde{\bar{a}}\rangle_{ss} = |\langle\bar{a}\rangle|_{ss}e^{i\phi}, \tag{92a}$$

$$\langle\tilde{\bar{J}}_-\rangle_{ss} = \frac{\wp}{1+|\langle\bar{a}\rangle|^2_{ss}}|\langle\bar{a}\rangle|_{ss}e^{i\phi}, \tag{92b}$$

$$\langle\bar{J}_z\rangle_{ss} = \frac{\wp}{1+|\langle\bar{a}\rangle|^2_{ss}}, \tag{92c}$$

where the mean field amplitude obeys the quadratic equation

$$|\langle\bar{a}\rangle|_{ss}\left(1-\wp+|\langle\bar{a}\rangle|^2_{ss}\right) = 0, \tag{93}$$

and the phase ϕ is arbitrary. Solutions to Eq. (93) reproduce the laser threshold behavior illustrated in Fig. 2. Below threshold $|\langle\bar{a}\rangle|_{ss} = 0$. The Fokker-Planck equation describing fluctuations about this state is separable. Writing

$$\bar{P}\Big(z(\tilde{z},t),z^*(\tilde{z}^*,t),\nu(\tilde{\nu},t),\nu^*(\tilde{\nu}^*,t),\mu,t\Big) = \tilde{\bar{X}}(\tilde{z}_1,\tilde{\nu}_1,t)\tilde{\bar{Y}}(\tilde{z}_2,\tilde{\nu}_2,t)\bar{M}(\mu,t), \tag{94}$$

with

$$z = e^{-i\omega_C t}(\tilde{z}_1+i\tilde{z}_2), \qquad \nu = e^{-i\omega_C t}(\tilde{\nu}_1+i\tilde{\nu}_2), \tag{95}$$

the *laser Fokker-Planck equation below threshold without adiabatic elimination of the polarization* is given by the three equations:

$$\frac{\partial\tilde{\bar{X}}}{\partial t} = \left[\kappa\frac{\partial}{\partial\tilde{z}_1}(\tilde{z}_1-\tilde{\nu}_1)+\frac{\gamma_h}{2}\frac{\partial}{\partial\tilde{\nu}_1}(\tilde{\nu}_1-\wp\tilde{z}_1)+\frac{1}{2}\kappa\bar{n}\frac{\partial^2}{\partial\tilde{z}_1^2}+\frac{1}{2}\xi^{-1}C\gamma_\uparrow\frac{\gamma_h^2}{(\gamma_\uparrow+\gamma_\downarrow)^2}\frac{\partial^2}{\partial\tilde{\nu}_1^2}\right]\tilde{\bar{X}}, \tag{96a}$$

$$\frac{\partial\tilde{\bar{Y}}}{\partial t} = \left[\kappa\frac{\partial}{\partial\tilde{z}_2}(\tilde{z}_2-\tilde{\nu}_2)+\frac{\gamma_h}{2}\frac{\partial}{\partial\tilde{\nu}_2}(\tilde{\nu}_2-\wp\tilde{z}_2)+\frac{1}{2}\kappa\bar{n}\frac{\partial^2}{\partial\tilde{z}_2^2}+\frac{1}{2}\xi^{-1}C\gamma_\uparrow\frac{\gamma_h^2}{(\gamma_\uparrow+\gamma_\downarrow)^2}\frac{\partial^2}{\partial\tilde{\nu}_2^2}\right]\tilde{\bar{Y}}, \tag{96b}$$

$$\frac{\partial\bar{M}}{\partial t} = (\gamma_\uparrow+\gamma_\downarrow)\left[\frac{\partial}{\partial\mu}\mu+\xi^{-1}C\left(1-\frac{\wp^2}{4C^2}\right)\frac{\partial^2}{\partial\mu^2}\right]\bar{M}. \tag{96c}$$

Equations (96a) and (96b) determine the average photon number below threshold:

$$\begin{aligned}\langle a^\dagger a\rangle_< \equiv \langle a^\dagger a\rangle^<_{ss} &= \left(\overline{\tilde{z}_1\tilde{z}_1}\right)_{\tilde{\bar{X}}_{ss}}+\left(\overline{\tilde{z}_2\tilde{z}_2}\right)_{\tilde{\bar{Y}}_{ss}}\\ &= \frac{\bar{n}[2\kappa(1-\wp)+\gamma_h](\gamma_\uparrow+\gamma_\downarrow)+2C\gamma_\uparrow\gamma_h}{(1-\wp)(2\kappa+\gamma_h)(\gamma_\uparrow+\gamma_\downarrow)}.\end{aligned} \tag{97}$$

The result given by the rate equation theory [Eq. (40)] is recovered when the laser medium relaxes much faster than the field [for $\xi = 2\kappa/(\gamma_\uparrow + \gamma_\downarrow) \ll 1$]. Then

$$\begin{aligned}\langle a^\dagger a\rangle_< &= \frac{\bar{n}(\gamma_\uparrow + \gamma_\downarrow) + C(\gamma_\uparrow + \gamma_\downarrow) + C(\gamma_\uparrow - \gamma_\downarrow)}{(1-\wp)(\gamma_\uparrow + \gamma_\downarrow)} \\ &= \frac{\bar{n} + C + \frac{1}{2}\wp}{1-\wp} \\ &= \frac{\bar{n} + n_{spon}}{1-\wp}. \end{aligned} \tag{98}$$

The comparison between Eqs. (97) and (98) illustrates the way in which the stochastic model deduced from photon number rate equations [Eq. (58)] is limited by the implicit adiabatic elimination of the polarization. Let us now see how to eliminate the polarization from Eqs. (96a) and (96b) and arrive at the linearized version of Eq. (58).

3.2 Adiabatic elimination of the polarization and laser linewidth. For any Fokker-Planck equation there exists an equivalent set of Ito stochastic differential equations.[16] Stochastic differential equations are much like the ordinary differential equations that generate phase space trajectories in classical mechanics. The difference is that stochastic equations include noise terms that cause a random wandering about otherwise deterministic trajectories. Averages calculated from an (infinite) ensemble of stochastic trajectories agree with those calculated from the equivalent Fokker-Planck equation.

We adiabatically eliminated the polarization using the Ito stochastic differential equations equivalent to Eqs. (96a) and (96b). These read

$$d\tilde{z}_i = -\kappa(\tilde{z}_i - \tilde{\nu}_i)dt + \sqrt{\kappa\bar{n}}\,dW_z^i, \tag{99a}$$

$$d\tilde{\nu}_i = -\frac{\gamma_h}{2}(\tilde{\nu}_i - \wp\tilde{z}_i)dt + \frac{\gamma_h}{\gamma_\uparrow + \gamma_\downarrow}\sqrt{\xi^{-1}C\gamma_\uparrow}\,dW_\nu^i, \tag{99b}$$

where dW_z^i and dW_ν^i are independent Weiner incrementa, and $i = 1,2$. For $\xi \ll 1$ we set $d\tilde{\nu}_i = 0$ on the left-hand side of Eq. (99b) and solve for $\tilde{\nu}_i dt$ in terms of $\tilde{z}_i dt$ and dW_ν^i. Substituting the result into Eq. (99a) and combining the noise terms proportional to dW_z^i and dW_ν^i into a single noise source, we find

$$d\tilde{z}_i = -\kappa(1-\wp)\tilde{z}_i dt + \sqrt{\kappa(\bar{n} + n_{spon})}\,dW_i. \tag{100}$$

The equivalent Fokker-Planck equations are

$$\kappa^{-1}\frac{\partial\tilde{\tilde{X}}}{\partial t} = \left[(1-\wp)\frac{\partial}{\partial\tilde{z}_1}\tilde{z}_1 + \frac{1}{2}(\bar{n} + n_{spon})\frac{\partial^2}{\partial\tilde{z}_1^2}\right]\tilde{\tilde{X}}, \tag{101a}$$

$$\kappa^{-1}\frac{\partial\tilde{\tilde{Y}}}{\partial t} = \left[(1-\wp)\frac{\partial}{\partial\tilde{z}_2}\tilde{z}_2 + \frac{1}{2}(\bar{n} + n_{spon})\frac{\partial^2}{\partial\tilde{z}_2^2}\right]\tilde{\tilde{Y}}. \tag{101b}$$

Thus, the *laser Fokker-Planck equation below threshold* is given by

$$\kappa^{-1}\frac{\partial\tilde{\bar{P}}}{\partial t}=\left[(1-\wp)\left(\frac{\partial}{\partial\tilde{z}}\tilde{z}+\frac{\partial}{\partial\tilde{z}^*}\tilde{z}^*\right)+2(\bar{n}+n_{spon})\frac{\partial^2}{\partial\tilde{z}\partial\tilde{z}^*}\right]\tilde{\bar{P}}. \tag{102}$$

This is the equation obtained by linearizing Eq. (58) about the steady state $\langle\tilde{\bar{a}}\rangle_< = 0$.

The spectrum of the laser field is the Fourier transform of the field autocorrelation function. This correlation function is given by a two-time operator average. In the representation we are using, two-time averages for phase space variables correspond to operator averages written in normal and time order. The field autocorrelation function can therefore be calculated from the Green function solutions to Eqs. (101a) and (101b):

$$\begin{aligned}g_<^{(1)}(\tau)&=\left(\langle a^\dagger a\rangle_<\right)^{-1}\left[\lim_{t\to\infty}\langle a^\dagger(t)a(t+\tau)\rangle\right]\\&=\left(\langle a^\dagger a\rangle_<\right)^{-1}e^{-i\omega_C\tau}\left[\lim_{t\to\infty}\left(\overline{\tilde{z}_1(t)\tilde{z}_1(t+\tau)}\right)_{\tilde{X}}+\lim_{t\to\infty}\left(\overline{\tilde{z}_2(t)\tilde{z}_2(t+\tau)}\right)_{\tilde{Y}}\right]\\&=e^{-i\omega_C\tau}e^{-\kappa(1-\wp)|\tau|}.\end{aligned}\tag{103}$$

The Fourier transform is a Lorentzian, and the *laser linewidth below threshold* (half width at half maximum) is

$$\left(\Delta\omega\right)_<=\kappa(1-\wp)=\kappa\frac{\bar{n}+n_{spon}}{\langle a^\dagger a\rangle_<}=2\kappa^2\frac{1}{\hbar\omega_C}\frac{\bar{n}+n_{spon}}{P_<}, \tag{104}$$

where $P_< = 2\kappa\langle a^\dagger a\rangle_<$ is the output power. The intensity correlation function can also be calculated from Eqs. (101a) and (101b):

$$\begin{aligned}g_<^{(2)}(\tau)&=\left(\langle a^\dagger a\rangle_<\right)^{-1}\left[\lim_{t\to\infty}\langle a^\dagger(t)a^\dagger(t+\tau)a(t+\tau)a(t)\rangle\right]\\&=\frac{1}{2}+\left(\langle a^\dagger a\rangle_<\right)^{-1}\left[\lim_{t\to\infty}\left(\overline{\tilde{z}_1^2(t)\tilde{z}_1^2(t+\tau)}\right)_{\tilde{X}}+\lim_{t\to\infty}\left(\overline{\tilde{z}_2^2(t)\tilde{z}_2^2(t+\tau)}\right)_{\tilde{Y}}\right]\\&=1+e^{-2\kappa(1-\wp)|\tau|}.\end{aligned}\tag{105}$$

This result shows the photon bunching, or Hanbury-Brown-Twiss effect,[24] of a thermal field, which is not surprising; recall that the photon number distribution below threshold is that of a "thermal" field [Eq. (50a)].

4. Laser Fokker-Planck Equation at Threshold

The linearized treatment of fluctuations breaks down at the laser threshold. For $\wp = 1$ the drift term in Eq. (102) vanishes. Then there is no restoring force to prevent the fluctuations from growing without bound — the mean photon number, for example, diverges at threshold [Eqs. (97) nd (98)]. Such divergences arise in nonlinear systems at any critical bifurcation point where one of the eigenvalues of the linearized deterministic dynamics vanishes. What has happened physically in the laser case is that the linearization neglects gain saturation. The system remains stable so long as the gain is less than the loss. At threshold the gain equals the loss, and the energy deposited in the laser mode by thermal photons and spontaneous emission grows indefinitely if gain saturation is not included. Mathematically, what has happened is that the coefficient of one of the dominant order terms in the system size expansion vanishes for $\wp = 1$. Under these conditions higher order terms must be included. This has two effects. It means that nonlinear terms (gain saturation terms) will appear in the Fokker-Planck equation. It also means that different scaling relations are needed for a self-consistent expansion.

4.1 System size expansion and adiabatic elimination of atomic variables. We need not know the new scaling relationship in advance. In place of Eq. (86) we now write

$$\bar{\alpha} = e^{-i\omega_C t}\langle\bar{\tilde{a}}\rangle_{thr} + n_{sat}^{-q} z, \qquad \bar{v} = e^{-i\omega_C t}\langle\bar{\tilde{J}}_-\rangle_{thr} + n_{sat}^{-q}\nu, \qquad \bar{m} = \langle\bar{J}_z\rangle_{thr} + n_{sat}^{-q}\mu, \quad (106)$$

where $\langle\bar{\tilde{a}}\rangle_{thr} = \langle\bar{\tilde{J}}_-\rangle_{thr} = 0$, and $\langle\bar{J}_z\rangle_{thr} = 1$, and leave the parameter q to be determined. The distribution in scaled variables is defined by

$$\bar{P}(z, z^*, \nu, \nu^*, \mu, t) = N^3 n_{sat}^{1-5q}\frac{1}{16C^3}P\big(\alpha(z), \alpha^*(z^*), v(\nu), v^*(\nu^*), m(\mu), t\big). \quad (107)$$

The Fokker-Planck equation obtained after expanding Eq. (81) and retaining lowest order terms has equivalent Ito stochastic differential equations (in a rotating frame)

$$d\tilde{z} = -\kappa(\tilde{z} - \tilde{\nu})dt + n_{sat}^{q-\frac{1}{2}}\sqrt{\kappa\bar{n}}\,(dW_z^1 + idW_z^2), \quad (108a)$$

$$d\tilde{\nu} = -\frac{\gamma_h}{2}(\tilde{\nu} - \tilde{z} - n_{sat}^{-q}\mu\tilde{z})dt + n_{sat}^{q-\frac{1}{2}}\frac{\gamma_h}{\gamma_\uparrow + \gamma_\downarrow}\sqrt{\xi^{-1}C\gamma_\uparrow}\,(dW_\nu^1 + idW_\nu^2), \quad (108b)$$

$$\begin{aligned} d\mu = &-(\gamma_\uparrow + \gamma_\downarrow)\big[\mu + n_{sat}^{-q}\tfrac{1}{2}(\tilde{\nu}^*\tilde{z} + \tilde{\nu}\tilde{z}^*)\big]dt \\ &+ n_{sat}^{q-\frac{1}{2}}\sqrt{\xi^{-1}2C(\gamma_\uparrow + \gamma_\downarrow)(1 - 1/4C^2)}\,dW_\mu. \end{aligned} \quad (108c)$$

The adiabatic elimination of variables $\tilde{\nu}$ and μ gives

$$d\tilde{z} = -\kappa n_{sat}^{-2q}|\tilde{z}|^2\tilde{z}dt + n_{sat}^{q-\frac{1}{2}}\sqrt{\kappa(\bar{n} + n_{spon})}\,(dW_1 + idW_2), \quad (109a)$$

with the equivalent Fokker-Planck equation

$$\kappa^{-1}\frac{\partial\bar{\tilde{P}}}{\partial t} = \left[n_{sat}^{-2q}\left(\frac{\partial}{\partial\tilde{z}}\tilde{z} + \frac{\partial}{\partial\tilde{z}^*}\tilde{z}^*\right)|\tilde{z}|^2 + n_{sat}^{2q-1}2(\bar{n} + n_{spon})\frac{\partial^2}{\partial\tilde{z}\partial\tilde{z}^*}\right]\bar{\tilde{P}}. \quad (109b)$$

We now determine q by requiring the terms in Eq. (109b) to be of the same order in n_{sat}. Thus, $q = \frac{1}{4}$. The *laser Fokker-Planck equation at threshold* reads

$$n_{sat}^{\frac{1}{2}}\kappa^{-1}\frac{\partial\bar{\tilde{P}}}{\partial t} = \left[\left(\frac{\partial}{\partial\tilde{z}}\tilde{z} + \frac{\partial}{\partial\tilde{z}^*}\tilde{z}^*\right)|\tilde{z}|^2 + 2(\bar{n} + n_{spon})\frac{\partial^2}{\partial\tilde{z}\partial\tilde{z}^*}\right]\bar{\tilde{P}}. \quad (110)$$

The fluctuations at threshold scale as $n_{sat}^{-1/4}$ rather than $n_{sat}^{-1/2}$. They are therefore much larger than the fluctuations below threshold. Note also that the timescale in Eq. (110) is set by $n_{sat}^{1/2}\kappa^{-1}$ rather than κ^{-1}. This, together with the slow relaxation generated by the nonlinear drift, is responsible for the well known *slowing down of fluctuations at laser threshold*. Equation (110) is the same as the Fokker-Planck equation derived from Eq. (58) by setting $\wp = 1$ and $\tilde{\alpha} = n_{sat}^{1/2}\tilde{\tilde{\alpha}} = n_{sat}^{1/4}\tilde{z}$.

4.2 Steady-state solution and threshold photon number. Because of the nonlinearity it is quite difficult to find time-dependent solutions to Eq. (110). The methods used to obtain such quantities as $g_{thr}^{(1)}(\tau)$ and $g_{thr}^{(2)}(\tau)$ are reviewed by Haken.[6] Risken's book on Fokker-Planck equations is also a good source of information on this subject.[17] On the other hand, the steady-state solution is easily obtained. From this we can calculate the average photon number

$$\langle a^\dagger a\rangle_{thr} \equiv \langle a^\dagger a\rangle_{ss}^{thr} = n_{sat}^{\frac{1}{2}}\left(\overline{\tilde{z}^*\tilde{z}}\right)_{\tilde{\tilde{P}}_{ss}} \tag{111}$$

for comparison with the result given by rate equations.

Eq. (110) is phase independent. We therefore transform to amplitude and phase variables r and ψ, with

$$\bar{\bar{P}}(r,\psi,t) = r\tilde{\tilde{P}}(re^{i\psi}, re^{-i\psi}, t). \tag{112}$$

Equation (110) now reads

$$n_{sat}^{\frac{1}{2}}\kappa^{-1}\frac{\partial\bar{\bar{P}}}{\partial t} = \left\{\frac{\partial}{\partial r}\left[r^3 - (\bar{n} + n_{spon})\frac{1}{2r}\right] + \frac{1}{2}(\bar{n} + n_{spon})\left(\frac{\partial^2}{\partial r^2} + \frac{1}{r^2}\frac{\partial^2}{\partial\psi^2}\right)\right\}\bar{\bar{P}}, \tag{113}$$

and has the steady-state solution

$$\bar{\bar{P}}_{ss}(r,\psi) = \frac{1}{2\pi}\sqrt{\frac{2}{\pi}}\frac{2}{\sqrt{\bar{n} + n_{spon}}}r\exp\left(-\frac{1}{2}\frac{r^4}{\bar{n} + n_{spon}}\right). \tag{114}$$

The *average photon number at threshold* is given by

$$\begin{aligned}\langle a^\dagger a\rangle_{thr} &= n_{sat}^{\frac{1}{2}}\sqrt{\frac{2}{\pi}}\frac{2}{\sqrt{\bar{n} + n_{sat}}}\int_0^\infty dr\, r^3\exp\left(-\frac{1}{2}\frac{r^4}{\bar{n} + n_{spon}}\right)\\ &= \sqrt{\frac{2}{\pi}}\sqrt{n_{sat}(\bar{n} + n_{spon})}.\end{aligned} \tag{115}$$

This result differs from that given by the rate equation theory [Eq. (41)] by the factor $\sqrt{2/\pi}$. It agrees with the result [Eq. (51b)] obtained from the birth-death equation constructed in section 1.3. In section 1.3 we saw that the photon number rate equation is obtained from the probabilistic birth-death, or stochastic, models via the factorization $\langle n^2\rangle \to \langle n\rangle^2$, or $\langle|\tilde{\alpha}|^4\rangle \to (\langle|\tilde{\alpha}|^2\rangle)^2$. This factorization is unimportant below threshold, since there the nonlinearity is negligible; we therefore obtained the average photon number below threshold exactly using rate equations. However, at threshold the nonlinearity is important. The disagreement between Eqs. (115) and (41) results from the factorization assumed by the photon number rate equation.

5. Laser Fokker-Planck Equation Above Threshold

Sufficiently far above threshold the nonlinear drift will again be negligible as it was below threshold. Gain saturation must certainly be included; but only saturation by the mean coherent intensity $|\langle\bar{a}\rangle|^2_> = \wp - 1$ [Eq. (93)] that is important. We do not need to include the small correction to the saturating intensity contributed by the fluctuations. We might expect to return to the treatment of fluctuations described in sections 3.1 and 3.2. Unfortunately, the treatment of fluctuations above threshold is not this simple. Equation (86) expands the fluctuations about a macroscopic field and polarization with defined amplitude *and* phase. Above threshold the macroscopic equations fix the steady-state amplitudes of the laser field and medium polarization [Eqs. (92)]. They also require the phases of the field and polarization to be locked. But the common phase ϕ for the field and polarization is left undetermined. This means that fluctuations in ϕ are free to grow without bound, producing a phase-symmetric steady-state with zero mean field and polarization amplitudes. Clearly the scaling defined by Eq. (86) is inadequate to treat such phase-symmetric fluctuations. Our treatment of fluctuations above threshold must be based on an expansion in amplitude and phase variables.

5.1 System size expansion and adiabatic elimination above threshold. The system size expansion in amplitude and phase variables is, in principle, no more complicated than the expansion outlined in section 3.1; although, the algebra gets a little confusing because the change of variables is a nonlinear transformation. There are now two sources of nonlinearity: the change of variables and nonlinear terms in the phase space equation of motion. Particular care must be taken in the treatment of the phases. Without fluctuations the phases of the field and polarization are locked. In the presence of fluctuations small differences between these phases arise. These fluctuations in the phase difference must be retained to give a correct treatment of the much larger fluctuations that develop in the phase sum. Fluctuations in the phase sum are responsible for the laser linewidth above threshold.

I will outline the main steps in the calculation. The details are given in Ref. 7. In place of Eq. (86) we write

$$\bar{\alpha} = e^{i\phi}\left(\bar{A}(t) + n_{sat}^{-\frac{1}{2}} z\right), \qquad \bar{v} = e^{i\theta}\left(\bar{J}(t) + n_{sat}^{-\frac{1}{2}} \nu\right), \qquad \bar{m} = \langle \bar{J}_z(t)\rangle + n_{sat}^{-\frac{1}{2}} \mu, \tag{116}$$

and we also introduce scaling relationships for the phase sum and difference:

$$\phi + \theta = \Psi(t) + n_{sat}^{-\frac{1}{2}} \psi, \qquad \phi - \theta = \Delta(t) + n_{sat}^{-\frac{1}{2}} \delta. \tag{117}$$

In Eq. (116) z and ν now represent *real* rather than complex amplitude fluctuations. These fluctuations can be both positive and negative, but must fall within the ranges $-n_{sat}^{1/2}\bar{A}(t) \leq z < \infty$ and $-n_{sat}^{1/2}\bar{J}(t) \leq \nu < \infty$, bounded below, since $\left(\bar{A}(t) + n_{sat}^{-1/2} z\right)$ and $\left(\bar{J}(t) + n_{sat}^{-1/2}\nu\right)$ must be positive. The phase space distribution in scaled variables is defined by

$$\bar{P}(z,,\nu,\psi,\delta,\mu,t) = N^3 n_{sat}^{-\frac{5}{2}} \frac{1}{32C^3}\left(\bar{A}(t) + n_{sat}^{-\frac{1}{2}} z\right)\left(\bar{J}(t) + n_{sat}^{-\frac{1}{2}}\nu\right)$$
$$\times\, P\Big(\alpha(z,\psi,\delta,t), \alpha^*(z,\psi,\delta,t), v(\nu,\psi,\delta,t), v^*(\nu,\psi,\delta,t), m(\mu,t), t\Big). \tag{118}$$

After making the change of variables in Eq. (81) the terms of order $n_{sat}^{1/2}$ determine the macroscopic equations. They give the *laser equations without fluctuations in amplitude and phase variables*:

$$\kappa^{-1}\frac{d\bar{A}}{dt} = -\bar{A} + \bar{J}\cos\Delta, \tag{119a}$$

$$(\gamma_h/2)^{-1}\frac{d\bar{J}}{dt} = -\bar{J} + \langle\bar{J}_z\rangle\bar{A}\cos\Delta, \tag{119b}$$

$$(\gamma_\uparrow + \gamma_\downarrow)^{-1}\frac{d\langle\bar{J}_z\rangle}{dt} = -\langle\bar{J}_z\rangle + \wp - \bar{J}\bar{A}\cos\Delta, \tag{119c}$$

$$\frac{d\Psi}{dt} = -2\omega_C - \left(\kappa\frac{\bar{J}}{\bar{A}} - \frac{\gamma_h}{2}\langle\bar{J}_z\rangle\frac{\bar{A}}{\bar{J}}\right)\sin\Delta, \tag{119d}$$

$$\frac{d\Delta}{dt} = -\left(\kappa\frac{\bar{J}}{\bar{A}} + \frac{\gamma_h}{2}\langle\bar{J}_z\rangle\frac{\bar{A}}{\bar{J}}\right)\sin\Delta. \tag{119e}$$

The steady-state solution to these equations above threshold is

$$\bar{A}_> \equiv \bar{A}_{ss}^> = \sqrt{\wp - 1}, \tag{120a}$$

$$\bar{J}_> \equiv \bar{J}_{ss}^> = \sqrt{\wp - 1}, \tag{120b}$$

$$\langle\bar{J}_z\rangle_> \equiv \langle\bar{J}_z\rangle_{ss}^> = 1, \tag{120c}$$

$$\Psi_> \equiv \Psi_{ss}^> = 2\omega_C t + \tilde{\Psi}, \tag{120d}$$

$$\Delta_> \equiv \Delta_{ss}^> = 0, \tag{120e}$$

where the phase $\tilde{\Psi}$ is arbitrary. The Fokker-Planck equation describing fluctuations about this state separates into an equation for the field amplitude, the polarization amplitude, and the inversion, and an equation describing phase fluctuations. We write the distribution $\bar{P}$ as the product

$$\bar{P}(z,\nu,\psi,\delta,\mu,t) = \bar{A}(z,\nu,\mu,t)\bar{\Phi}(\psi,\delta,t). \tag{121}$$

Then the Fokker-Planck equationsatisfied by $\bar{A}$ has equivalent stochastic differential equations

$$dz = -\kappa(z-\nu)dt + \sqrt{\kappa\bar{n}}\,dW_z, \tag{122a}$$

$$d\nu = -\frac{\gamma_h}{2}\left(\nu - z - \sqrt{\wp - 1}\,\mu\right)dt + \frac{\gamma_h}{\gamma_\uparrow + \gamma_\downarrow}\sqrt{\xi^{-1}C\gamma_\uparrow}\,dW_\nu, \tag{122b}$$

$$\begin{aligned} d\mu = {}& -(\gamma_\uparrow + \gamma_\downarrow)\left(\mu + \sqrt{\wp - 1}\,z + \sqrt{\wp - 1}\,\nu\right)dt \\ & + \sqrt{\xi^{-1}2C(\gamma_\uparrow + \gamma_\downarrow)(1 - 1/4C^2)}\,dW_\mu. \end{aligned} \tag{122c}$$

The Fokker-Planck equation satisfied $\bar{\Phi}$ has equivalent stochastic differential equations

$$d\psi = -\left(\kappa - \frac{\gamma_h}{2}\right)\delta\, dt + \frac{1}{\sqrt{\wp - 1}}\left(\sqrt{\kappa\bar{n}}\, dW_1 + \frac{\gamma_h}{\gamma_\uparrow + \gamma_\downarrow}\sqrt{\xi^{-1}C\gamma_\uparrow}\, dW_2\right), \tag{123a}$$

$$d\delta = -\left(\kappa + \frac{\gamma_h}{2}\right)\delta\, dt + \frac{1}{\sqrt{\wp - 1}}\left(\sqrt{\kappa\bar{n}}\, dW_1 - \frac{\gamma_h}{\gamma_\uparrow + \gamma_\downarrow}\sqrt{\xi^{-1}C\gamma_\uparrow}\, dW_2\right); \tag{123b}$$

dW_z, dW_ν, dW_μ, dW_1, and dW_2 are independent Weiner increments.

Adiabatically eliminating the polarization amplitude and inversion from Eqs. (122) gives the stochastic differential equation

$$dz = -\kappa 2(\wp - 1) z\, dt + \sqrt{\kappa(\bar{n} + n_{spon})}\, dW. \tag{124a}$$

for the amplitude of the laser field. Thus, the *laser Fokker-Planck equation for field amplitude fluctuations above threshold* is

$$\kappa^{-1}\frac{\partial \bar{A}}{\partial t} = \left[2(\wp - 1)\frac{\partial}{\partial z}z + \frac{1}{2}(\bar{n} + n_{spon})\frac{\partial^2}{\partial z^2}\right]\bar{A}. \tag{124b}$$

Adiabatically eliminating the phase difference δ from Eqs. (123) gives

$$d\psi = \frac{2}{\sqrt{\wp - 1}}\left(1 + \frac{2\kappa}{\gamma_h}\right)^{-1}\sqrt{\kappa(\bar{n} + n_{spon})}\, dW, \tag{125}$$

which describes a Brownian diffusion of the phase sum, with variance increasing proportional to t. Since the fluctuations in *delta* have constant variance, we can neglect δ compared to ψ and write

$$\phi = \tfrac{1}{2}(\Psi_> + \Delta_>) + n_{sat}^{-\frac{1}{2}}\tfrac{1}{2}(\psi + \delta) = \omega_C t + n_{sat}^{-\frac{1}{2}}\tfrac{1}{2}\psi. \tag{126}$$

Then, defining

$$\Phi(\phi, t) = \tilde{\Phi}(\tilde{\phi}, t) = n_{sat}^{-\frac{1}{2}}\tfrac{1}{2}\bar{\Phi}(\psi, t), \tag{127a}$$

with

$$\phi - \omega_C t = \tilde{\phi} = n_{sat}^{-\frac{1}{2}}\tfrac{1}{2}\psi, \tag{127b}$$

the laser *Fokker-Planck equation for phase fluctuations above threshold* is given by

$$\kappa^{-1}\frac{\partial \tilde{\Phi}}{\partial t} = \frac{1}{2}\left(1 + \frac{\kappa}{\gamma_h/2}\right)^{-2}\frac{\bar{n} + n_{spon}}{n_{sat}(\wp - 1)}\frac{\partial^2}{\partial \tilde{\phi}^2}\tilde{\Phi}. \tag{128}$$

5.2 Quantum fluctuations above threshold. The phase diffusion described by Eq. (128) determines the laser linewidth above threshold. The Green function solution to this equation — with $\tilde{\phi}$ distributed in the range $0 \le \tilde{\phi} < 2\pi$ — is

$$\tilde{\Phi}(\tilde{\phi}, t|\tilde{\phi}_0, 0) = \frac{1}{2\pi} \sum_{m=-\infty}^{\infty} \exp\left[im(\tilde{\phi} - \tilde{\phi}_0) - \frac{1}{2} \frac{\bar{n} + n_{spon}}{n_{sat}(\wp - 1)} m^2 t\right]. \tag{129}$$

Then

$$\begin{aligned} g_{>}^{(1)}(\tau) &= \left(\langle a^\dagger a\rangle_{>}\right)^{-1} \left[\lim_{t\to\infty} \langle a^\dagger(t) a(t+\tau)\rangle\right] \\ &= e^{-i\omega_C t} \lim_{t\to\infty} \left(\overline{e^{-i\tilde{\phi}(t)} e^{i\tilde{\phi}(t+\tau)}}\right)_{\tilde{\Phi}} \\ &= e^{-i\omega_C t} \sum_{m=-\infty}^{\infty} \exp\left[-\kappa \frac{1}{2} \frac{\bar{n} + n_{spon}}{n_{sat}(\wp - 1)} m^2 |\tau|\right] \left(\frac{1}{2\pi} \int_0^{2\pi} d\tilde{\phi}_0 \, e^{-i(m+1)\tilde{\phi}_0}\right) \\ &\quad \times \left(\frac{1}{2\pi} \int_0^{2\pi} d\tilde{\phi} \, e^{i(m+1)\tilde{\phi}}\right) \\ &= e^{-i\omega_C t} \exp\left[-\kappa \frac{1}{2} \frac{\bar{n} + n_{spon}}{n_{sat}(\wp - 1)} |\tau|\right]. \end{aligned} \tag{130}$$

The Fourier transform gives a Lorentzian line; the *laser linewidth above threshold* (half width at half maximum) is

$$(\Delta\omega)_{>} = \kappa \frac{1}{2} \frac{\bar{n} + n_{spon}}{n_{sat}(\wp - 1)} = \kappa \frac{1}{2} \frac{\bar{n} + n_{spon}}{\langle a^\dagger a\rangle_{>}} = \kappa^2 \frac{1}{\hbar\omega_C} \frac{\bar{n} + n_{spon}}{P_{>}}. \tag{131}$$

It is interesting to note that the expression in terms of the output power $P_{>}$ only differs from the corresponding expression below threshold [Eq. (104)] by a factor of two. Of course, the actual linewidth varies a great deal. The output power, or, alternatively, the mean intracavity photon number, increases by many orders of magnitude from below threshold to above threshold. Thus, a linewidth of the order of the cavity width κ below threshold is replaced by a very much narrower line above threshold: Using $\bar{n} + n_{spon} = 1$ and $n_{sat} = 10^8$ gives $(\Delta\omega)_{>}/\kappa \sim 10^{-6}$ for $\wp - 1 = 10^{-2}$.

From Eq. (124b) we can calculate the variance of the steady-state amplitude fluctuations:

$$\langle a^\dagger a\rangle_{>} - n_{sat}(\wp - 1) = \left(\overline{z^2}\right)_{\tilde{A}} = \frac{\bar{n} + n_{spon}}{4(\wp - 1)}. \tag{132}$$

This result differs from that obtained using the rate equation theory [Eq. (420] by the the factor of four in the denominator on the right-hand side. The difference arises from the factorization used to pass to a rate equation description from Eq. (55). Finally, the intensity correlation function above threshold is given by

$$g_{>}^{(2)}(\tau) = 1 + \frac{\bar{n} + n_{spon}}{n_{sat}(\wp - 1)^2} e^{-2\kappa(\wp-1)|\tau|} = 1 + \frac{\pi}{2} \left(\frac{\langle a^\dagger a\rangle_{thr}}{\langle a^\dagger a\rangle_{>}}\right)^2 e^{-2\kappa(\wp-1)|\tau|}. \tag{133}$$

Compare this with the "thermal" result below threshold [Eq. (105)]. For $\bar{n} + n_{spon} = 1$ and $n_{sat} = 10^8$, the difference between Eq. (133) and the correlation function $g^{(2)}(\tau) = 1$ for coherent light is $\sim 10^{-4}$ ($\wp - 1 = 10^{-2}$).

This completes my series of lectures. I just have enough time for a few concluding remarks. It is clearly an understatement to say that the laser revolutionized modern optics. The laser did not just revolutionize a field, it defined a new field; indeed it defined a whole collection of new fields. The most visible signs of the revolution are seen in the laser's scientific and commercial applications. most of these do not rely in any essential way on the quantum statistical theory we have developed in these lectures — at least, not beyond the statement that the laser is a "thermal" source below threshold and a good approximation to a coherent source above threshold.. As I mentioned in my introduction, the approximation to coherence above threshold is generally determined by engineering, not quantum mechanical, considerations. However, as the uses of lasers become more sophisticated — in information processing, perhaps, and in high precision measurement — this will change. Small devices and small powers will require that attention be payed to questions of quantum statistics, and there will always be a few individuals who carry the challenges of stabilization to their ultimate conclusion where quantum fluctuations set the fundamental limits. In a recent development we are now seeing an interest in new sources of light with exotic statistical properties such as antibunched or sub-Poissonian photons, and "squeezed" quantum fluctuations. Perhaps these will find wide application, perhaps not. Whatever the outcome, it is certainly a fascinating theoretical enterprise to study the "engineering" of quantum fluctuations in optical systems. Many of the tools used in this enterprise are variations and developments of those you have met during these lectures.

Acknowledgements

This work was supported by the National Science Foundation under Grant No. PHY-88-10502. I would like to thank Prof. H. J. Kimble and those working in the Quantum Optics Laboratory at the California Institute of Technology for their hospitality.

References

† On leave from the Physics Department, University of Arkansas, Fayetteville, AR 72701.

1. A. L. Schawlow and C. H. Townes, *Phys. Rev.* **112**, 1940 (1958).
2. M. O. Scully and W. E. Lamb Jr., *Phys. Rev. Lett.* **16**, 853 (1966); *Phys. Rev.* **159**, 208 (1967); **166**, 246 (1968).
3. W. Weidlich, H. Risken, and H. Haken, *Z. Physik* **201**, 396 (1967).
4. M. Lax, *Phys. Rev.* **157**, 213 (1967).
5. J. P. Gordon, *Phys. Rev.* **161**, 367 (1967).
6. H. Haken, *Handbuch der Physik*, Vol. XXV/2c, ed. by L. Genzel, Springer: Berlin, 1970.
7. H. J. Carmichael, "Quantum Statistical Methods in Quantum Optics," *Lecture Notes in Physics*, Springer: Berlin, to be published.

8. W. E. Lamb Jr., *Phys. Rev.* **134**, A1429 (1964).
9. M. Sargent III, M. O. Scully, and W. E. Lamb Jr., *Laser Physics*, Addison-Wesley: Reading, Massachusetts, 1974, Chapt. VIII.
10. Reference 9, pp. 21ff.
11. Reference 9, pp. 104,203.
12. W. H. Louisell, *Quantum Statistical Properties of Radiation* Wiley: New York, 1973, p 288; Reference 9, p. 22.
13. M. R. Young and S. Singh, *Phys. Rev.* A **35**, 1453 (1987).
14. Reference 16, equation (10.4.4).
15. I. S. Gradshteyn and I. M. Ryźhik, *Tables of Integrals Series and Products*, Academic Press: New York, 1965, p. 930.
16. C. W. Gardiner, *Handbook of Stochastic Methods*, Springer: Berlin (1983).
17. H. Risken, *The Fokker-Planck Equation*, Springer: Berlin (1984).
18. A. Yariv, *Introduction to Optical Electronics*, Holt, Rinehart and Wilson: New York, 1976, pp. 118ff.
19. Reference 6,pp. 51ff; Reference 12, Chapt. 6.
20. M. Hillery, R. F. O'Connell, M. O. Scully, E. P. Wigner, *Physics Rep.* **106**, 121 (1984).
21. R. J. Glauber, *Phys. Rev.* **131**, 2766 (1963).
22. E. C. G. Sudarshan, *Phys. Rev. Lett.* **10**, 277 (1963).
23. N. G. van Kampen, *Can. J. Phys.*, **39**, 551 (1961); N. G. van Kampen, *Stochastic Processes in Physics and Chemistry*, North Holland: Amsterdam, 1981, Chapts. IX – XI.
24. R. Hanbury-Brown and R. Q. Twiss, *Nature* **177**, 27 (1956); **178**, 1046 (1956); *Proc. R. Soc. Lond.* A **242**, 300 (1957); **243**, 291 (1957).

QUANTUM STATISTICS OF THE ELECTROMAGNETIC FIELD

Vittorio Degiorgio
Dipartimento di Elettronica-Sezione di Fisica Applicata,
Università di Pavia, 27100 Pavia, Italy

Contents:

1. Coherence and Correlations

1.1 First-order coherence.

The electric field associated with an electromagnetic wave having central frequency ω_0 is, in complex notation,:

$$E(\rho,t) = A \exp[-i(\omega_0 t - k \cdot \rho + \phi)]$$

where A and ϕ are real quantities. If A and ϕ do not depend on ρ and t, (1) is a monochomatic plane wave, fully coherent in the sense that, given $E(\rho,t)$ at ρ,t, $E(\rho+\varrho,t+\tau)$ is completely predictable for any point ϱ,τ. If A and ϕ are random functions of t (intrinsic noise or external disturbance), the field is partially coherent because predictability of E(t) is limited to a time interval τ_c around t. A similar consideration can be applied to the space dependent properties.

Quantitative measurement of coherence is based on the use of **correlation functions.** The simplest is the field correlation function:

$$G_1(t,\tau) = \langle E^*(t)E(t+\tau)\rangle$$

<...> means ensemble average (classically: average over a probability density, quantum mechanically : Trace over a density matrix).

In many practical cases, E(t) is a stationary and ergodic process. If it is so, G_1 depends only on τ and ensemble average can be substituted by time average (much simpler from an experimental point of view):

$$G_1(\tau) = \lim \frac{1}{2T}\int_{-T}^{T} E^*(t)E(t+\tau)dt$$

Monochromatic wave:

$$G_1(\tau) = I \exp(i\omega_0\tau), \quad \text{where } I = E^*(t)E(t).$$

If A and ϕ fluctuate with t :

$$G_1(\tau) = \exp(i\omega_0\tau) \langle A(t)A(t+\tau) \exp\{i[\phi(t)-\phi(t+\tau)]\}\rangle = \langle I\rangle \exp(i\omega_0\tau)\, g(\tau)$$

$g(0) = 1$; $g(\infty) = 0$. $|g|$ is the degree of temporal coherence.

If $|g(\tau)| = \exp(-\tau/\tau_c)$, we can call τ_c the coherence time of the optical field.

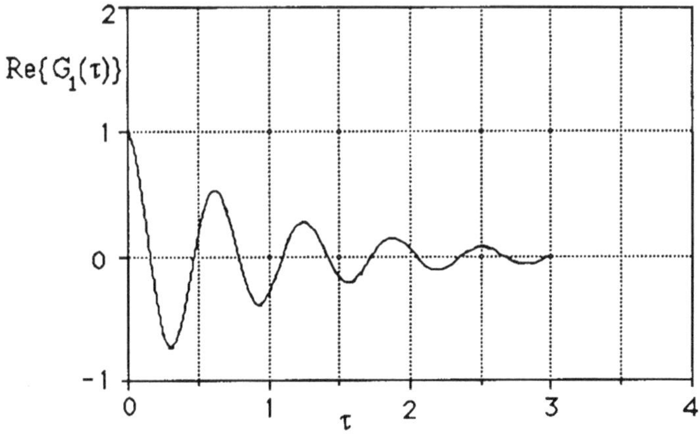

Fig.1. A typical behavior of the real part of the field correlation function.

If E(t) is expressed in terms of a Fourier integral:

$$E(t) = \int E(\omega) \exp(-i\omega t)\, d\omega ,$$

we can define the optical spectrum as: $S(\omega) = \langle |E(\omega)|^2 \rangle$. It can be easily shown that:

$$G_1(\tau) = \int S(\omega) \exp(-i\omega\tau)\, d\omega$$

Since $\langle I \rangle = \int S(\omega)\, d\omega$, we define the spectral density $s(\omega)$ as: $s(\omega) = \dfrac{S(\omega)}{\langle I \rangle}$. Clearly,

$$g(\tau) = \int s(\omega) \exp[-i(\omega-\omega_0)\tau]\, d\omega$$

If $s(\omega)$ is symmetrical around ω_0, $g(\tau)$ is real. The optical spectrum can be measured by a Fabry-Perot interferometer (see Fig.2) which is essentially a narrow-band periodic filter.

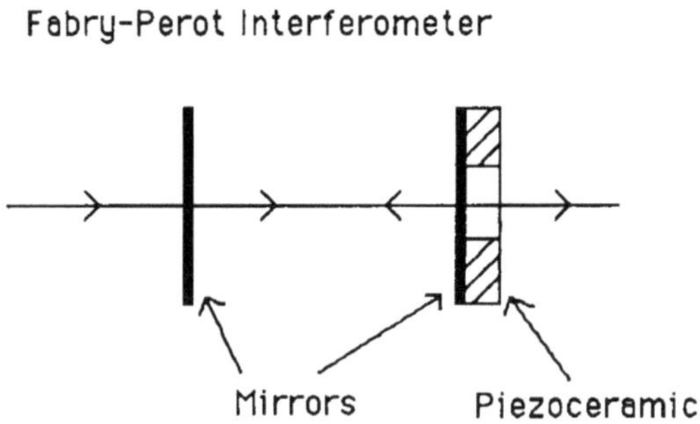

Figure 2. Fabry-Perot interferometer.

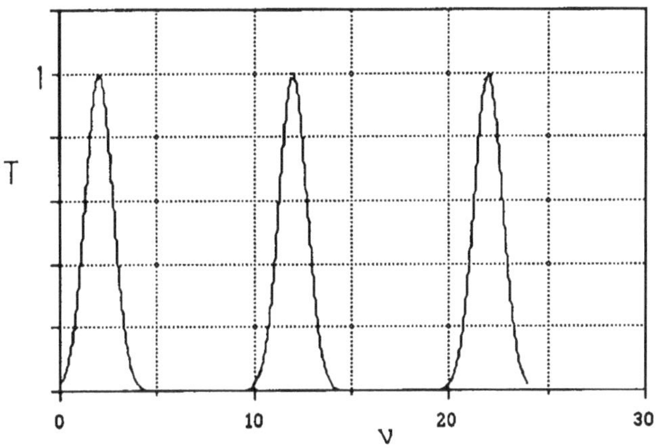

Figure 3. Transmission of a Fabry-Perot versus the optical frequency ν.

Example: Lorentzian Spectrum $s(\omega) = \frac{\Gamma/\pi}{\Gamma^2+(\omega-\omega_0)^2}$

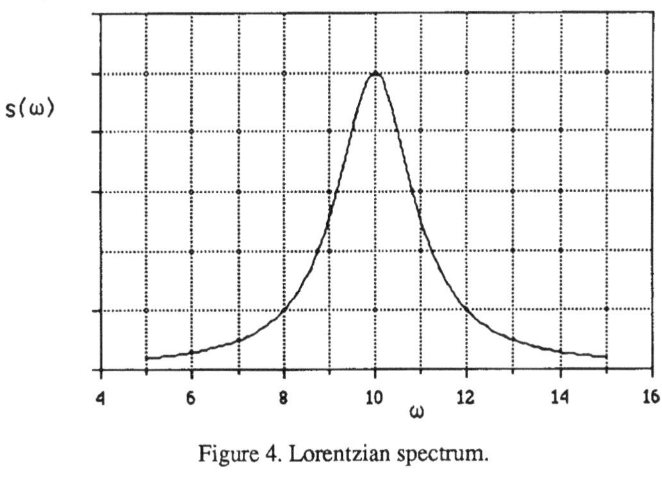

Figure 4. Lorentzian spectrum.

We find that: $g(\tau) = \exp(-\Gamma|\tau|)$, where $\tau_c = \Gamma^{-1}$ is the coherence time. $\Gamma = 0$ corresponds to a monochromatic field, $\Gamma \rightarrow \infty$ to a white spectrum.

Measurement of $g(\tau)$: Michelson interferometer

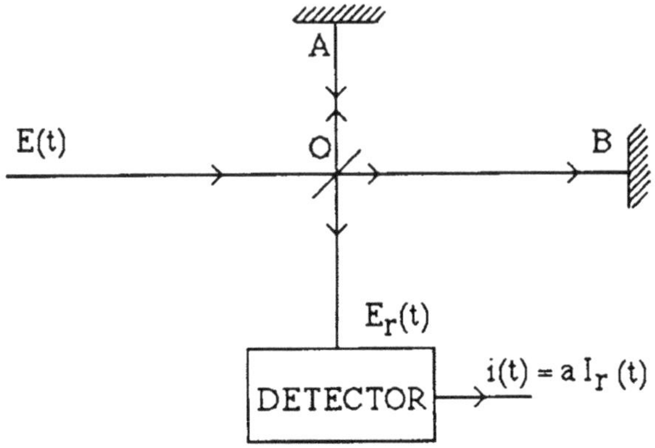

Figure 5. Michelson interferometer.

$E_R(t,\tau) = 1/2\ [E(t) + E(t+\tau)]$; $\tau = 2(OB\text{-}OA)/c$

$I_R = |E_R|^2 = 1/4\ [I(t)+I(t+\tau)+E(t)E^*(t+\tau) + c.c.]$

$\langle i(t)\rangle = a\langle I_R\rangle = a/2\ [\langle I\rangle + \mathrm{Re}\{G_1(\tau)\}]$

Note that $|g(\tau)|$ o 0 for τ o ∞ because spectrum is continuous. If we take instead:

$E(t) = E_1 \exp[-i(\omega_1 t+\phi_1)] + E_2 \exp[-i(\omega_2 t+\phi_2)]$,

we find, under the hypothesis of independent random phases,:

$G_1(\tau) = E_1{}^2 \exp(i\omega_1\tau) + E_2{}^2 \exp(i\omega_2\tau)$

If $E_1 = E_2$; $|\omega_1-\omega_2| << \omega_1$:

$\mathrm{Re}\{G_1(\tau)\} = E_1{}^2 \cos[(\omega_1+\omega_2)\tau/2]\cos[(\omega_1-\omega_2)\tau/2]$

In this case it not so easy to define a coherence time.

Space dependence : one can make a similar discussion, and define a coherence area A_c.. By combining spatial and temporal coherence properties, one can define a coherence volume V_c (cylinder with base A_c and height $c\tau_c$).

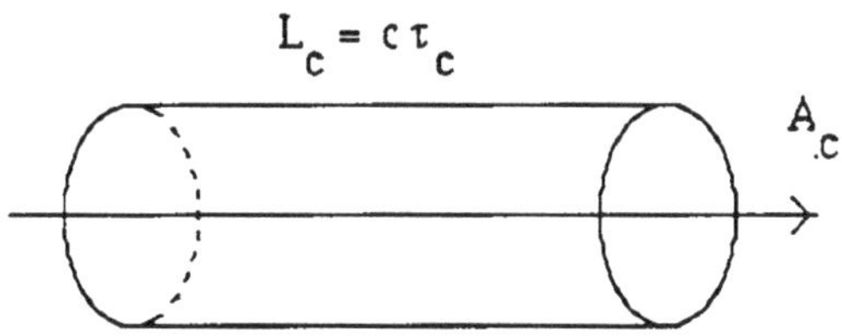

Figure 6. Coherence volume.

We have discussed sofar only first-order correlations. Why do we need higher order correlations?

-to characterize the **laser field**, that is, the field generated by an active nonlinear device.

-to overcome technical difficulties in measuring G_1: **intensity correlations** applied in stellar interferometry and in dynamic light scattering.

-ultrashort pulses.

-nonclassical fields.

1.2 Random Variables and Processes.

A continuous variable x is characterized by a probability density W(x). W(x) is normalized : $\int W(x)dx = 1$. The average of f(x) is defined as:

$$\langle f(x)\rangle = \int f(x)W(x)dx$$

The moments are: $M_k = \langle x^k\rangle = \int x^k W(x)dx$.

In particular, the variance: $v = \langle x^2\rangle - \langle x\rangle^2 = \langle \Delta x^2\rangle$, and the standard deviation: $\sigma = v^{1/2}$.

The characteristic function is:

$$Q(\lambda) = \langle e^{-i\lambda x}\rangle = \int e^{-i\lambda x}\, W(x)dx \quad = \Sigma_k(-1)^k(i\lambda)^k(k!)^{-1}M_k$$

The cumulants K_n are defined through : $\ln Q(\lambda) = \Sigma_n\, (-i\lambda)^n(n!)^{-1}K_n$

$$K_1 = M_1$$

$$K_2 = M_2 - M_1{}^2 \;=\; \langle(\Delta x)^2\rangle = \text{ variance}$$

$$K_3 = M_3 - 3M_1M_2 \;= \langle(\Delta x)^3\rangle = \text{ skewness}$$

$$K_4 = \;\langle(\Delta x)^4\rangle - 3\langle(\Delta x)^2\rangle^2$$

If W(x) is Gaussian, Q(λ) is also Gaussian

$Q(\lambda) = \exp(-i\lambda K_1 - \lambda^2 K_2/2)$, with $K_n = 0$ for $n > 2$.

Central limit theorem: given N random variables x_s ($s = 1,2,\ldots,N$), the variable $x = \frac{1}{N} \Sigma_s(x_s - \langle x_s \rangle)$ is Gaussian for large N.

The demonstration is straightforward, if one uses the cumulant expansion: W(x) is the convolution of the $W(x_s)$, therefore $Q(\lambda)$ is the product of the $Q(\lambda_s)$:

$$Q(\lambda) = \exp[-\frac{\lambda^2}{2} \sum_{s=1}^{N} K_{2s} - \frac{i\lambda^3}{6N^{1/2}} \sum_{s=1}^{N} K_{3s} + \ldots]$$

Clearly, $Q(\lambda)$ becomes Gaaussian when N goes to infinity.

A discrete variable n is characterized by a probability distribution p(n), which is normalized: $\Sigma p(n) = 1$.

The moments are: $M_k = \langle n^k \rangle = \Sigma n^k p(n)$.

The factorial moments: $F_k = \langle n(n-1)\ldots(n-k+1) \rangle$

A multidimensional continuous variable $x = (x_1,\ldots,x_N)$ is characterized by a joint probability density $W_N(x_1,\ldots,x_N)$. The electric field is a bidimensional variable , $x = (x_1,x_2)$, where x_1 and x_2 are amplitude and phase (or real and imaginary part).

Conditional probability , $W_2(x_1/x_2)$, is the probability to find x_2 given x_1.

$W_c(x_1/x_2) = W_2(x_1,x_2)/W_1(x_1)$

For statistically independent variables :

$W_2(x_1,x_2) = W_1(x_1)W_1(x_2)$, that is, $W_c(x_1/x_2) = W_1(x_2)$

For a bidimensional variable, one can define joint moments :

$M_{kj} = \langle x_1^k x_2^j \rangle = \iint x_1^k x_2^j W_2 dx_1 dx_2$

$M_{10} = \langle x_1 \rangle$; $M_{01} = \langle x_2 \rangle$; $M_{11} = \langle x_1 x_2 \rangle$, this is the lowest order correlation function.

Complex statistical variable : $z = x+iy$

$W(z)d^2z = W_2(x,y)dxdy$

$\langle z^{*k} z^h \rangle = \iint z^{*k} z^h W(z) d^2z$

If W(z) depends only on $|z|$ (circular simmetry), $\langle z^{*k} z^h \rangle = 0$ for $k \neq h$.

Let us present some examples.

Deterministic variable : $W(x) = \delta(x-x_o)$

$\langle x \rangle = x_o$; $\langle x^k \rangle = x_o^k$; $v = 0$.

Gaussian variable:

$$W(x) = \frac{1}{\sigma\sqrt{2\pi}} \exp[-\frac{(x-x_o)^2}{2\sigma^2}]$$

$\langle x \rangle = x_o$; $\langle (x-x_o)^{2k+1} \rangle = 0$; $v = \sigma^2$.

Poisson distribution:

$p(n) = (n!)^{-1} x_o^n \, e^{-x_o}$ n integer ≥ 0

$<x> = x_o$; $v = x_o$; $F_k = x_o^k$

Bose-Einstein distribution:

$p(n) = \frac{x_o^n}{(1+x_o)^{n+1}}$ n integer ≥ 0

$<x> = x_o$; $v = x_o + x_o^2$; $F_k = k! x_o^k$

Let us now consider a random process x(t). If the statistical properties do not depend on t : stationary process. Random process is specified by joint probability densities:

$W_2(x(t_1),x(t_2))$

$W_3(x(t_1),x(t_2),x(t_3))$

...

W_N

Given W_N, all W_k with $k \leq N$ are given.

Time correlation function : $<x(t_1)x(t_2)> = \iint dx_1 dx_2 x_1 x_2 W_2(x_1,x_2)$

Markov processes (short memory) are specified only by W_2:

$W_c(x(t_1),x(t_2),...,x(t_{n-1})/x(t_n)) = W_c(x_{n-1}/x_n)$

Gaussian process :W_n is a n-dimensional Gaussian function. A Gaussian process is specified only by $<x_1 x_2>$.

The electric field associated with the optical wave is a complex random process. The general expression for its correlation functions is:

$G_{nm} = <E^*(t_1)...E^*(t_n)E(t_1')...E(t_m')>$

However, since detectors of optical signals are square-law detectors (energy detectors), n is always equal to m. We can use the notation: $G_{nn} = G_n$.

Besides the field correlation function, we consider in this article second- and third-order intensity correlations, such as:

$G_2 = <E^*(t_1)E^*(t_2)E(t_2)E(t_1)> = <I(t_1)I(t_2)>$.

Up to now, we have considered the electric field as a classical variable. Let us now recall the quantum definitions: the field operator E can be written as

$\mathbf{E} = \mathbf{E}^+ + \mathbf{E}^-$, where $\mathbf{E}^+ = \Sigma_k \mathbf{a}_k \exp(-i\omega_k t)$ and $\mathbf{E}^- = \Sigma_k \mathbf{a}_k^+ \exp(i\omega_k t)$

a : annihilation operator

a$^+$: creation operator

$[\mathbf{a},\mathbf{a}^+] = \mathbf{a}\mathbf{a}^+ - \mathbf{a}^+\mathbf{a} = 1$

Coherent states are eigenstates of **a**:

$a|\alpha\rangle = \alpha|\alpha\rangle$ where α is a c-number

$$|\alpha\rangle = \exp[-\frac{|\alpha|^2}{2}] \Sigma_n \alpha^n \frac{1}{\sqrt{n!}} |n\rangle$$

Coherent state is a superposition of $|n\rangle$ states

$$|\langle n|\alpha\rangle|^2 = p(n) = \exp(-|\alpha|^2)\frac{|\alpha|^{2n}}{n!} \quad \text{(Poisson)}$$

α states are complete, but not orthogonal.

Diagonal representation of the density matrix:

$$\rho = \int d^2\alpha \, P(\alpha) \, |\alpha\rangle\langle\alpha|$$

$P(\alpha) \geq 0$ is a quasi-probability density

$$G_n = \mathrm{Tr}\{\rho \mathbf{E}^-(x_1)...\mathbf{E}^+(x_n')\}$$

Normal ordered correlation functions can be computed by using $P(\alpha)$ representation [1].

Note that $\langle a^+aa^+a\rangle = \langle n^2\rangle = \langle a^+a^+aa\rangle + \langle a^+a\rangle = \langle n(n-1)\rangle + \langle n\rangle = F_2 + M_1$

The difference between $\langle a^+aa^+a\rangle$ and $\langle a^+a^+aa\rangle$ is $O(1/n)$: many-photon field is classical.

Since, as we will discuss later on, what is actually measured is a photon number and not a field amplitude, it is important to know the relation between $P(\alpha)$ and the photon probability distribution $p(n)$. The relation is:

$$p(n) = \iint d^2\alpha \, P(\alpha) \exp(-|\alpha|^2)\frac{|\alpha|^{2n}}{n!}$$

If $P(\alpha)$ is $\delta(\alpha-\alpha_o)$, $p(n)$ is a Poisson. If $P(\alpha)$ is Gaussian, $p(n)$ is a Bose-Einstein.

1.3 Coherence

The Glauber definition [1,2] is the following: a field is coherent to the n-th order if G_n obeys the factorization relation:

$$|G_n(x_1,...,x_n;x_1',...,x_n')| = [\Pi_k G_1(x_k;x_k) \, \Pi_k G_1(x_k';x_k')]^{1/2}$$

where x_k is a space-time point.

Since $G_1(x_k;x_k) = \langle E^*(x_k)E(x_k)\rangle = \langle I(x_k)\rangle$,

the definition can be written as :

$$|g_n| = \frac{|G_n|}{\{\Pi_k\langle I(x_k)\rangle^{1/2}\Pi_k\langle I(x_k')\rangle^{1/2}\}} = 1$$

Note that Glauber definition applies also to fields which are non-stationary in time and (or) space. For a stationary field : $\langle I(x_k)\rangle = \langle I\rangle$. Therefore :

$$|g_n| = \frac{|G_n|}{<I>^n} = 1$$

First order coherence for a stationary field: $|G_1| = <I>$.

Since $G_1(\tau) = <I> \exp(i\omega_0\tau)g(\tau)$, coherence means : $|g(\tau)| = 1$, in agreement with classical definition.

Second order coherence for a stationary field: $G_2 = <I>^2$.

For a stationary field, first-order coherence implies monochromaticity, second-order coherence implies no amplitude fluctuations.

Meaning of the general definition of coherence: **no fluctuations** (in amplitude and phase).

For a normal-ordered correlation function,

$$G_k(0) = F_k$$

where F_k is the k-th factorial moment of the photon probability distribution p(n).

Coherence implies : $F_k = <n>^k$. This is the property of the Poisson distribution (coherent state).

Conventional optical sources (lamps) have only first-order coherence, whereas an ideal laser beam is coherent to all orders.

1.4 Chaotic Source.

A chotic source, such as a conventional lamp, is a collection of many independent emitters.

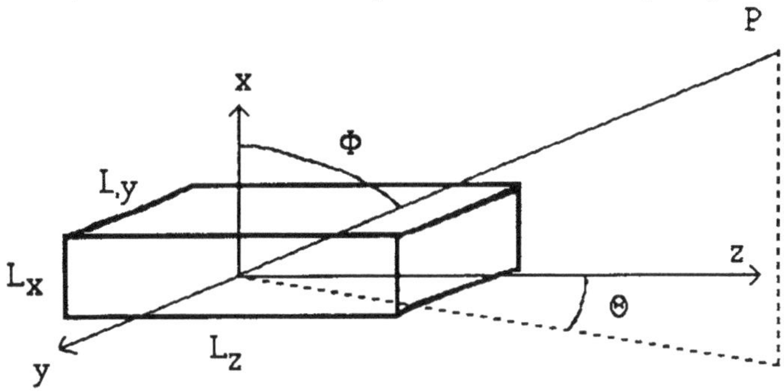

Figure 7. Chaotic source.

Let us assume that the shape of the source is that of a rectangular box. The box contains N identical atoms which are excited and emit independently one from the other. The field observed at point P is the superposition of N uncorrelated fields:

$$E(t) = \sum_{j=1}^{N} E_j(t)$$

We discuss first the temporal coherence of E(t). The field correlation function is:

$$G_1(\tau) = \langle E^*(t)E(t+\tau)\rangle = \sum_{jk=1}^{N} \langle E_j^*(t)\, E_k(t+\tau)\rangle)$$

Since $\langle E_j^*(t)\, E_k(t+\tau)\rangle = 0$ for $j \neq k$, we see immediately that

$$G_1(\tau) = \sum_{k=1}^{N} G_{1k}(\tau) = N\, G_{1k}(\tau)\, ,$$

where $G_{1k}(\tau) = \langle E_k^*(t)\, E_k(t+\tau)\rangle$. A similar calculation shows that the intensity correlation function can be written as

$$G_2(\tau) = \sum_{jklm=1}^{N} \langle E_j(t)E_k(t)E_l^*(t+\tau)E_m(t+\tau)\rangle = N\, G_{2k}(\tau) + N(N-1)\, \{G_{1k}^2(0) + |\, G_{1k}(\tau)|^2\, \}$$

When N is large, the first term at right-hand-side can be neglected, and we recover the relation expected for a Gaussian field. This is not surprising, because the emitted field is, for large N, the superposition of many independent contribution, and must, therefore, present a Gaussian probability density W(E).

Let us now discuss spatial coherence. Whereas temporal coherence depends on atomic dynamics, spatial coherence is only related to the geometry of the source. If the observation point P is in the far field, only one spatial mode will contribute to the field in P.

Elementary calculation of coherence area: take a linear source of length d.

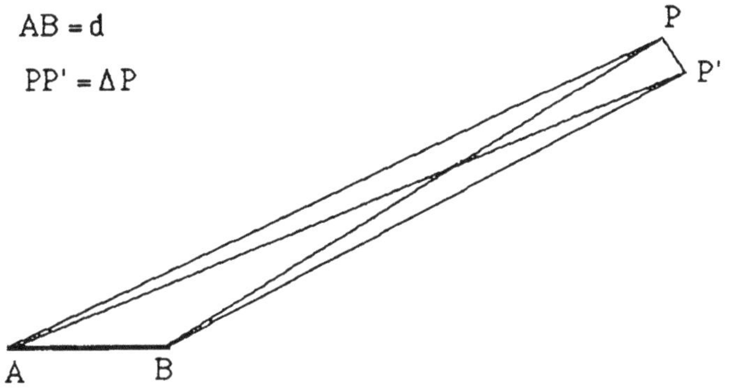

Figure 8. Spatial coherence of a linear source AB.

Let us assume that the field emitted by A and B yields constructive interference in P. Now we move from P to P'. The phase difference between the two emitted fields is, in P',:
$\Delta\phi = (2\pi/\lambda)(AP'-AP-BP'+BP) = (2\pi/\lambda)\Delta P d\sin\theta/R$
where ΔP is the distance PP' and Θ is the angle between AB and PP'. The displacement ΔP which produces a phase difference $\Delta\phi = \pi$ is:

$$\Delta P = \frac{\lambda R}{2d\sin\theta}$$

The general expression for the solid coherence angle of a rectangular source is:

$$\Omega_c(\theta,\phi) = \frac{\lambda^2}{L_yL_z\sin\theta\cos\phi+L_xL_z\sin\theta\sin\phi+L_xL_y\cos\theta}$$

The field in P is the linear superposition of many independent emission events (two-dimensional random walk) , and has, therefore, a Gaussian probability density.

$$W(E) = \frac{1}{\pi\langle I\rangle}\exp\left(-\frac{|E|^2}{\langle I\rangle}\right)$$

$$G_1(\tau) = \langle I\rangle \exp(i\omega_o\tau)\exp(-\tau/\tau_c)$$

Exponential correlation o Lorentzian linewidth (typical of natural or collisional broadening).

$$G_2(\tau) = G_1^2(0) + |G_1(\tau)|^2$$

This a very important relation, valid only for Gaussian fields. Everything is known, once first-order correlation properties are known.

$$G_2(\tau) = \langle I\rangle^2[1 + \exp(-\frac{2\tau}{\tau_c})]$$

The correlation time for intensity fluctuations is simply 1/2 of the correlation time for field fluctuations.

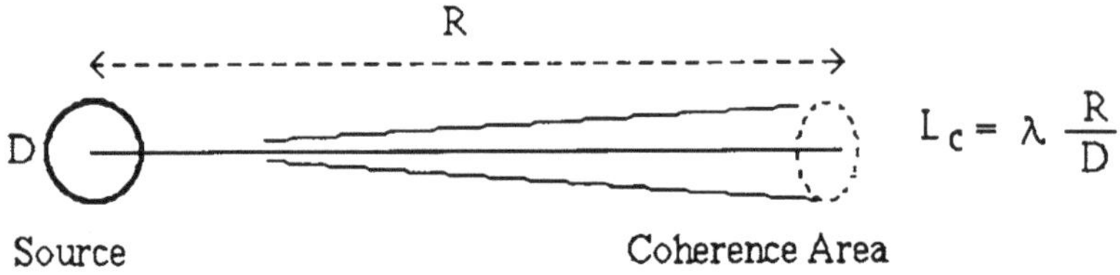

Figure 9. Stellar interferometer.

Measurement of L_c (size of coherence area) gives R/D or D (if R is known).The source may be a star.

SUN $\quad R = 150\ 10^6$ Km

$\qquad D = 1.4\ 10^6$ Km

$\qquad$ take $\lambda = 0.5\ \mu m$ o $L_c \approx 50\ \mu m$

PROXIMA CENTAURI

$\qquad R = 4.2$ light-years

$\qquad \theta = D/R \approx 0.01''$ o $L_c \approx 10$ m

If $L_c > 10$ m, it is difficult to eliminate air turbulence effects, it is better to use intensity interferometer (Hanbury-Brown & Twiss [3]).

2. Experimental Techniques

2.1 Photon Counting.

Detectors of electromagnetic waves at optical frequencies work through the absorption of e.m. energy in quantized steps (absorption of photons) and the consequent excitation (or emission) of electrons.Photon counting techniques rely on a photoelectric detector, the photomultiplier tube (PMT), which is able to give at the output a measurable electric pulse after the absorption of a single photon. If the optical signal is so weak that it is very unlikely to detect more than one photon within the response time of the photodetector, the output electric current consists of a train of randomly occurring and nonoverlapping photoelectron pulses.

Photoelectron pulses are not all identical because the PMT is a statistical device. The experimental apparatus includes a discriminator (to eliminate small spurious pulses) and a standardizer (to present pulses of standard shape to the counter). The output of the counter is sent to a multichannel analyzer, if the aim of the experiment is to measure a probability distribution, or to a correlator, if the aim is to get a correlation function [4].

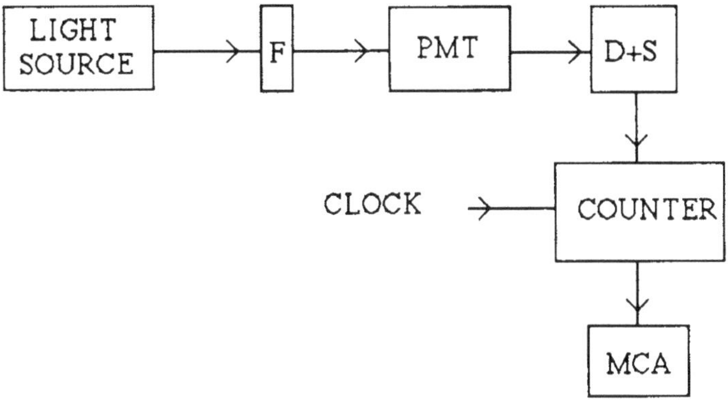

Figure 10. Photon counting experiment. F: optical filter, PMT: photomultiplier, D+S: discriminator and pulse shaper, MCA: multichannel analyzer.

Let us call p(m) the measured photoelectron probability distribution. The first question is: how is p(m) related to the true photon probability distribution p(n)? If h is the quantum efficiency of the detector (probability of obtaining a useful photoelectron as a consequence of a photon absorption process), the relation is

$$p(m) = \Sigma_n\, p(n) \binom{n}{m} \eta^n (1-\eta)^{n-m}$$

Note, incidentally, that the same relation occurs when the light beam passes through a semitransparent mirror having transmission η. Which are the properties of the transformation?

The factorial moment F_k scales with η^k:

$$F_k^m = \eta^k F_k^n$$

In particular, if p(n) is a Poisson (a Bose-Einstein), also p(m) is a Poisson (a Bose-Einstein) with an average <m> = η<n>.

Experimental problems:

1) the useful area of the detector must be much smaller than the coherence area of the field
2) the sampling interval T should be chosen as: $\tau_D \ll T \ll \tau_{c2}$, where τ_D is the dead time of the apparatus and τ_{c2} is the correlation time for intensity fluctuations.
3) subtraction of uncorrelated noise: stray light, spurious photoelectron pulses,...

Joint photocount distributions: similar measurement technique, but a multidimensional memory is needed!

2.2 Digital Correlator.

Measurements of the intensity correlation function $G_2(\tau)$ are performed by digital correlators. The scheme of the experiment is the same as that of Fig.10, with the digital correlator substituting the MCA. A block diagram of the correlator is shown in Fig.11 [5].

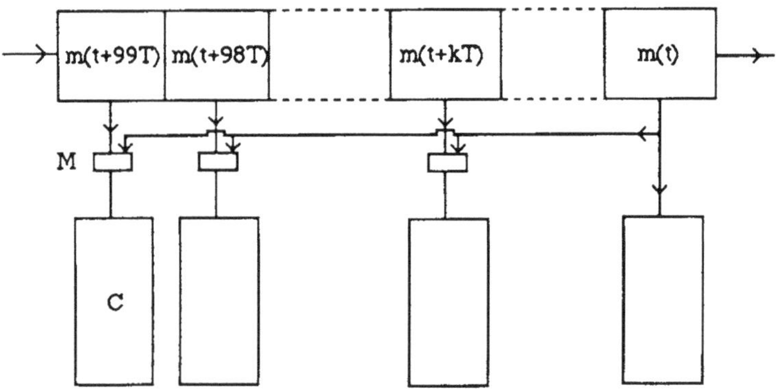

Figure 11. Scheme of a digital correlator. M: multiplier, C: counter. The shift-register has 100 cells. m(t), before being kicked out by the sample m(t+100T), is multiplied by all the samples m(t+kT), with k=1,...,99. The result of each multiplication is stored into the appropriate counter. Present state of the art: 4 bits, 20 MHz.

Call m(t,T) the number of photoelectron pulses counted in an interval T centered around time t. The autocorrelation function of the random variable m is

$$R(\tau,T) = \langle m(t,T)m(t+\tau,T)\rangle$$

If $T \ll \tau_{c2}$, then

$$R(\tau,T) \div G_2(\tau)$$

The maximum number of photoelectron pulses m recorded in the interval T is limited by the dead time of the apparatus: $m_{max} \leq T/\tau_D$. If real time operation is desired at high sampling frequency, another limit is set by the fact that large numbers require long multiplication and registration times.

In the low-frequency range, it is possible to build "software correlators" by feeding the signal directly into a computer without intermediate hardware.

The technique can be extended to higher order intensity correlations. Third-order correlations have been measured on laser light near threshold [6].

In some cases, it is possible to measure G_n directly by using multiphoton absorption or harmonic generation (picosecond-pulse techniques, see later).

2.3 Reference-beam technique (heterodyne)

Intensity correlations can be exploited to measure $G_1(\tau)$ by using the reference-beam technique. The method does not rely on the assumption that the field is Gaussian.

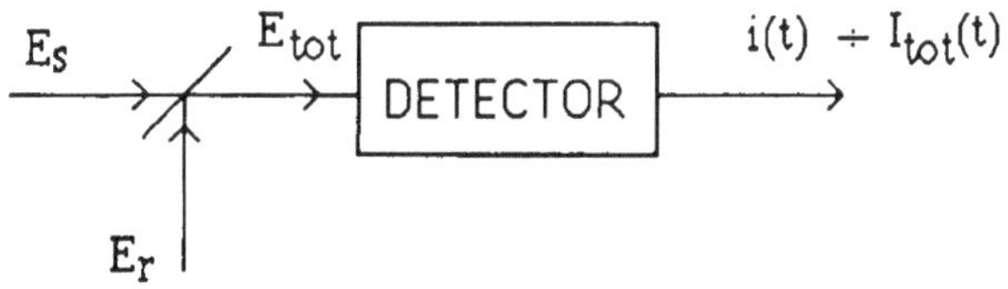

Figure 12. Heterodyne technique.

E_s is the field we want to study, E_r is a reference field. $E_{tot}(t) = E_s(t) + E_r(t)$

$I_{tot} = |E_{tot}|^2 = I_s(t) + I_r(t) + 2\,\mathrm{Re}\{E_s(t)E_r^*(t)\}$

$$G_{2tot} = G_{2s} + G_{2r} + 2\,\mathrm{Re}\{G_{1s}G_{1r}^*\}$$

Put: $E_s(t) = E_{os}(t)\exp[i(\omega_{os}t+\phi_s(t))]$

$$E_r(t) = E_{or}\exp[i(\omega_{or}t+\phi_r)]$$

with E_{os} and ϕ_s random variables, E_{or} and ϕ_r deterministic. We can write:

$$G_{2tot}(\tau)= \langle I_s\rangle^2(1+g_{2s}(\tau)) + I_r + 2I_r\langle I_s\rangle\cos[(\omega_{os}-\omega_{or})\tau]\, g_{1s}(\tau)$$

If $<I_s> \ll I_r$, the first term at r.h.s. can be neglected, and g_{1s} can be easily derived from G_{2tot}.

3. Scattered Fields.

A chaotic field can be generated by a conventional lamp or by a scattering experiment. Lamps have very short coherence times (typically $\tau_c \leq 1$ns) and very few photons per mode. It is therefore very difficult to verify that p(m) is a Bose-Einstein or to measure G_2 with a lamp. However, one can make a pseudo-chaotic source by using a coherent source (laser) plus a random scatterer (a ground-glass disk, in the first experiments by Martienssen and Spiller and by Arecchi [3]). In such a case, τ_{c2} is determined by the rotation velocity of the disk and can be controlled at will.

We will now shortly discuss the coherence properties of the field scattered by a dispersion of particles undergoing Brownian motion.

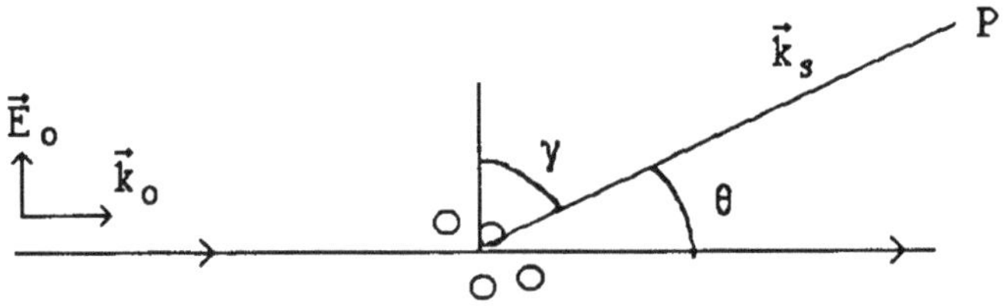

Figure 13. Light scattering experiment. P is the observation point, the incident beam is linearly polarized in the vertical direction .

A basic quantity is the scattering vector $\vec{k} = \vec{k}_s - \vec{k}_o$, which is the difference between the wave vectors of the scattered and incident beam. Since $|k_s| = |k_o| = k_o$, $k = 2k_o\sin(\theta/2)$, where θ is the scattering angle.

A scattering particle having linear size a behaves as a point scatterer if $ka \ll 1$.

Field scattered by a small particle [7]:

$$E_s(R,t)=\hat{k}_s\cdot(\hat{k}_s\cdot\hat{E}_o)\frac{1}{4\pi\varepsilon R}\,\alpha\exp[i(\omega_o t-\vec{k}_s\cdot\vec{\rho})]e^{i\vec{k}\cdot\vec{\rho}}$$

where α is the polarizability of the particle, $\alpha = (n^2-n_o^2)V$. n and n_o are the indices of refraction of the particle and of the solvent, V is the particle volume.

The scattered intensity:

$$I_s = I_o \,(k_o/n_o)^4 (n^2-n_o^2)^2 V^2 \frac{\sin^2\gamma}{16\pi^2 R^2}$$

If the particle is moving, $\mathbf{r}$ is a function of time, and therefore the phase of the scattered field, ϕ_s, will take a temporal behavior which contains information on particle dynamics. There are two important cases:

a) particle with uniform deterministic velocity:

$$\mathbf{r} = \mathbf{r}_o + \mathbf{v}t$$

$$\phi_s = \omega_o t - \mathbf{k}_s \cdot \mathbf{R} + \mathbf{k} \cdot \mathbf{v}t$$

Scattered field is monochromatic, but it is frequency shifted (Doppler effect).

Example: $\mathbf{v}$ parallel to $\mathbf{k}$, $\lambda = 500$ nm (green light), $\theta = \pi$, o $\Delta\nu_D = 40$ kHz

b) particle undergoing Brownian motion: $\mathbf{r}$ is a random variable, the displacement $\Delta\mathbf{r} = \mathbf{r} - \mathbf{r}_o$ is described by a tridimensional Gaussian probability distribution with variance

$\langle \Delta\mathbf{r}^2 \rangle = 6Dt$, where D is the translational diffusion coefficient of the particle.

The scattered field is randomly phase modulated.

$$G_1(\tau) = I_s \exp(i\omega_o\tau) \langle \exp\{ik \cdot [\mathbf{r}(t) - \mathbf{r}(t+\tau)]\}\rangle$$

The exponential can be written as a series expansion:

$$\exp\{ik[r_k(t) - r_k(t+\tau)]\} = \exp(ik\Delta r_k(\tau) = \Sigma_j (ik)^j \frac{[\Delta r_k(\tau)]^j}{(j!)^{-1}}$$

The moments of the Gaussian variable Δr_k are

$$\langle \Delta r_k^{2j+i} \rangle = 0$$

$$\langle \Delta r_k^{2j} \rangle = (2j-1)(2j-3)\ldots\langle \Delta r_k^2 \rangle^j$$

Recalling that $\langle \Delta r_k^2 \rangle = 2Dt$, we obtain

$$G_1(\tau) = I_s \exp(i\omega_o\tau) \exp(-Dk^2\tau)$$

The spectrum is:

$$S(\omega) = I_s \frac{(Dk^2/\pi)}{(Dk^2)^2 + (\omega - \omega_o)^2}$$

For a spherical particle, the Einstein-Stokes relation holds:

$$D = \frac{k_B T}{6\pi\eta a}$$

where k_B is the Boltzmann constant, T the absolute temperature, and η the viscosity of the solvent

Example: a = 2.5 nm (protein, micelle), $\eta = 1$ cp, $n_o = 1.33$ (water), T = 293°K, $\lambda = 0.5$ µm,

$\theta = 90°$. We obtain: $D = 10^{-6}\ cm^2/s$, and $k^2 = 5\ 10^{10}\ cm^{-2}$. The linewidth is:

$$\Delta\nu = \frac{Dk^2}{2\pi} = 8kHz$$

Both $\Delta\nu_D$ and $\Delta\nu$ are very small in comparison with the optical frequency, and are, in any case, too small to be measured with optical techniques. In fact :

a) no optical instrument has the adequate resolution: a Fabry-Perot (or a Michelson) interferometer have, at best, a resolution of a few MHz.

b) the laser linewidth itself is much larger than the frequency shift (or broadening) to be seen.

If the scattering volume contains many particles, the scattered field is the superposition of many scattering events (all independent, if particles are not interacting), and is, therefore, a Gaussian process. We can obtain G_1 through a measurement of the intensity correlation G_2 , as first shown by Benedek and Ford, and by Cummins et al. in 1965 [5,7].

4. Single-mode Steady-state Laser

I shortly recall the treatment of the statistical behavior of a single-mode laser in the threshold region by using the Langevin equation approach and the notation first used by Risken [8].

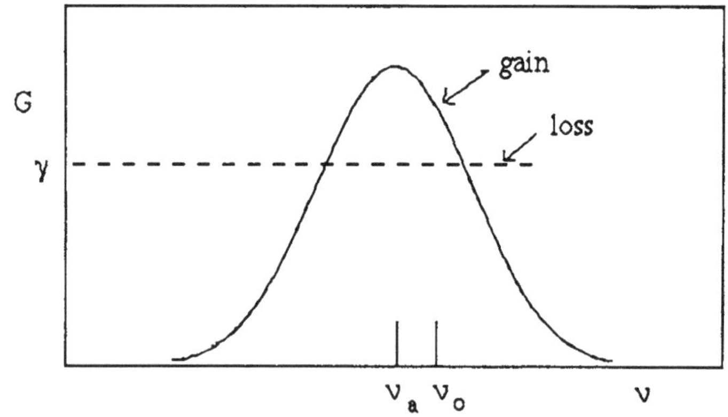

Figure 14. Laser gain profile. γ is the loss level, ν_a is the central atomic frequency and ν_o is the frequency of the lasing mode.

The electric field of the lasing mode is $E(t) = \alpha(t) \exp(-i\omega_0 t)$, where a(t) is a slowly varying amplitude.

We first recall the damped-harmonic-oscillator treatment: in terms of a Langevin equation:

$$\dot{\alpha} + \gamma\alpha = \Gamma(t)$$

$\Gamma(t)$: noise source, complex Gaussian process. $<\Gamma(t)> = 0$; $<\Gamma^*(t)\Gamma(t+\tau)> = Q\delta(\tau)$

Assume that $\alpha(t)$ is a Markov process, fully characterized by the conditional probability $W_c(\alpha_0,0/\alpha,t)$.

From the Langevin equation one can derive a Fokker-Planck equation for W_c:

$$\frac{\partial W_c}{\partial t} - \gamma \mathrm{div}_\alpha(\alpha W_c) = \frac{Q}{4}\Delta W_c$$

Solution: $W_c = \frac{1}{\pi\sigma^2(t)} \exp[-\frac{|\alpha-\alpha_0\exp(-\gamma t)|^2}{\sigma^2(t)}]$

This is a Gaussian probability density with a variance $\sigma^2(t) = \frac{Q}{2\gamma}[1-\exp(-2\gamma t)]$.

Stationary solution: $W(\alpha) = \frac{2\gamma}{\pi Q} \exp[-\frac{|a|^2}{(Q/2\gamma)}]$

$\alpha(t)$ may represent a single mode of a chaotic field with a Lorentz spectrum.

In the case of the single mode laser, we add a saturable gain $G = g_0 - \beta|\alpha|^2$

$$\dot{\alpha} + (\gamma - G)\alpha = \Gamma(t) \qquad \text{or} \qquad \dot{\alpha} - (d - |\alpha|^2)\alpha = \Gamma(t)$$

where $d = \frac{g_0-\gamma}{\beta}$.

The associated Fokker-Planck equation is:

$$\frac{\partial W_c}{\partial t} + \beta\, \mathrm{div}_\alpha[(d-|\alpha|^2)\alpha W_c] = \frac{Q}{4}\Delta W_c$$

Put $\alpha = re^{i\phi}$,

$$\frac{\partial W_c}{\partial t} + \frac{\beta}{r}\frac{\partial}{\partial r}[(d-r^2)rW_c] = \frac{Q}{4}[\frac{1}{r}\frac{\partial}{\partial r}(r\frac{\partial W_c}{\partial r}) + \frac{1}{r^2}\frac{\partial^2 W_c}{\partial\phi^2}]$$

Stationary solution:

$$W(r,\phi) = N \exp[-(\beta/Q)r^4 + (2\beta d/Q)r^2]$$

Far below threshold ($d \ll -1$):

$$W(r) = N \exp[-(2\beta|d|/Q)r^2]$$

This is a Gaussian distribution with variance $<r^2> = \frac{Q}{2\beta|d|}$. The associatedp(n) will be a Bose-Einstein.

Far above threshold ($d \gg 1$):

$$W(r) = N \exp[-\frac{4\beta d}{Q}(r-\sqrt{d})^2]$$

This is a displaced Gaussian. The associated p(n) will become close to a Poisson far from threshold.

At threshold: intermediate situation.

The correlation functions far below threshold are:

$$|G_1(\tau)| = \frac{Q}{2\beta|d|} e^{-\beta|d|\tau}$$

$$G_2(\tau) = (\frac{Q}{2\beta|d|})^2 [1 + e^{-2\beta|d|\tau}]$$

Far above threshold:

$$|G_1(\tau)| = de^{-(Q/4d)\tau} + \frac{Q}{8\beta d} e^{-(2\beta d+Q/4d)\tau}$$

$$G_2(\tau) = d^2 + \frac{Q}{2\beta} e^{-2\beta d\tau}$$

Note that above threshold intensity fluctuations are very small, the main noise is phase diffusion noise.

The behavior of the laser near threshold is similar to that of a system in thermal equilibrium near a critical point [9]. At the basis of this analogy is the fact that one can define a laser "potential" or "free energy" as

$$\dot{\alpha} + \frac{\partial V}{\partial \alpha} = \Gamma(t)$$

The theoretical predictions have been verified in detail by many experiments, mostly performed on single-mode He-Ne lasers. A description of the earlier measurements can be found, for instance, in Ref. 4. I will only mention here a less well-known experiment which involves third-order intensity correlations [6]. Such an experiment was performed by using a specially built correlator which can measure in a single experimental run $g_3(\tau,\tau')$ for fixed τ' and for 100 values of τ. The function g_3 is defined as:

$$g_3(\tau,\tau') = \frac{\langle I(t)I(t+\tau)I(t+\tau')\rangle}{\langle I\rangle^3}$$

From the measured g_3 and g_2 we derive the normalized cumulant function

$$K_3(\tau,\tau') = \frac{\langle \Delta I(t)\Delta I(t+\tau)\Delta I(t+\tau')\rangle}{\langle I\rangle^3} = g_3(\tau,\tau') + 2 - g_2(\tau) - g_2(\tau') - g_2(\tau-\tau')$$

which is reported in Figs. 15 and 16 for two distinct values of the threshold parameter $a = \beta d/Q$.

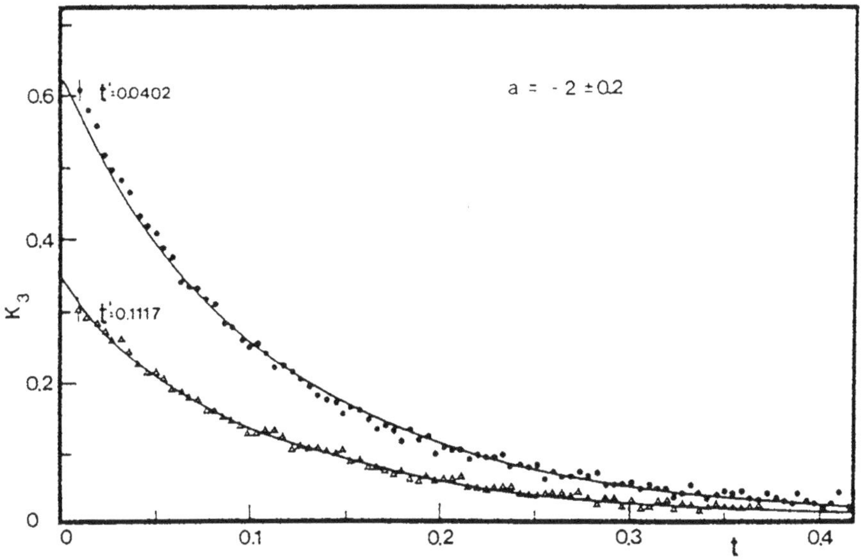

Fig.15. The third order cumulant $K_3(t,t')$ versus t for two distinct values of t'. The laser is operating below threshold. The delays t and t' are expressed in dimensionless form, as explained in the text. (from Ref.6)

Full lines in the figures represent theoretical predictions.

There are lasers, such as the dye laser, for which the simple Langevin-equation model is not adequate to describe experimental results. A possible complication of the model consists in the introduction of multiplicative noise. The Langevin equation is transformed into the following:

$$\dot{\alpha} - (d - |\alpha|^2)\alpha = \Gamma(t) + \xi(t)\alpha(t)$$

where both $\Gamma(t)$ and $\xi(t)$ are complex white-noise processes with zero average:

$$\langle\xi(t)\rangle = 0; \quad \langle\xi^*(t)\xi(t')\rangle = 4P\delta(t-t').$$

Recently, a model experiment was performed by using a single-mode laser with an externally added source of multiplicative noise [10]. I show in Fig.17 some of the reported results. The lower curve corresponds to zero multiplicative noise, that is, to the situation of a regular laser. We see that the addition of multiplicative noise has a dramatic effect near threshold (the laser threshold corresponds to $\langle I\rangle = 1$ for the plot of Fig.17).

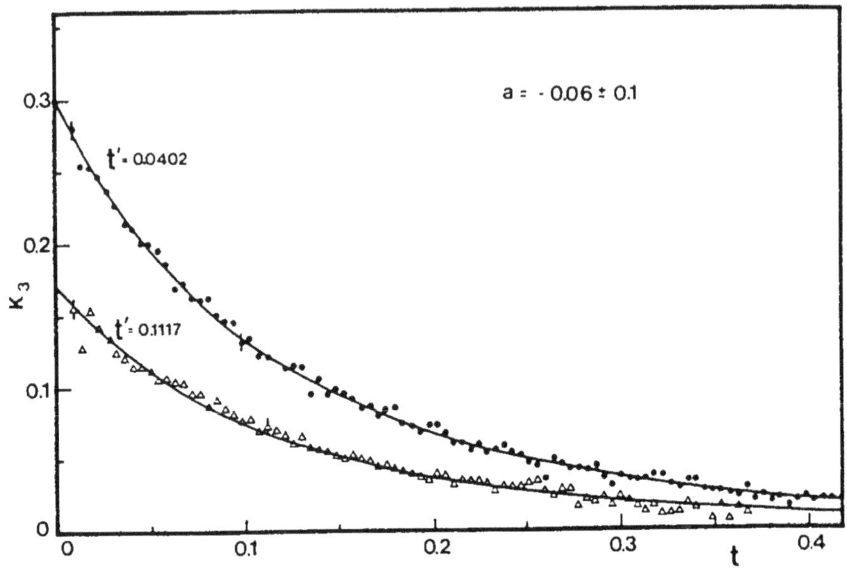

Fig.16. The third order cumulant $K_3(t,t')$ versus t for two distinct values of t'. The laser is operating at threshold. The delays t and t' are expressed in dimensionless form, as explained in the text. (from Ref.6)

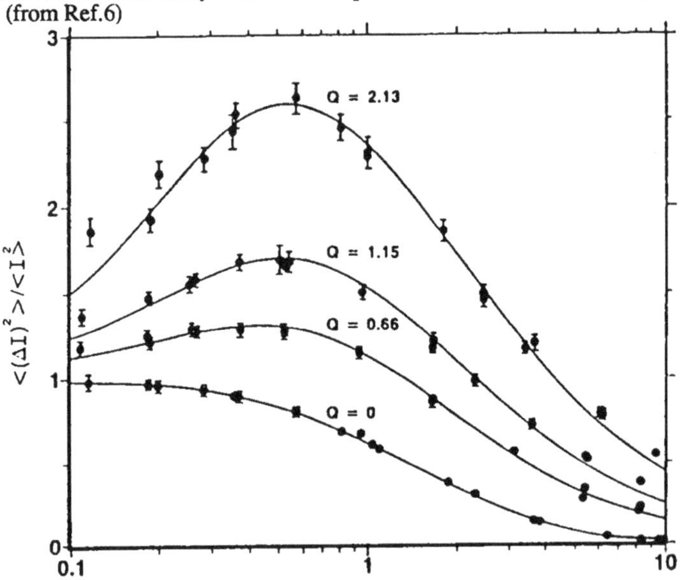

Fig.17.Comparison of theoretical and measured normalized variance. Mean intensity is in dimensionless units.Statistical uncertainties, whenever are larger than the dot size, are shown

5. Laser Transients.

5.1 Q-switched Gas Laser

Statistical properties of a single-mode transient: fluctuations are large even far away from threshold. It is interesting from the general point of view: decay of an unstable state in absence of systematic forces. The decay is initiated by a fluctuation. The evolution of the average intensity: exponential growth (linear amplification regime) for short times, saturation becomes important around the inflection point. The evolution of the variance: huge peak in correspondence with the inflection point. A detailed description of the results can be found in Ref.4. I will simply recall here that the behavior of the system is very well described by a simple phenomenological model which assumes a deterministic evolution from a statistically defined initial condition. The model was later extended to superradiant pulses and to Q-switched solid-state lasers.

The deterministic equation describing the transient intensity of a single-mode gas laser is:

$$\dot{I} = aI - 2\beta I^2 \qquad \text{with the initial condition: } I(0) = I_0.$$

The solution is:

$$I(t,I_0) = \frac{I_0 I_s}{I_s e^{-at} + I_0(1-e^{-at})}$$

where $I_s = \dfrac{a}{2\beta}$.

I_0 is a statistical variable with a probability density

$$p(I_0) = \frac{1}{\langle I_0 \rangle} \exp\left(-\frac{I_0}{\langle I_0 \rangle}\right)$$

(this is like a Bose-Einstein with large $\langle n \rangle$).

Time evolution of the moments of I(t):

$$\langle I^k(t) \rangle = \int_0^{\infty} I^k(t,I_0)\, p(I_0)\, dI_0$$

As an example, the relative variance

$$v(t) = \frac{\langle \Delta I^2(t) \rangle}{\langle I^2(t) \rangle}$$

can be expressed as an universal function of the scaled variable z

$$z = \frac{I_s e^{-at}}{I_0(1-e^{-at})}$$

$$v(z) = \frac{z[1-H(z)]-H^2(z)}{[1-H(z)]^2} \quad , \quad \text{where} \quad H(z) = ze^z \int_z^1 \frac{e^{-y}}{y} dy \quad .$$

The general conclusions concerning the treatment of superradiant or Q-switched pulses are: if the laser is single-mode, and is sufficiently above threshold, the intrinsic fluctuations affect negligibly the shape and area of the pulse. The only macroscopic effect is a fluctuation in the time position of the pulse with respect to the switching instant. Such a time jitter is due to the fact that the initial photon number at time $t = 0$ is a random variable with a broad distribution (Bose-Einstein). The time jitter is comparable to the pulse duration.

5.2 Short Pulses.

The measurement of the duration of ultrashort pulses ($\tau \leq 100$ ps) is performed by some nonlinear technique, such as second harmonic generation or multiphoton fluorescence, which yelds an intensity correlation function of the optical field. I show in Fig.18 a scheme of a two-photon fluorescence experiment.

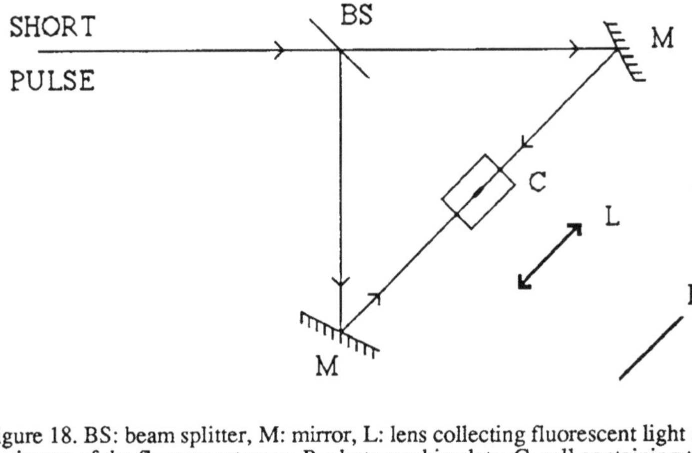

Figure 18. BS: beam splitter, M: mirror, L: lens collecting fluorescent light and creating on P the image of the fluorescent zone, P: photographic plate, C: cell containing the solution of the two-photon-fluorescence dye.

Fluorescence signal:

$$F(\tau) = A \int |E_1(t)+E_2(t+\tau)|^4 dt = A\{G_2(0) + 2G_2(\tau) + \text{rapidly varying terms}\}$$

where $G_2(\tau) = \langle I(t)I(t+\tau)\rangle$. The behavior of $F(\tau)$ is shown in the upper part of Fig.19.

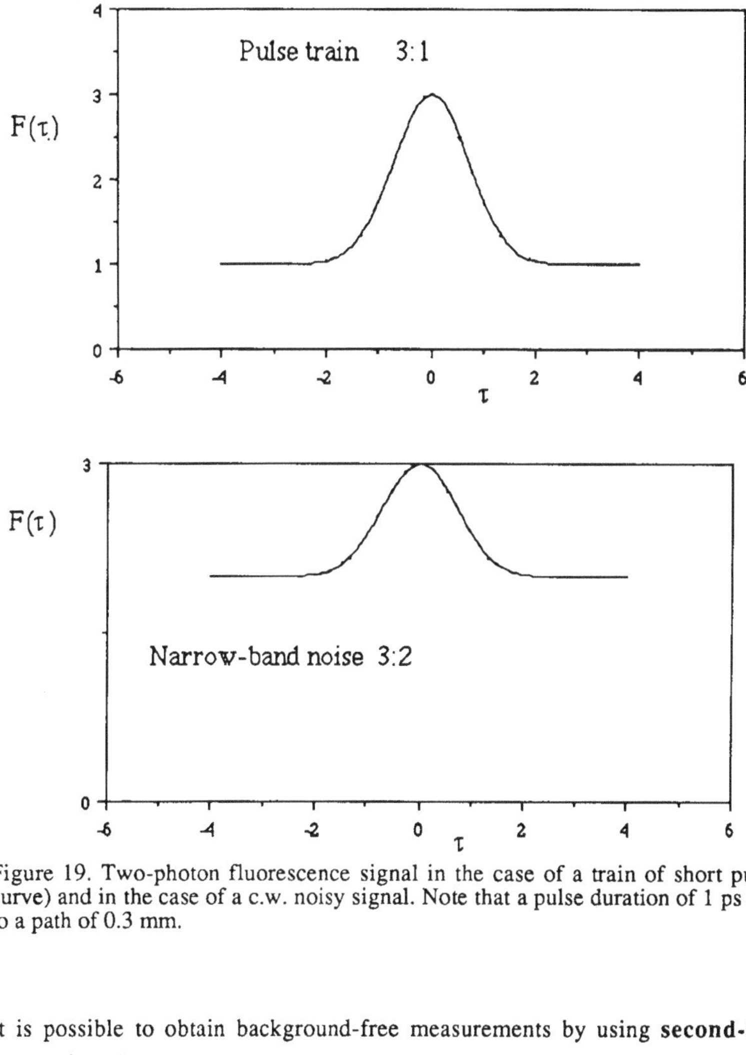

Figure 19. Two-photon fluorescence signal in the case of a train of short pulses (upper curve) and in the case of a c.w. noisy signal. Note that a pulse duration of 1 ps corresponds to a path of 0.3 mm.

It is possible to obtain background-free measurements by using **second-harmonic-generation** (SHG), as shown schematically in Fig.20.

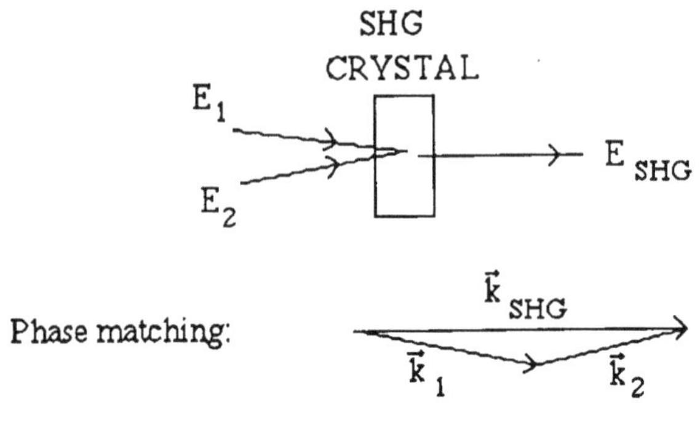

Figure 20. Measurement of pulse duration by SHG. The correlation function is obtained by varying the delay between E_1 and E_2. E_{SHG} is $\neq 0$ only if E_1 and E_2 are simultaneously present:

$$F_{SH}(\tau) = \int |E_1(t)E_2(t+\tau)|^2 dt = \{G_2(\tau) + \text{rapidly varying terms}\}$$

There are commercial optical autocorrelators using SHG. It can be used only with a long train of pulses (in order to scan a wide range of delays τ), whereas TPF works with a single pulse.
Typically, the translation stage : 2 μm per step, that is, 12 fs per step (1 fs = 10^{-15} s).

References.

1. R.J. Glauber, in "Quantum Optics and Electronics", edited by C. De Witt et al., Gordon and Breach, New York, 1965.
2. R. Loudon "The Quantum Theory of Light", Clarendon Press, Oxford, 1985.
3. R. Hanbury-Brown "The Intensity Interferometer", Taylor and Francis, London, 1974.
4. F.T. Arecchi and V. Degiorgio, in "Laser Handbook", edited by F.T. Arecchi and E.O. Schulz-DuBois, North-Holland, Amsterdam, 1972, p. 191. This article contains the relevant references also to the important work done by Armstrong and Smith, Freed and Haus, Mandel and coworkers, and others.
5. H.Z. Cummins and E.R. Pike Eds., "Photon Correlation and Light Beating Spectroscopy", Plenum Press, New York, 1974.
6. M. Corti and V. Degiorgio, Phys. Rev. A **14**, 1475 (1976)

7. B.J. Berne and R. Pecora "Dynamic Light Scattering", Wiley, New York, 1976.

8. H. Risken, Progress in Optics **8**, 241 (1970)

9. V. Degiorgio and M.O. Scully, Phys. Rev. A **2**, 1170 (1970); R. Graham and H. Haken, Z. Phys. **237**, 31 (1970).

10. M.R. Young and S. Singh, Optics Lett. **13**, 21 (1988). This paper contains the relevant references about theoretical and experimental work on the statistical properties of dye lasers by Roy and Mandel, Schenzle and Brand, and others.

AN INTRODUCTION TO LASER INSTABILITIES, TO THE LASER-LORENZ MODEL and TO THE BEHAVIOR OF A SINGLE MODE FIR AMMONIA LASER

N.B. Abraham*
Laboratoire de Spectroscopie Hertzienne, CNRS
Université de Sciences et Techniques de Lille, Flandres Artois
U.F.R. de Physique, Bâtiment P5
F-59655 Villeneuve d'Ascq, Cedex, FRANCE

*permanent address: Department of Physics, Bryn Mawr College,
Bryn Mawr, PA 19010-2899 USA
NEAL@BRYNMAWR.bitnet

Introduction

One of the simplest models for spontaneous pulsations in a single mode laser is that involving interaction with a homogeneously broadened, two-level medium with the laser in a unidirectional ring configuration. We review the characteristics and signatures of the instability thresholds and of the phase diagrams for different types of pulsations. We then present the experimental results from studies of an optically-pumped Far-Infrared laser with ammonia as the active medium. We show that in many respects this laser closely emulates the model though puzzles remain as to exactly how to justify the description of this laser by the simple set of five equations used in the model.

The structure of this chapter is as follows: It begins with a brief overview of the early history of the study of laser instabilities and a survey of the collections of recent research in the field. Then the classic model for "coherent" single mode laser instabilites the "Maxwell Bloch" or "laser-Lorenz" model is reviewed in detail because of its role as a paradigm for single mode laser instabilities over more than thirty years of maser and laser research. For comparison, the experimental characteristics of a Far-Infrared (FIR) ammonia laser that closely emulates the laser-Lorenz model are presented and discussed.

History and collections of recent research

As a maturing field, optical instabilities now offers a variety of weighty references which provide entry points for experts and beginners /1-19,24/. These reflect a series of international schools and conferences held over the last three years which have focussed on these subjects, the emergence of many of these subjects from the shadow of a foster guardian, optical bistability (which has also nutured independent branches on photonic switching and optical computing), and by the appearance of monographs, reviews, special issues of journals and collections of research articles.

For summaries of the three optical instabilities conferences (June 1985, June 1986, and July 1987) there are proceedings volumes for the first two /2,3/, a detailed summary of the third included in the lecture notes volume of a preceeding school /5/, and technical digests are still available for the first /16/ and third /17/. *Optical Instabilities* /2/ is a product of the first international conference on the subject, held in Rochester, NY, USA in June 1985, and is notable for its inclusion of 16 review articles written by speakers who presented tutorial overviews. It also includes 53 three-page summaries of a selection of the contributed papers. *Optical Chaos* /3/ is the proceedings of a smaller meeting held the following year in Quebec, CANADA. Already in this second volume one sees the broadening of the field and a breaking down of barriers in classifications. The next major international conference on Optical Instabilities and related subjects (in this case combined with Optical Bistability) will be held in June 1990 in Afton, OK in the USA and the proceedings will be published by the Optical Society of America.

Collections of research papers include two special issues of the Journal of the Optical Society of America B (January 1985 and May 1988) which managed to capture a major percentage of the ongoing research work as of their deadlines (September 1984 and October 1987, respectively). *Instabilities and Chaos in Quantum Optics* /4/ is a collection of 12 chapters written by researchers in particular fields and provides both tutorial reviews and research progress in the specific fields as of the closing dates of the articles which ranged from 1985 to 1986.

There have now been three schools substantially or entirely devoted to this subject for which there have been or will soon be publications of the lecture notes or proceedings. *Laser Physics and Laser Instabilities* /11/ presents updated and revised lecture notes from a school organized at Jilin University in Changchun, China in August of 1985. *Instabilities and Chaos in Quantum Optics II* /5/ is a volume in the NATO Advanced Study Institutes Program, with its first 23 chapters representing the lecture notes from a school held in Il Ciocco, Italy in July 1987. The last chapter offers a topical overview of the three-day international research workshop (the second large Optical Instabilities meeting) that followed the school and it also provides a set of references updated through December 1987 on various fields not fully covered by the lectures in the school. *Laser Physics and Quantum Optics* /15/ (this present volume) represents the proceedings of a Latin American School in August 1988 which was substantially devoted to this field. *Dynamics of Nonlinear Optical Systems* are the proceedings of a gathering in Santander, SPAIN, in October 1988 at which the sometimes disparate fields of study of Instabilities, noise driven systems and squeezing were shown to be inexorably intertwined /15a/.

Other notable reviews of the field include articles in journals /6,7/, review chapters /17/, reviews in the *Progress in Optics* series /10,18/ and books titled *Chaos in Laser Matter Interactions* /8/, *Laser Instabilities* /12/, and others that are provisionally expected from such authors as Khanin /14/ and Casperson /13/.

In a rapidly moving field, proceedings volumes and reviews often quickly become outdated, and special issues of journals and collections of research chapters fare little better than providing an historical snapshot of the state of the field at the time the material was prepared. Often by the time the material is published the field has moved to more interesting subjects or there have been important new discoveries which clarify or refute the published claims. While these almost universal maxims necessarily apply to this field as well, the five major references /1,2,4,5/ continue to serve the field as important resources that merit their acquisition by libraries and researchers alike. Not the least of these values is the remarkable concentration in these five works of both review material and the latest of research progress. A researcher entering the field or changing topic within the field can find discussions of most issues and nearly complete references in this select group of sources. One expects that in a few yeears one might turn instead to textbooks and recent reviews such as /8,10-14/ but the critical judgement on these volumes is not yet available.

The fact that so much information on this active field is concentrated in a few volumes may be due in part to a dormancy period in the study of laser instabilities. The field made great progress from the earliest days of lasers and masers in describing potential instabilities, but the divergence between the instabilities found in simple laser models and the instabilities arising from more complicated dynamics or technical designs left much of the study of nonlinear optical dynamics to an academic exercise. However, there was a rich literature and tradition available to be assembled when Haken in 1975 /20/ observed that the laser had the nonlinear dynamics and chaos of the better studied Lorenz equations /21/ of convective hydrodynamics, when Bonifacio, Lugiato, McCall and others in 1978 /22/ began a systematic study of instabilities in models of optical bistability (OB), when Ikeda in 1979 /23/ discovered chaos in delay equations used to model some OB systems (see also the wealth of material in OB since then /24,18,19/), when Casperson in 1978 /25/ systematically unmasked the origin of dynamical instabilities in inhomogeneously broadened lasers through his studies of the Lamb equations and the behavior of xenon lasers and when Arecchi and coworkers in 1982 /26/ found and explained chaotic behavior in a modulated CO_2 laser. In fact, it seems that it was the merger of the nonlinear dynamical studies of laser pulsations and mode interactions with the more popular studies of chaos, period doublings, fractals, and strange attractors that welded the field around a new focus and signalled its emergence as a significant subfield.

Classifications

Most overviews of laser problems begin with a sorting of the many different topics. Unfortunately no classification is entirely satisfactory and so one must be prepared to disrupt the neat order of the arrangement of topics almost as quickly as it has been prepared. More subtle is the fact that classifications which are appropriate to the steady state and single-mode stability analyses may break down when the dynamical pulsations carry the system far from equilibrium or when intermode beat frequencies excite rapid time variations. Nevertheless, and not only because the literature is replete with these classifications, it is useful to begin by looking at some of them.

Among the various ways of sorting problems and issues are:

a) coherent equations versus rate equations versus field equations,

b) field-matter interactions in a cavity versus interacting beams,

c) vectorial versus scalar fields,

d) optically thin versus optically thick media,

e) plane wave versus transverse profiles,

f) ring cavities versus Fabry-Perot cavities,

g) 'active media' versus 'passive media', and

h) single mode versus multimode.

Some deails of the meanings of these different classifications are presented below.

a) Coherent equations versus rate equations versus field equations.

The field-matter interaction may be described by a "coherent" interaction involving the off-diagonal elements of the material density matrix, that is, involving the material polarization in a dynamical way, or it may be described in terms of a reduced set of variables. This is usually justified by an adiabatic elimination of the variable(s) that relaxes rapidly (because of a large decay rate) to a kind of adiabatic equilibrium with the more slowly evolving variables. The details of the justification of such an adiabatic elimination are nontrivial and are addressed in such references as /27/. The reduced set of equations describing only the intensity of the electromagnetic field and the population difference in the medium on the two levels interacting with the field is called the "rate equation approximation" and appears to adequately describe many phenomena in a rather large class of lasers such as ruby, YAG, other solid-state lasers, and CO_2 and other molecular gas lasers at moderate pressures. For the inclusion of detuning of the laser cavity from resonance, such lasers sometimes need to be described by a field amplitude equation (a complex variable) and the population difference equation and in this form the behavior of semiconductor lasers is also usually modelled, though with phenomenologically modified equations. Finally, under conditions in which the material variables can all be adiabatically eliminated with respect to a more slowly evolving field amplitude, one arrives at the field equations which are used to describe certain lasers, notably dye lasers, argon-ion lasers, and helium-neon red lasers. However, there are many complicating factors that merit special attention if one is trying to model specific laser systems rather than trying to study the model itself. The Doppler broadening in gas lasers can lead to unusual complications that may be sufficiently important that they override an otherwise acceptable adiabatic elimination. In addition, any form of multimode behavior of a laser generates a wide combination of frequencies at intermediate and relatively high frequencies. These rapid variations may be faster than the "fast" variables can follow and so the multimode behavior may excite important dynamics in these variables which were safely and satisfactorily adiabatically eliminated for description of single-mode operation.

The appropriate warning note is that while classifications based on the relative relaxation rates of the field, populations and material polarization are quite popular, they must be used cautiously and viewed as only one of many ways of classifying and simplifying the long list of separate laser problems.

b) Field-matter interactions in a cavity versus interacting beams.

It is sometimes said that nonlinear dynamics requires both nonlinear coupling and feedback. The importance of feedback was reinforced by Ikeda's /23a/ demonstration of the role of delayed feedback in obtaining chaos in a simple model for multimode optical bistability. In this sense, it is the round-trip time in the optical cavity which provides the delay and the cavity returns the field to the interaction region providing the feedback. There is a temptation to replace the round-trip time in optical cavities with delays and then to neglect the time derivatives with respect to the simple delay which gives a mapping. That this can lead to serious errors has been noted in an extensive literature that is discussed in /23b/. Note that the error continues to be made, as, for example, in ref. 23c. However, if one looks closely at the models one uses to describe most optical systems, the boundary conditions most often disappear into the specification of modal functions and the cavity losses and detunings which appear in the evolution equations for the modal amplitudes. Modal expansions of delay-differential equations require an infinite number of modes with frequencies differing by multiples of the inverse of the delay time. The resulting equations appear to describe cavity free interaction between field amplitudes and materials. This analogy is complete as many interesting temporal, transverse, and longitudinal dynamics and formation of patterns occur in the interaction of one or more beams with a suitably nonlinear medium. Hence such topics as four-wave mixing and phase conjugation stand on equal footing with dynamics in cavities in the present understanding of optical instabilities.

c) Vectorial versus scalar fields.

Most optical experiments use Brewster angle components in order to have lower loss conditions for one linear polarization of the electromagnetic field. Such components provide higher loss for the orthogonal linearly polarized field and so the operation and associated descriptions often deal only with a scalar (single polarization) field. However, where such losses are not so critical or where the material has a definite preference for certain eigenstates of the field (such as circular polarizations because of angular momentum selection rules) the vector character of the field can be activated. This may be described in the form of dynamics of the electric field vector or in the form of interaction between nearly independent fields of orthogonal polarizations. Sometimes one sees this in the literature referred to as the dynamics of "two-mode lasers" as, for example, in reference 67. Such dynamics have been seen in both bistable optical systems using atomic vapors and in lasers using atomic gas media.

d) Optically thin versus optically thick media

A common approximation in describing field-matter interactions in cavities is that the single pass interaction is extremely weak, that is, the material is optically thin. In that case, the field can be thought of as nearly uniform in the cavity and in the interaction region, with changes occuring only after many roundtrips or at the mirrors because of partial transmission. If the mirror transmission is reduced to zero as the interaction (optical thickness) is reduced to zero, then a "uniform field" approximation can be made that ignores longitudinal spatial evolution except for that generated by the overlap of modes of different wavelengths. Though many lasers and optical bistability systems do not satisfy these conditions, the approximation appears to be a robust one and does not seem to discard important physical phenomena in most cases.

e) Plane wave versus transverse profiles.

For mathematical simplicity, most problems are described by models that assume that the field is a plane wave, that is, the models do not take account of the finite transverse extent of the field or the medium that is necessarily present in any real experiment. The development of the transverse mode theory of lasers came early in the development of such devices because of the interest in beam quality for focusing and other applications. However, transverse effects have only rarely been included in the studies of laser dynamics. There are perhaps two such classes of problems. In one case the transverse modes have significantly different resonance frequencies, perhaps on the order of magnitude of the difference between longitudinal modes, in this case one gets the temporal modulations at the relatively high intermode frequency differences that are common of most multimode interactions as discussed below. However, for different cavity geometries or mirror curvatures, the transverse modes may have optical frequencies that are very similar and when they are close enough the interactions can lead to a variety of interesting behavior. Enforcing of a transverse profile can extinguish all hope of instabilities or, at least, simplify them (see articles by Lugiato and Sargent in /1/) Transverse pattern dynamics also arise from partially separated modes (see work by Fleck and Kidder, Shih and Moloney, as discussed in /6/ and /8/). In one case the mode amplitudes are modulated out of phase with respect to each other at the relatively low frequency of the single mode relaxation oscillation frequency leading to a pattern that varies in time as the total intensity is modulated /29/. In another dramatic case the modes may lock to a common optical frequency with the formation of new stationary, transverse patterns /29a,30,31/. To build a description of these modes requires a linear combination of the spatial parts of the modal eigenfunctions modulated by a single time-dependent amplitude which includes the frequency shift of the optical frequency. Similar behavior is observed and similar models are required in the case of optically thick media or cavities with large localized losses (such as a highly transmissive mirror). Whether these are most conveniently viewed as a "cooperatively frequency locked" /29/ condition of the empty cavity modes (in which each space-time mode has a time dependent amplitude of the right frequency to bring the total optical frequency of that mode to the common frequency) or whether the new spatial pattern with a constant intensity profile and a single optical carrier frequency should be viewed as a nonlinear mode of the nonlinear system /30/ is a matter of taste. I believe it helps to keep both descriptions in mind.

Only in the detailed engineering studies of semiconductor lasers were such interactions considered from the earliest period of laser dynamics /29a/. Independently Lugiato and Narducci /30/ and coworkers have developed this study more formally and their language has been used in the description of experimental observations of changes in stationary transverse patterns and in the observation of transverse pattern dynamics in CO_2 /32/ and HeNe red lasers /33/. Similar phenomena were recently found in the theory and experiments on Nd:YAG lasers by Hollinger and coworkers /31/.

f) Ring cavities versus Fabry-Perot cavities.

Much is often made of the differences found in standing wave cavities in comparison with the phenomena found in ring cavities. The most important differences are:

- that the ring cavity supports two different modes of the same wavelength which travel in opposite directions in the laser cavity;
- that oppositely travelling waves in either case create a spatial interference pattern that causes longitudinal spatial hole-burning and which can change mode interaction coefficients or which can provide opportunities for other modes (which would have been suppressed in a case of uniform saturation of the medium) to oscillate;

- that oppositely travelling waves of the same optical frequency in Doppler broadened media interact with different velocity classes of the material depending on their detuning and thus they may have a strong interaction only if they both interact with the zero-velocity class.

The theorists' "toy model" is usually the unidirectional ring laser because of the uniformity of the gain saturation and the resulting simplicity of the equations. While such lasers have been built in a few cases /34-36/, the majority of lasers are of the bidirectional ring or the standing wave types. Most of the subtleties of these differences have only begun to be addressed, but fortunately a wide class of phenomena seem to be relatively insensitive to the exact cavity geometry. For example, the standing wave Doppler broadened laser behaves like the unidirectional ring laser model as soon as the laser is detuned sufficiently so that the counterpropagating portions of the standing wave interact with different atoms.

g) 'Active media' versus 'passive media'

As a general classification it is often helpful to begin by dividing materials into those which are prepared in the absence of the field in an absorbing state and those which are prepared in an amplifying state -- passive versus active. However, most of the interesting dynamical behavior involves the strong driving of the initial atomic state by the field, causing stimulated emission and absorption and turning the amplifying medium into an absorber or vice versa. Instabilties in steady state behavior can be viewed as the initiation of new optical fields where the nonlinear interaction of the strong field and the medium have created gain by multiwave interactions, Raman effects, Rabi sidebands; or other nonlinear processes. Hence Gibbs and Khitrova /37/ have argued that the distinction is somewhat artificial. That is probably the case in a quite broad view of the field, but there are important distinctions between those "passive" systems that require external driving from a field to get the nonlinear process started and those "active" systems which can begin spontaneously from intrinsic noise. The differences include the relative importance of the reference amplitude, phase, and polarization of the driving field in the "passive" case and the absence of such an absolute reference in many "active cases". This is the general basis for separating the dynamics of "optical bistability" (passive devices) from "laser instabilities" (active devices), but there is merit in keeping in mind many similarities rather than enforcing the separation.

h) Single mode versus multimode.

When one is dealing with a single mode optical field interacting with the medium, one has reduced the complexity of the field description to that of a single complex amplitude which then must be coupled to the appropriate number of atomic variables. The dynamical interactions then rely heavily on the nonlinear field-matter couplings. When more than one mode of the optical field is present, then the dynamics involve mode-mode coupling which arises primarily from the interaction of the modes with the same material.

The modes typically have different spatial patterns and different optical frequencies. The mere presence of two such modes in a linear combination leads to spatial and temporal interference and the nonlinear field-matter interactions then create corrections to the linear combination. This point is worth making clearly: that a linear combination of single mode solutions of a nonlinear problem is not a solution. However, sometimes there is a solution that is nearly that of the linear combination. That is, the corrections are small. These may involve either the generation of very small amplitudes of new optical frequencies at combination tones (or harmonics) or shifts in the frequencies of the modes themselves (an effect called "mode-pushing" or "mode-pulling"). In such cases the dominant feature one would observe is temporal oscillations in the output intensity at the difference between the optical frequencies of the two modes.

When three or more modes are present, there are four important kinds of interactions. Consider $\omega_1 < \omega_2 < \omega_3$ to be the modal frequencies. In the first case there is simply the beating (intensity modulation at the output) at all of the frequency differences between the modes (e.g., $\Delta_{12} = \omega_1 - \omega_2$; $\Delta_{23} = \omega_2 - \omega_3$; $\Delta_{13} = \omega_1 - \omega_3$). In the second case the presence of the other modes distorts the gain and dispersion of the medium so that each mode is modified in its amplitude and in its optical frequency (the frequency shifts are called mode pushing and mode pulling). In the third case the nonlinear interactions within the laser medium lead to all rational combinations of the optical frequencies (with an odd number of terms) which are sometimes called combination tones (e.g., $\omega_i + \omega_j - \omega_k$) and these result in lower frequency intensity modulations at $\Delta_{12} - \Delta_{23}$ which means that one has more complex quasiperiodic (two or more characteristic frequencies) modulation of the intensity /38-41,31/. Finally, three important forms of mode-locked (fixed relative phases) can also develop:

- Mode locking -- when the modes lock to equally spaced frequencies and generate a regular train of pulses ($\Delta_{12} = \Delta_{23}$) [This can be resonantly enhanced if this common frequency spacing matches the Rabi frequency or the relaxation oscillation frequency.];

- Combination tone locking -- when the difference between two intermode beat frequencies locks onto an internal resonance such as the Rabi frequency or the relaxation oscillation frequency or to a subharmonic of the intermode beat frequency;
- Cooperative frequency locking -- when the modes lock to the same optical frequency with the result that there are no intensity pulsations.

The first has been observed in the interaction of both longitudinal and transverse modes in laser experiments, the third has been seen in cases of transverse profile dynamics, and all three have also been seen in various models for such lasers, but a unified picture and description has yet to be fully developed. This is in part because the changes from one type of behavior to another usually involve nontrivial corrections to the original modal functions making mode identification difficult.

4. The "Laser-Lorenz" or "Maxwell-Bloch" model

This model is one of the simplest possible and relative to the classifications mentioned above it corresponds to the assumptions of a) coherent interactions (including the atomic polarization dynamics as well as those of the field and the population inversion), b) a uniform field (optically thin medium) approximation, c) single mode (other transverse and longitudinal modes are eliminated), d) scalar optical field (polarization selective optics are assumed), e) unidirectional ring cavity (directional selectivity by nonreciprocal elements are assumed), f) the plane wave approximation.

One begins with Maxwell's wave equation for the scalar electromagnetic field and with the Schrödinger equation for the material, namely,

$$\nabla^2 E_0 - (1/c^2)(\partial^2 E_0/\partial t^2) = \mu_0 P_0 \tag{1}$$

and the equations for a two-level system for the material with an interaction Hamiltonian given by

$$H = e\,\mathbf{r} \circ \mathbf{E_0}\,. \tag{2}$$

We then make the following approximations:

- the paraxial approximation (that variations in the transverse directions are much slower than variations in the longitudinal direction) and
- the slowly varying amplitude approximation

$$E_0 = (1/2)\{E(r,t)\exp(i(kz-\omega t)) + c.c.\} \quad (3a)$$

and

$$P_0 = (1/2)\{P(r,t)\exp(i(kz-\omega t)) + c.c.\}, \quad (3b)$$

where the longitudinal and transverse variations remaining in E and P are small on the distance of a wavelength and the temporal variations are small on the order of an optical period.

Substituting eqs. (3) into the Maxwell wave equation (1) and taking the approximations mentioned we arrive at the basic transverse wave equation

$$\nabla_t^2 E + 2ik(\partial E/\partial z) + (2i\omega/c^2)E + [k^2 - (\omega^2/c^2)]E = -\mu_0\omega^2 P \quad (4)$$

and by taking the travelling wave to satisfy $ck = \omega$ and rearranging we arrive at

$$-(ic^2/\omega)\nabla_t^2 E + c(\partial E/\partial z) + \partial E/\partial t = -\mu_0\omega c^2 P, \quad (5)$$

which is a basic travelling wave equation for the slowly-varying amplitude of the field with its slow longitudinal and transverse variations driven by the polarization which comes from the collective action of the dipole moments of the material variables described by the Schrödinger equation. Making a plane wave assumption we can also eliminate the transverse Laplacian in eq. (5).

5. Material equations from the two-level atom picture

In further modification of the equations, the boundary conditions of the laser cavity are converted into distributed losses and phase shifts corresponding to detunings of the carrier frequency from the cavity resonance frequency. This is most straight forwardly done in the uniform field limit which we will take for the following discussion. The procedure is carefully outlined in /10/.

If the dynamics of the two-level material polarization (P) and the population difference (D) variables are retained together with the field amplitude dynamics, then one has for a two-level medium the following equations,

$$E \;= -\kappa E + i\delta_{AC}E - \kappa AP \qquad (16a)$$

$$D \;= -\gamma_{\parallel}(D-1) + (\gamma_{\parallel}/2)(P^*E + PE^*) \qquad (16b)$$

$$P \;= -\gamma_{\perp}P - \gamma_{\perp}ED, \qquad (16c)$$

where

$$\delta_{AC} = \omega_A - \omega_C, \qquad (16d)$$

(the difference between the material frequency and the cavity frequency) and the variables have been suitably rescaled for convenience (see /10/).

These are the famous "Maxwell-Bloch Laser equations" (because of their orign in the Maxwell wave equation and the Bloch equations for the evolution of a two-level quantum system driven by an electromagnetic field), where the slowly varying amplitudes P and E are taken in the rotating reference frame of the atomic resonance frequency. (These equations are also known as the laser-Lorenz equations, or the Haken equations, because the resonant case is ismorphic to the Lorenz equations used to describe convective hydrodynamics since 1963 as first noticed by Haken in 1975 who had popularized the use of the Maxwell-Bloch form of these equations in the study of lasers.). The equations were first studied for lasers by Oraevskiy and Uspenskiy /21,48/. Spontaneous pulsations are possible and were noted as early as 1962 but the pulsations described by the model did not correspond to the types of pulsations nor to the operating conditions that plagued early devices.

The steady state solutions are given by:

$$E = (I_{ss})^{1/2} \exp(-i\Delta t); \qquad (17a)$$

$$I_{ss} = A-1-\Delta^2 \;\; ; \text{ and} \qquad (17b)$$

$$\Delta = -\delta_{AC} / (1 + (\kappa/\gamma_{\perp})) \,. \qquad (17c)$$

There is also a "trivial" (zero-intensity) solution which is stable for $A < 1+\Delta^2$. The non-trivial solutions are physical only for positive intensities ($A > 1+\Delta^2$) and then the intensity of these solutions grows linearly with the pump parameter A.

When the non-trivial steady state solutions are analyzed for their stability, it is found that if $\kappa > \gamma_{\perp} + \gamma_{\parallel}$, then there exists a value of A above which the steady states are unstable with respect to small perturbations. The fact that the cavity linewidth must be larger than the sum of the material linewidths both makes it difficulty to find lasers that satisfy this criterion and leads to the terminology that this is a "bad cavity" instability. This stability boundary in the parameter space is shown in Fig. 1a,b and sample pulsatons are depicted in Fig. 2. Below the stability boundary shown in Fig. 1 there may be a region of coexistence in which there are both stable steady state solutions and stable pulsing solutions.

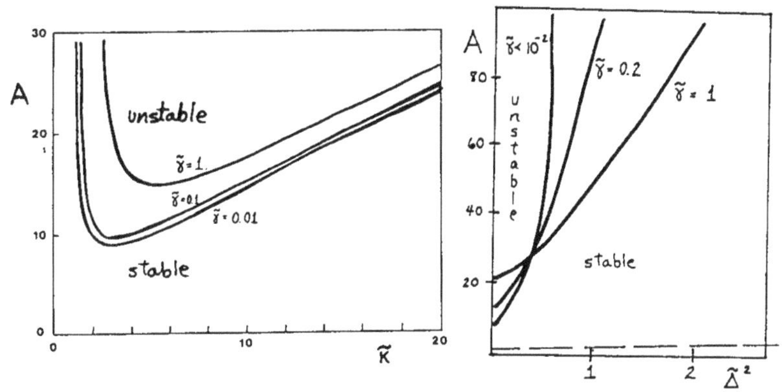

Fig. 1 Stability boundary in resonance (a) versus cavity decay rate and with detuning (b). Parameters marked with tilde's denote division by $\gamma_\perp$/10/.

The classic "Lorenz chaos" such as is shown in Figures 2 and 3 (found in the single-mode laser model in the resonant case above the instability threshold) involves alternate spiraling (of chaotically varying spiral lengths) about the two unstable fixed points.

Upon detuning the laser, more periodic behavior is discovered as indicated in Figure 3.

The first such "Lorenz-like" chaotic solutions to this laser model were reported by Grasiuk and Oraevskii in 1962 /21/. However, the Lorenz equations were more thorougly studied and characterized by those interested in fluids and the phenomena of nonlinear dynamics, and so Haken's discovery of the isomorphism endowed the laser equations with a substantial phenomenology /20/. Hence one should have expected to see

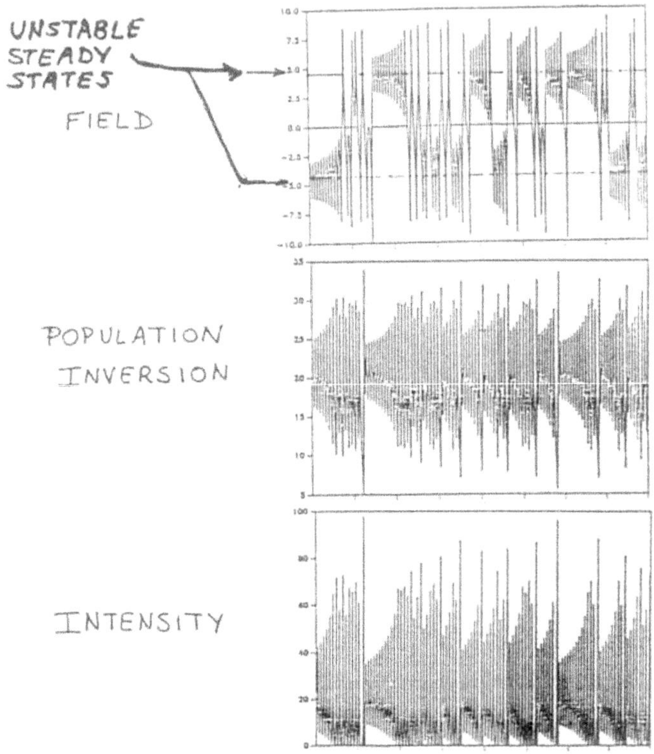

Fig. 2 Sample of chaotic "spiral" intensity and field pulsations in the unstable region of the laser operation.

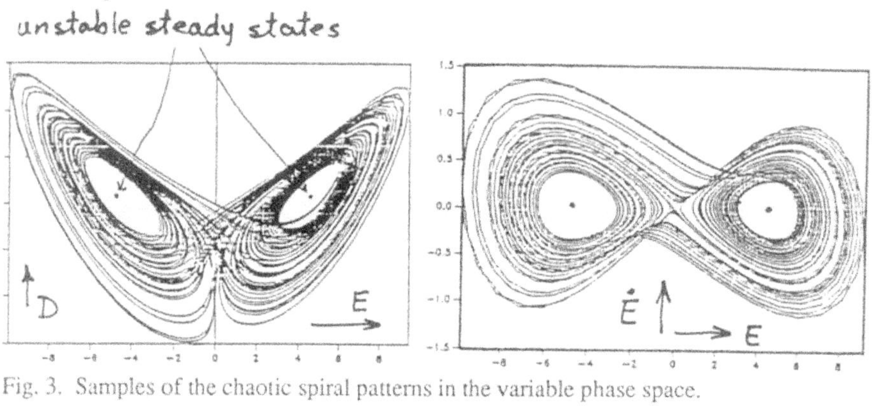

Fig. 3. Samples of the chaotic spiral patterns in the variable phase space.

the richness of the Lorenz type chaos in the appropriate laser system. This richness includes not only the heteroclinic spirally about two unstable fixed points, but a variety of symmetric and asymmetric (periodic and chaotic) solutions that can be so classified because the equations have a basic inversion symmetry that says that any solution (E,P,D) has a corresponding solution given by (-E,-P,D).

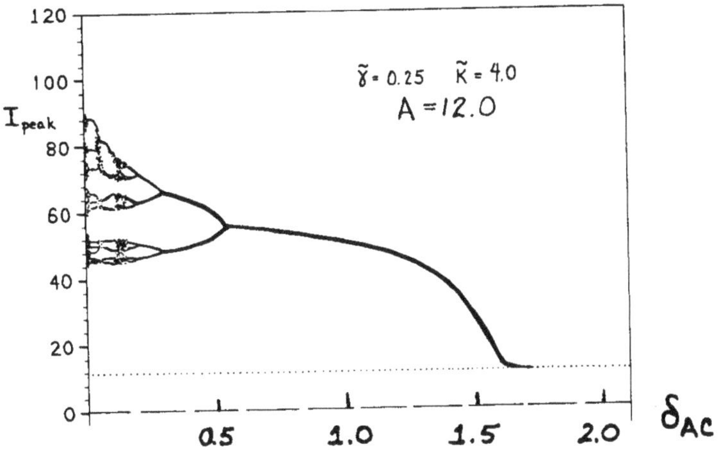

Fig. 4 Bifurcation diagram showing the peak intensity of pulsations for different detunings /10/.

Indeed, the laser-Lorenz model is even richer than the Lorenz model because there are real parameters in the Lorenz model which are naturally made complex in the laser model simply by detuning the laser cavity from resonance with the medium. This breaks the symmetries and permits a rich family of period doublings and period doubling chaos as well /48/.

The connection between the period-doubling behavior out of resonance and the spiralling intermittent chaos close to resonance has yet to be fully explored. It is worth noting that in both regions the behavior of successive peak intensity values (or the time intervals between peaks) are almost completely given by one-dimensional maps. In the period doubling region the map has a quadratic maximum while in the spiral-itermittency region it has a cusp maximum. How the topological deformation takes place in the multiparameter space merits further study.

With detuning there is also time dependent evolution of the phase of the laser field. Not only is this a frequency shift of the optical carrier for the pulsing solution with respect to the stable solution, but one finds dynamical frequency and phase pulsations that are driven by the intensity pulsations. Samples of this are seen in Fig. 6

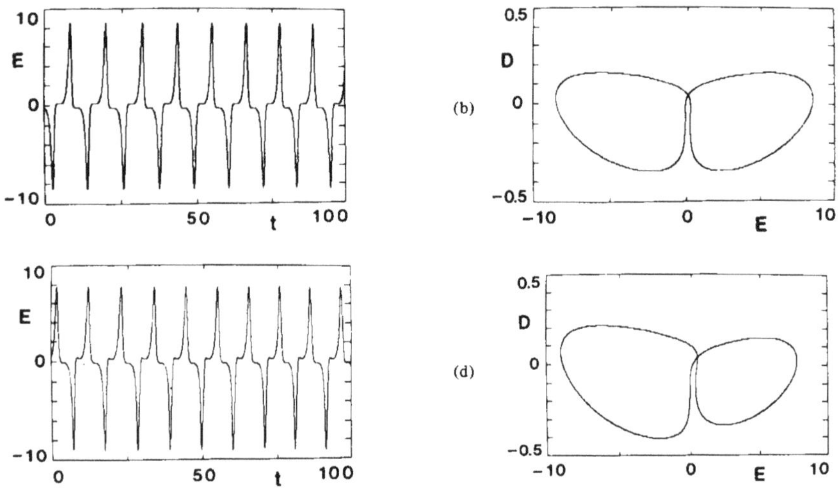

Fig. 5. Samples of symmetric and asymmetric periodic solutions for the laser operating on resonance /Narducci, *et al.*, Opt. Commun. **55**, 370 (1985)/.

One sees that the phase is time dependent. At the peak of each pulse the instantaneous laser frequency is pulled a little closer to resonance with the atoms and then the frequency moves well beyond even the empty cavity frequency as the intensity passes near zero. New studies show that the pulsing detuned laser is best described in a rotating reference frame at a different frequency than that of the steady state laser solution for the same parameter values /48 (1988)/. In this reference frame the plot of the complex field has the same topological structure as one finds in plotting the intensity versus the population inversion. In other reference frames the plot of the complex field trajectory appears unnecessarily complicated and quasiperiodic because of the rotation in the complex plane induced by the frequency offset. The frequency shift of the optimal reference frame perhaps serves as evidence of an additional phase shift for each trajectory in the five-dimensional variable space of the system, as a generalization of the phase shifts associated with topological phases and Berry's phase in quantum mechanics (I thank R. Chiao for this suggestion - Ref.: Optics News, December 1988).

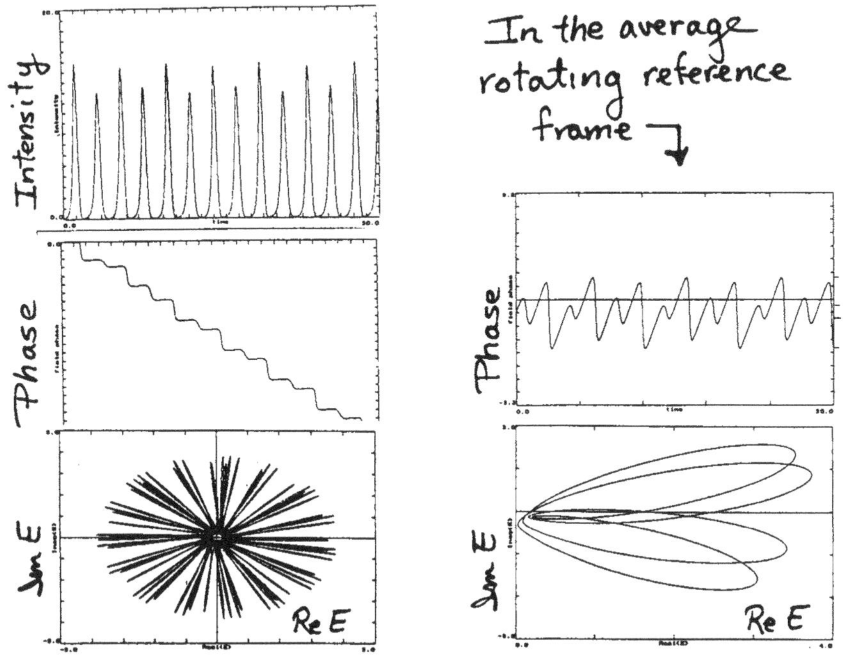

Fig. 6. Intensity and phase pulsations for a detuned laser (Zeghlache *et al.*, /48/).

6. Experimental characteristics of the FIR ammonia lasers

Unfortunately for the study of nonlinear dynamics real experimental devices corresponding to these equations were nonexistent for the first twenty-five years of laser studies. However, recent work by Weiss and collaborators has demonstrated that the FIR ammonia lasers behave almost identically to the laser-Lorenz equations in a certain region of operating parameters.

The experimental design is shown in Figure 7.

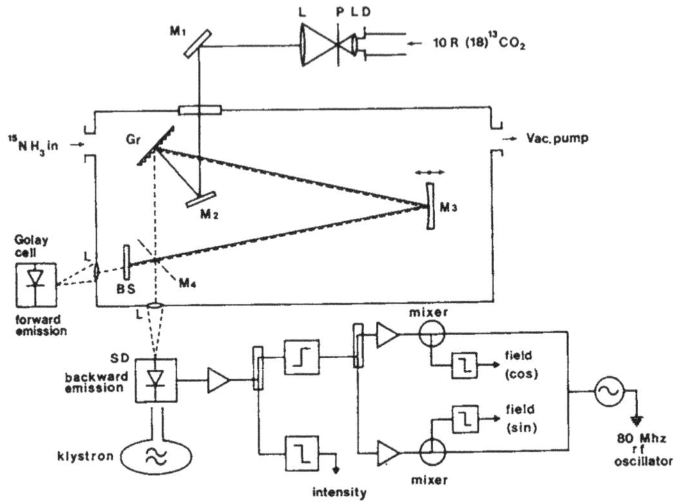

Fig. 7. Experiemental set-up of an optically pumped FIR laser and the detection electronics for heterodyne detection of the phase evolution.

Several features of this laser help to make it more nearly equivalent to the unidirectional ring laser with a two-level medium than would first seem possible. The Doppler broadening of the ammonia gas is used together with selective optical pumping of a particular velocity class of the molecules to separate the forward and backward gain profiles. Then by proper tuning of the laser it can be made to lase on the backward gain only. This is helpful because the coherent optical pumping causes Rabi splitting of the pump transition and coherences in the three relevant levels of the pump and emission transitions that would seem to forbid description by a simple two-level model. Pressure broadening, multiple magnetic sublevels and crossed polarizations of the pump and emission lines, and backward emission all reduce the coherent effects.

The features of the laser-Lorenz model first observed were the appropriate instability threshold, the period doublings upon detuning and the spiralling chaos /35/.

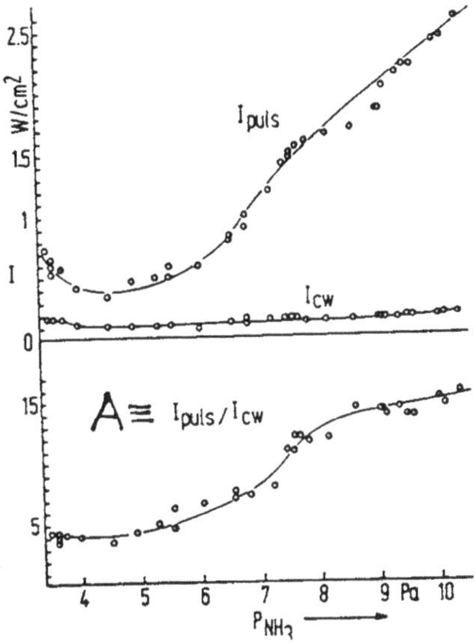

Fig. 8 Experimental measurements of the instability threshold /35/.

Clear evidence has also been obtained for the sequences of intensity pulsations both in resonance and with detuning that are expected from the model.

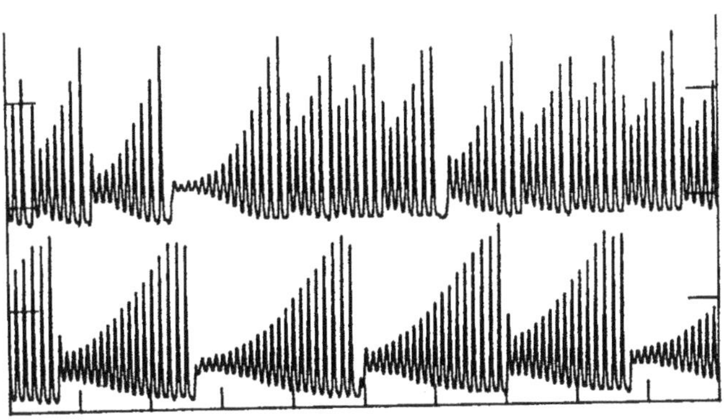

Fig. 9. Chaotic spiral intensity pulsations /11/.

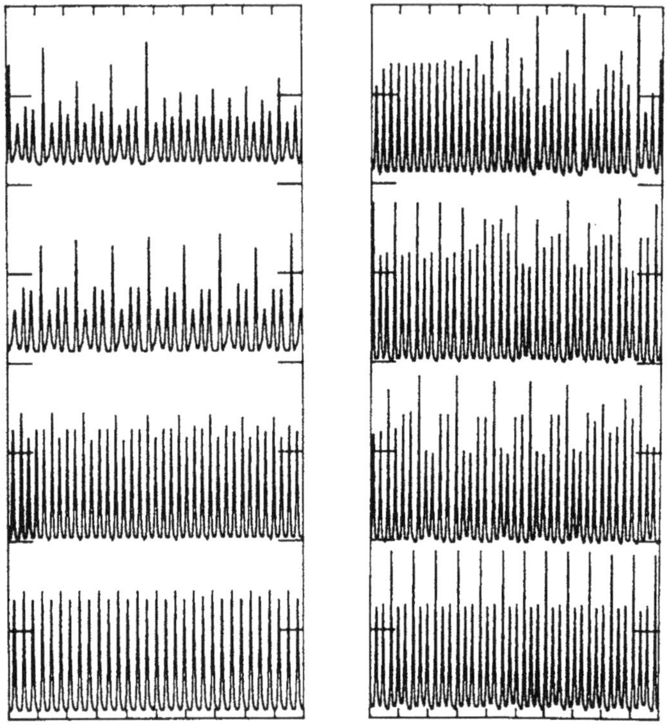

Fig. 10. Chaotic and periodic pulsations of the laser upon detuning /11,35,68/.

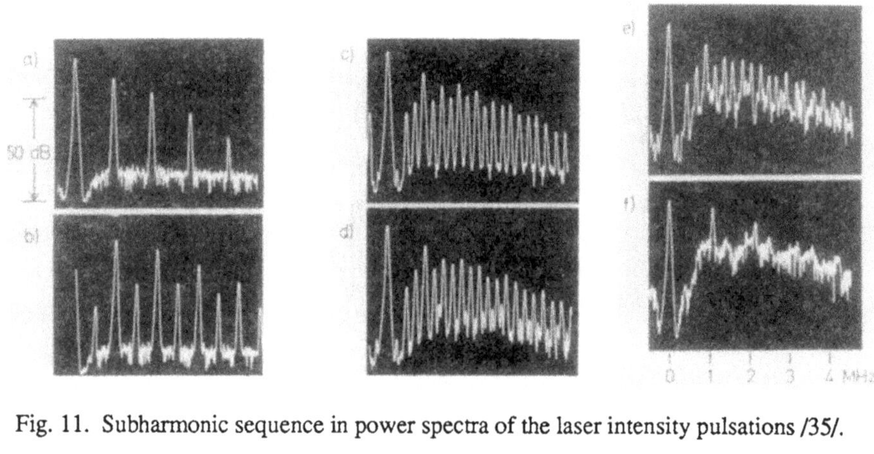

Fig. 11. Subharmonic sequence in power spectra of the laser intensity pulsations /35/.

However, the measured variable was the laser intensity which corresponds to the square of one of the inversion symmetric variables, hence the heteroclinic nature of the Lorenz-like chaos could only be inferred. New heterodyne measurements /50/ reveal the behavior of the optical electric field and show that in certain parameter regions the electric field changes sign at the end of each spiral indicating that the attractors involve alternation between the neighborhoods of two unstable fixed points.

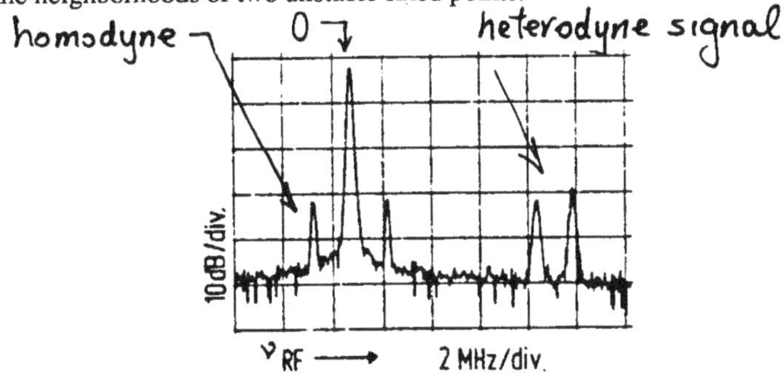

Fig. 12 Heterodyne spectrum showing symmetry for spirals /9/.

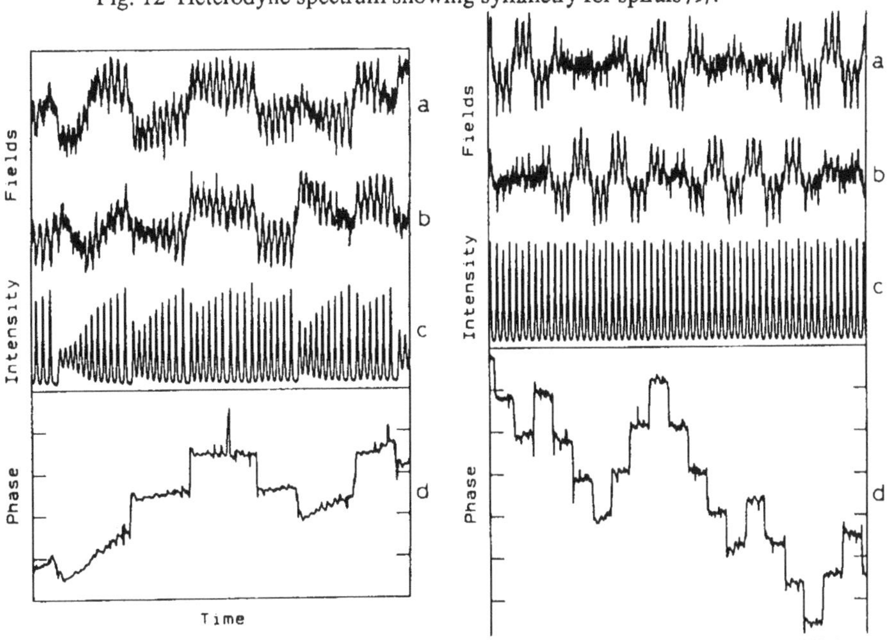

Fig. 13 Heterodyne signals with intensity and phase for spirals showing pi jumps /50/.

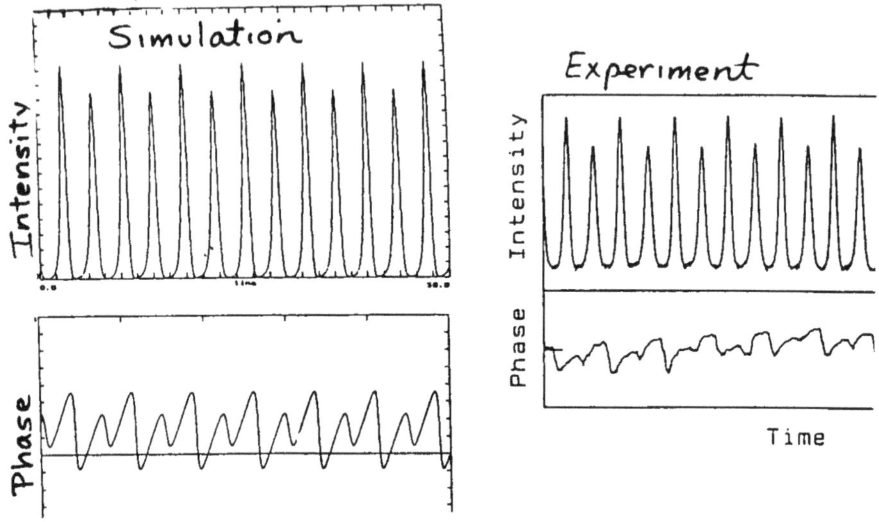

Fig. 14. Heterodyne signals with intensity pulsations for periodic pulsations (/50, and Zeghlache /48/).

7. Dimensions and entropies

The chaotic pulsations have also been characterized for their dimensions and entropies as a comparison between the theoretical model and the experimental data. The methods are those described elsewhere in the chapter by Tredicce and Abraham in this volume. These are summarized in ref. 68. All evidence found to date strongly supports the conclusion that the two-level Lorenz model is sufficient to describe the experimental behavior. This issue has also been addressed in numerical simulations that incorporate all of the relevant physical processes and effects leading to hundreds of equations. The result is that when enough of the physics has been included then the simple experimental behavior is recovered /69/.

8. Symmetry breaking

The laser-Lorenz model in resonance has three variables and a particular kind of symmetry such that if (E(t),P(t),D(t)) is a solution, then (-E(t),-P(t),D(t)) is also a solution. If these two solutions have the same attracting set in phase space they are called "symmetric" and if they have different attracting sets they are called "asymmetric". There are special features and problems associated with these solutions. Symmetric solutions cannot period double /54/ and so can only break their symmetry or make abrupt (crisis induced) transitions to new types of solutions. In contrast, as a control parameter is changed, asymmetric solutions may period-double or they may become attached to their partner asymmetric solution in a symmetry restoring "glueing bifurcation" /55/. The characteristics of signals necessary to distinguish these various types of evolution are concealed by measurements of the intensity. Symmetry broken solutions appear as period doublings of the intensity. Symmetry-restoring, glueing bifurcations may be missed altogether. Hence this interesting family of characteristics requires more subtle measurement techniques, typically heterodyne experiments to record the behavior of the field variable as well as the intensity.

The new heterodyne measurements that monitored the behavior of the FIR laser field /50/ have been used to identify symmetric and asymmetric solutions. Pulsing that might have been classified on the basis of the intensity periodicities can now be more subtlely analyzed for proper underlying symmetries as well. Trajectories in the phase space with n orbits around one fixed point and m orbits around the other appear as (m+n)-periodic intensity pulsations but are in reality symmetric or asymmetric orbits of (m+n) pulses in a single path around the attractor. Such orbits are better described by Sparrow's /21,65/ notation $x^n y^m$ but when n=m one needs to add a C or S or symmetric or A symmetric possibilities!.

9. Homoclinic and heteroclinic ("Shil'nikov") chaos

Recently two special cases of chaotic pulsations have been found that differ from the classic routes to chaos of period doubling and quasi-periodicity /51/. These are instead special classes of what has been called the intermittency route to chaos /51/. In three experiments /56-58/, lasers have been shown to have chaotic pulses which originate near a homoclinic orbit involving a single unstable fixed point which successively attracts and repels trajectories /59/. In the Lorenz-like FIR laser case the chaos may be related to either a heteroclinic cycle as in the Lorenz model /21/ or to a homoclinic chaos.

The classic "Lorenz chaos" (found in the laser model in the resonant case above an excitation threshold if $\kappa > \gamma_\parallel + \gamma_\perp$) involves alternate spiraling (of chaotically varying spiral lengths) about the two unstable fixed points and is closely related in parameter space to a heteroclinic cycle that connects the two unstable points. The fixed points are called saddle foci as trajectories are attracted along a one-dimensional stable manifold and then are repelled in a spiral on the unstable two-dimensional manifold.

Other laser models have recently been studied which have similar inversion symmetries and which give rise to similar spiral chaos. Several of them have been proposed as more appropriate complex models for the three coherently coupled levels of the optically pumped FIR laser systems /60,61/. However, the FIR laser experiments are even more complicated, involving Doppler broadenings and detuning of the pump laser which breaks the symmetries of these more complex models. Solutions of a model which includes these complications show only homoclinic type chaos described below /61/. It is believed that the experiments /35,50/ include a further symmetry restoring effect based on optical pumping of the magnetic sublevels of the various transitions of the three levels by the orthogonal linear polarizations of the pump and FIR lasers and then a scrambling of these many coherences by a sufficiently large collision rate within the gaseous laser medium and the transverse profiles of the pump and lasing beams. It is a case in which a problem of extreme microscopic complexity (perhaps requiring thousands of equations) together with dissipative effects and stochastic collisions which behaves in a simplified way with collective variables permitting a satisfactory description by the simple five-equation model for the laser-Lorenz system.

In contrast, homoclinic chaos arises when there is a single unstable fixed point with a homoclinic orbit that connects this point to itself. Chaotic orbits can appear in parameter space on certain conditions and involve trajectories which depart from this point along the unstable manifold and which are then reinjected near the the point being attracted towards it along the stable manifold. When the unstable point is a saddle focus this is called a "Shil'nikov instability" and gives rise to "homoclinic chaos" /59,62/ or "Shil'nikov chaos" /63,64/. When the fixed point is an attracting focus and an unstable node (reversing the stable and unstable manifolds) is is called an "inverse Shil'nikov instability" /64/. The two types and their homoclinic orbits are related by simple time reversal /64/. Such orbits have been seen in the Belousov-Zhabotinski chemical reaction models and experiments /62/.

In cases of both heteroclinic chaos and homoclinic chaos there can be a very dense topological structure in the parameter space with an infinite number of homoclinic orbits and of periodic orbits. The countably infinite number of periodic orbits was identified by Shil'nikov /63/ and this structure has been studied and identified in the Lorenz equations /65/.

Homoclinic chaos has been seen experimentally in three laser systems: a laser with feedback of its output intensity to control the intracavity losses /56/; lasers with saturable absorbers /57,58/; and the FIR laser operated at low pressure /50/. Experimentally it is difficult to demonstrate the existence of the homoclinic orbit which generates the chaos because the homoclinic trajectory requires an infinite time. The models for each of these experimental situations do contain such homoclinic orbits near the parameter regions where the "quasi-homoclinic" chaos is seen. A notable feature of these trajectories is that the phase portrait for each burst of spirals is nearly identical, with the primary difference being long and chaotically varying lethargy near the fixed point which depends on how the signal is injected near the stable manifold of the unstable point which determines when it "discovers" the unstable direction. Such systems are often better measured for their chaotic properties by return maps of the times between successive pulses /59,62/ (this method has been developed and applied in the laser case in ref. /56/) rather than by attractor dimensions /52/ which would have to resolve very fine structure, requiring resolution not normally achievable in a practical experimental system. Alternative tests yet to be tried are discussed in /66/. These methods have been used to confirm Shil'nikov-type chaos in lasers with saturable absorbers/70/, though the fact that there are both converging and diverging spirals from the fixed-point indicates that this is a higher (than three) dimensional case of the homoclinic-induced chaotic pulsations.

10. Conclusion

Laser instbilities is a rich field for investigation. Many effects which have been observed by many experimentalist have been fought and suppressed because they were undesireable sources of noise. It now seems that we can use nonlinear dynamics to understand the fundamental processes of field matter interactions and that by understanding them we may gain a better idea of how to control and modify them.

Acknowledgements

I am grateful to my many collaborators over the last five years who have shared with me their insights and who have worked together with me on many of the issues covered in this review. My thanks particularly go to A.M. Albano, F.T. Arecchi, L.W. Casperson, R.S. Gioggia, P. Glorieux, U. Hübner, W. Lange, N. Lawandy, G.L. Lippi,

L.A. Lugiato, L.M. Narducci, G.L. Oppo, P. Mandel, M.F.H. Tarroja, J.R.Tredicce, C.O.Weiss, and H. Zeghlache.

References

1. Special Issues of the Journal of the Optical Society of America B, **2**, January 1985; **5**, May 1988.
2. Optical Instabilities, R.W. Boyd, M.G. Raymer, and L.M. Narducci, eds., (Cambridge U. Press, Cambridge, 1986).
3. Optical Chaos, J. Chrostowski and N.B. Abraham, eds., SPIE Proceedings volume 667, (SPIE, Bellingham, 1986).
4. Instabilities and Chaos in Quantum Optics, F.T. Arecchi and R.G. Harrison, eds., (Springer, Berlin, 1987).
5. Instabilities and Chaos in Quantum Optics II, N.B. Abraham, F.T. Arecchi and L.A. Lugiato, eds., (Plenum, NY, 1988).
6. R.G. Harrison and D.J. Biswas, Prog. Quantum Electron., **10**, 127 (1985).
7. J.R. Ackerhalt, P.W. Milonni and M.-L. Shih, Physics Reports (Physics Letters C), **128,** 205 (1985).
8. P.W. Milonni, J.R. Ackerhalt and M.-L. Shih, Chaos in Laser-Matter Interactions (World Scientific, Singapore, 1987).
9. C.O. Weiss, Opt. and Quant. Electron. **20**, 1 (1988).
10. N.B. Abraham, P. Mandel, and L.M. Narducci, in Progress in Optics, vol. XXV, E. Wolf, ed., (Elsevier, Amsterdam, 1988) p. 1.
11. L.M. Narducci and N.B. Abraham, Laser Physics and Laser Instabilities (World Scientific, Singapore, 1988).
12. C.O. Weiss and R. Vilaseca, Laser Dynamics (Physics-Verlag, Weinheim, manuscript under preparation for publication in 1989).
13. L.W. Casperson, manuscript in preparation for Cambridge University Press.
14. Ya. I. Khanin, manuscript in preparation for North Holland representing a translation and updating of his book in Russian (the third in a series that began with the two-volume Quantum Electronics first published with B. Fain).
15. Laser Physics and Quantum Optics, E. Quel, J.R. Tredicce and L.M. Narducci, eds., (World Scientific, Singapore, 1989).

15a. Dynamics of Nonlinear Optical Systems, F.J. Bermejo and L. Pesquera, eds., (World Scientific, Singapore, 1989).

16. Optical Instabilities, Technical Digest of the International Topical Meeting on Instabilities and Dynamics of Lasers and Nonlinear Optical Systems, Rochester, NY June 1985 (Optical Society of America).

17. Instabilities, Dynamics and Chaos of Nonlinear Optical Systems, Technical Digest of the International Workshop in Il Ciocco, Italy, July 1987 (available from Prof. E. Arimondo, Dipartimento di Fisica, Universita di Pisa, Piazza Torricelli 2, 56100 Pisa, Italy).

18. W. Firth, in Chaos, A. Holden, ed., (Adam Hilger, Bristol, 1986) p. 135.

19. J.C. Englund, R.R. Snapp, and W.C. Shieve, in Progress in Optics vol. XXI, ed. E. Wolf (North Holland, Amsterdam, 1984) p. 355.

20. H. Haken, Phys. Lett. **53A**, 77 (1975).

21. E.N. Lorenz, J. Atmos. Sci, **20**, 130 (1963); see also, C.T. Sparrow, The Lorenz Equations: Bifurcations, Chaos, Strange Attractors (Springer Verlag, Berlin, 1982). The analogy found by Haken is indicated by comparing Lorenz's results with the prior stability analysis and numerical solutions of the equations in the laser model by Grasyuk and Oraevsky as reported in two conferences in 1962 and 1963 in the Proceedings of the 4th International Congress on Microwave Tubes (Sheveningen, 1962), p. 446 and in Quantum Electronics and Coherent Light, Enrico Fermi School, August 1963, P.A. Miles, ed. (Academic Press, NY, 1964) p. 192 and in Rad. Eng. and Electron. Phys. (USSR) **9**, 424 (1964); and with the work of A.V. Uspenskiy, Sov. Phys. - Rad Eng. Electron. Phys. **8**, 1145 (1963) [Russian original in 1963] and extended by Korobkin and Uspenskiy, (1963), see Ref. 48. [If one is seeking the "first" presentations of the equations and of the chaotic behavior, Grasyuk and Oraevsky submitted their paper to the Sheveningen meeting in June 1962 and Uspenskiy submitted his first paper for publication also in June 1962 while Lorenz submitted his first paper in November 1962 -- and surely each has his story as to when he actually "discovered" particular features. It really does appear that these were rather independent efforts though in later work Uspenskiy acknowledges the precedence of Oraevskiy. This is reviewed from the point of view of one of the principals by A.N. Oraevskiy in Research on Laser Theory (Trans. of Proceedings of the Lebedev Physics Institute of the Academy of Sciences of the USSR, Vol. 171), A.N. Orayevskiy, ed., (Nova Science Pub., Commack, 1988) p. 1.]

22. R. Bonifacio and L.A. Lugiato, Lett. Nuovo Cim. **21**, 505 (1978); **21**, 510 (1978); S.L. McCall, Appl. Phys. Lett. **32**, 284 (1978); L.A. Lugiato, L.M. Narducci, D.K. Bandy and C.A. Pennise, Opt. Commun. **43**, 281 (1982); K. Ikeda and O. Akimoto, Phys. Rev. Lett. **48**, 617 (1982).
23. a) K. Ikeda, Opt. Commun. **30**, 257 (1979); b) N.B. Abraham in Methods and Applications of Nonlinear Dynamics, A. Saenz, ed., (World Scientific, Singapore, 1987); c) W. Lu and W. Tan, Opt. Commun. **61**, 271 (1987).
24. Conference proceedings: Optical Bistability (Plenum, 1981); Optical Bistability II (Plenum, 1984; Optical Bistability III (Springer, 1986); Optical Bistability IV (les Editions de Physique, 1988 -- J. Physique **49**, Colloque C2, Supplement n° 6); see also long reviews by L.A. Lugiato "The Theory of Optical Bistability" in Progress in Optics, Vol XXI, ed. E. Wolf (North-Holland, Amsterdam 1986) p. 69 and by H.M. Gibbs, Optical Bistability: Controlling Light by Light (Academic Press, NY, 1985).
25. L.W. Casperson, IEEE J. Quantum Electron., **QE-14**, 756 (1978).
26. F.T. Arecchi, R. Meucci, G.P. Puccioni and J.R. Tredicce, Phys. Rev. Lett., **49**, 1217 (1982).
27. H. Haken, Synergetics, 2nd Ed. (Springer, Berlin, 1977); L.A. Lugiato, P. Mandel, and L.M. Narducci, Phys. Rev. A, **29**, 1438 (1984); G.L. Oppo and A. Politi in Instabilities and Chaos in Quantum Optics II, N.B. Abraham, F.T. Arecchi and L.A. Lugiato, eds., (Plenum, NY, 1988) p. 363.
28. See, for example, Y. Silberberg and I. Bar-Joseph, Phys. Rev. Lett. **48**, 1541 (1982) and J. Opt. Soc. Am B **1**, 62 (1984); P. Narum, *et al.*, J. Opt. Soc. Am. B **5**, 623 (1988); G. Khitrova, *et al.*, Phys. Rev. Lett, **60**, 1126 (1988); D. Gauthier, et al., Phys. Rev. Lett, **58**, 1640 (1987) and Opt. Lett., **14**, 323 (1989), and Phys. Rev. Lett. **61**, 1827 (1988); R. Boyd, *et al.*, in Optical Chaos /12/; G. Grynberg, *et al.*, Opt. Commun. **67**, 363 (1988); *66*, 321 (1988); Europhys. Lett. **8**, 29 (1989); A. Rozanov, Sov. Phys. JETP **53**, 47 (1981); Firth and C. Pare, Opt. Lett. **13**, 1096 (1988); and references therein.
29. a) A.F. Suchkov, Sov. Phys. JETP, **22**, 1026 (1966) [Zh. Ek. T. Fiz., **49**, 1495 (1965)]; R.G. Allakhverdyan, A.N. Oraevskii, and A.F. Suchkov, Sov. Phys. Semiconductors, **4**, 227 (1970) [Fiz. Tekh. Polup., **4**, 341 (1970)]; b) See, for example, J. Buus, IEEE J. Quantum Electron., **QE-19**, 953 (1983) and refs. therein.
30. L.A. Lugiato, C. Oldano, and L.M. Narducci, J. Opt. Soc. Am. B, **5**, 879 (1988); L.A. Lugiato, F. Prati, L.M. Narducci, P. Ru, J.R. Tredicce, and D.K. Bandy, Phys.

Rev. A, **37**, 3847 (1988); L.A. Lugiato, G.-L. Oppo, M.A. Pernigo, J.R. Tredicce, L.M. Narducci, and D.K. Bandy, Opt. Commun., **68**, 63 (1988).

31. F. Hollinger, Chr. Jung, and H. Weber, private communication (1988).
32. J.R. Tredicce, E.J. Quel, A.M. Ghazzawi, C. Green, M.A. Pernigo, L.M. Narducci and L.A. Lugiato, Phys. Rev. Lett. **62**,1274 (1989)
33. Chr. Tamm, Phys. Rev. A, **38**, 5960 (1988); Phys. Rev. A (to be published 1989); see also his contribution to volume 15a.
34. L.M. Hoffer, T.H. Chyba, N.B. Abraham, J. Opt. Soc. Am. B, **2**, 102 (1985).
35. C.O. Weiss and J. Brock, Phys. Rev. Lett., **57**, 2804 (1986).
36. L.W. Hillman, J. Krasinski, R.W. Boyd and C.R. Stroud, Jr., Phys. Rev. Lett., **52**, 1605 (1984).
37. G. Khitrova, J.F. Valley, and H.M. Gibbs, Phys. Rev. Lett. **60**, 1126 (1988).
38. C.O. Weiss and H. King, Opt. Commun., **44**, 59 (1982); C.O. Weiss, A. Godone, and A. Olafsson, Phys. Rev. A, **28**, 892 (1983).
39. N.J. Halas, S.N. Liu, and N.B. Abraham, Phys. Rev. A, **28**, 2915 (1983).
40. D.J. Biswas and R.G. Harrison, Phys. Rev. A, **32**, 3835 (1985).
41. F. Hollinger and C. Jung, J. Opt. Soc. Am. B, **2**, 102 (1985); F. Hollinger, C. Jung, and H. Weber, in Optical Instabilities, R.W. Boyd, M.G. Raymer, and L.M. Narducci, eds. (Cambridge U. Press, Cambridge, 1986) p. 259; see also R. Hauck, F. Hollinger and H. Weber, Opt. Commun., **47**, 141 (1983).
42. L. Mandel, Optical Society of America Annual Meeting, October 1987.
43. M.M. Tehrani and L. Mandel, Phys. Rev. A, **17**, 677 (1978); W.R. Christian, T.H. Chyba and L. Mandel, Opt. Commun., **66**, 238 (1988).
44. I. McMackin, C. Radzewicz, M. Beck and M.G. Raymer, Phys. Rev. A, 38, 820 (1988); Yu.M. Ajvasjan, V.V. Ivanov, S.A. Kovalenko, V.M. Baev, E.A. Sviridenkov, H. Atmanspacher and H. Scheingraber, Appl. Phys. B, **46**, 175 (1988); C.R. Doering, J.N. Elgin, J.D. Gibbon and D.D. Holm, Phys. Lett., **129A,** 310 (1988); W. Brunner, R. Fischer and H. Paul, J. Opt. Soc. Am. B, **2,** 202 (1985); W. Brunner, R. Fischer and H. Paul, J. Opt. Soc. Am. B, **5**, 1139 (1988).
45. M.W. Phillips, H. Gong, A.I. Ferguson and D.C. Hanna, Opt. Commun., **61**, 215 (1987).
46. A.M. Ratner, Spectral, Spatial and Temporal Properties of Lasers, trans. P. Robeson, Jr., (Plenum, NY, 1972).

47. V.N. Morozov, A.B. Sergeev, and I.A. Skopin, IEEE J. Quantum Electron., **QE-19**, 1074 (1983). Rate equation formulations for multimode interactions often overlook the fourwave mixing processes by ignoring interference terms. Even in semiconductor lasers this is not a reasonable approach as shown by B. Thédrez, A. Jones and R. Frey, IEEE J. Quantum Electron., **24**, 1499 (1988). (Though the two-level equivalent model for such lasers is highly ad hoc, the experimental measurements and references cited in this article indicate that the multiwave mixing is clearly present.)

48. The equations with detuning were first analyzed for single mode instabilities by A.V. Uspenskiy, Sov. Phys. - Rad Eng. Electron. Phys., **8**, 1145 (1963) [Russian original in 1963]; **9**, 605 (1964) and extended by V.V. Korobkin and A.V. Uspenskiy, Sov. JETP, **18**, 693 (1964) [Russian original in 1963]. The equations with detuning were next studied by J. Zorell, Opt. Commun., **38**, 127 (1981) for the multimode problem (in extension of the work of reference 49) and he found that the threshold for the multimode amplitude instability increased with detuning. By the connection since proven between the single mode and multimode instabilities L.A. Lugiato and L.M. Narducci, Phys. Rev. A, **32**, 1576 (1985) we can say that Zorell's results indicate that the threshold for single mode instabilities also increases with detuning. The complex Lorenz equations suitably isomorphic to the laser equations given here with detuning were also derived and studied by A.C. Fowler, J.D. Gibbon and M.J. McGuinness, Physica, **4D**, 139 (1982). The single mode laser with detuning was reconsidered and more completely treated by Mandel and Zeghlache in: P. Mandel and H. Zeghlache, Opt. Commun., **47**, 146 (1983); H. Zeghlache and P. Mandel, J. Opt. Soc. Am. B, **2**, 18 (1985); H. Zeghlache, P. Mandel, N.B. Abraham, and C.O. Weiss, Phys. Rev. A, **38**, 3128 (1988); H. Zeghlache, Ph.D. thesis, U. Libre de Bruxelles, December 1988.

49. H. Risken and K. Nummedal, J. Appl. Phys. **39**, 4662 (1968); R. Graham and H. Haken, Z. Phys. **213**, 420 (1968); for a variety of extensions see references in /10/ and L.A. Lugiato, L.M. Narducci, E.V. Eschenazi, D.K. Bandy and N.B. Abraham, Phys. Rev. A, **32**, 1563 (1985) and references therein. To explain the bichromatic multimode dye laser behavior of /36/ ultimately required models not of the field equation type nor even of the Maxwell-Bloch coherent form but more complex equations given by Fu Hong and H. Haken, J. Opt. Soc. Am. B, **5**, 899 (1988).

50. C.O. Weiss, N.B. Abraham and U. Hübner, Phys. Rev. Lett., **61**, 1587 (1988).

51. J.P. Eckmann, Rev. Mod. Phys., **53**, 643 (1981); Y. Pomeau and P. Manneville, Commun. Math. Phys., **74**, 189 (1980).

52. P. Grassberger and I. Procaccia, Physica, **9D**, 189 (1983); Phys. Rev. A, **28**, 2591 (1983); Dimensions and Entropies, G. Mayer-Kress, ed. (Springer, Berlin, 1986); Chaos and Related Nonlinear Phenomena, I. Procaccia and M. Shapiro, eds., Physica Scripta, (to be published 1989); H. Shuster, Deterministic Chaos, 2nd Ed.(Physik Verlag, Weinheim, 1987); R. Badii and A. Politi in Instabilities and Chaos in Quantum Optics II, N.B. Abraham, F.T. Arecchi and L.A. Lugiato, eds., (Plenum, NY, 1988) p. 335; K. Pawelzik and H.G. Shuster, Phys. Rev. A, **35**, 481 (1987).

53. P. Atten, J.G. Caputo, B. Malraison and Y. Gagne, J. Mechanique Théorie et Applique, Numero speciale, 133 (1984); J.G. Caputo and P. Atten, Phys. Rev. A, **35**, 1311 (1987); R. Badii, G. Broggi, B. Derighetti, M. Ravani, S. Ciliberto, A. Politi, and M.A. Rubio, Phys. Rev. Lett., **60**, 979 (1988); F. Mitschke, M. Möller, and W. Lange, Phys. Rev. A, **37**, 4518 (1988); A.M. Albano, J. Muench, C. Schwartz, A.I. Mees and P.E. Rapp, Phys. Rev. A, **38**, 3017 (1988); H. Atmanspacher, H. Scheingraber and W. Voges, Phys. Rev. A, **37**, 1314 (1988); A. Destexhe, J.A. Sepulchre and A. Babloyantz, Phys. Lett., **132A**, 101 (1988).

54. J. Swift and K. Wiesenfeld, Phys. Rev. Lett. **52**, 705 (1984).

55. A. Arneodo, P. Coullet and C. Tresser, Phys. Lett., **81A**, 197 (1981); Y. Kuramoto and S. Koga, Phys. Lett., **92A**, 1 (1982); J. Gambaudo, P. Glendinning and C. Tresser, Nonlinearity, **1**, 203 (1988).

56. F.T.Arecchi, R. Meucci, W. Gadomski, Phys. Rev. Lett., **58**, 2205 (1987); F.T. Arecchi, A. Lapucci, R. Meucci; J.A. Roversi and P.H. Coullet, Europhys. Lett., **6**, 677 (1988); F.T. Arecchi, W. Gadomski, A. Lapucci, H. Mancini, R. Meucci, and J.A. Roversi, J. Opt. Soc. Am. B, **5**, 1153 (1988); see also the review by F.T. Arecchi in ref. 5 and in Nucl. Phys. B (Proc. Suppl.) **2**, 13 (1987).

57. D. Hennequin, F. de Tomasi, B. Zambon, and E. Arimondo, Phys. Rev. A, **37**, 2243 (1988); D. Hennequin, F. de Tomasi, L. Fronzoni, B. Zambon, and E. Arimondo, Opt. Commun. to be published; F. de Tomasi, D. Hennequin, B. Zambon and E. Arimondo, private communication 1988..

58. D. Dangoisse, A. Bekkali, F. Papoff, and P. Glorieux, Europhys. Lett., **6**, 335 (1988); A. Bekkali, F. Papoff, D. Dangoisse and P. Glorieux, J. Phys. Colloque C2 (Paris), Suppl. no. 6, **49**, C2-349 (1988).

59. O.E. Rössler, Phys. Lett. **71A**, 155 (1979); A. Arneodo, P. Coullet, C. Tresser, J. Stat. Phys. **27**, 171 (1982); P. Glendinning and C. Sparrow, J. Stat. Phys., **35**, 645 (1984); P. Gaspard, R. Kapral and G. Nicolis, J. Stat. Phys., **35**, 697 (1984); C.

Tresser, Ann. Inst. Henri Poincaré, **40**, 441 (1984); C. Tresser, J. Physique **45**, 837 (1984); P. Glendinning and C. Tresser, J. Phys. Lett. **46**, 347 (1985); A. Arneodo, P.H. Coullet, E.A. Spiegel, and C. Tresser, Physica, **14D**, 327 (1985); J.C. Antorantz and M.A. Rubio, J. Opt. Soc. Am. B, **5**, 1070 (1988).

60. J.S. Uppal, R.G. Harrison, and J.V. Moloney, Phys. Rev. A, **36**, 4823 (1987); Phys. Rev. Lett., **26**, 2868 (1987); P.A. Khandokhin, Ya. I. Khanin, and I.V. Koryukin, Opt. Commun., **65**, 367 (1988).
61. F. Laguarta, T. Pujol, R. Vilaseca, and R. Corbalan, J. Phys. Colloque **C2** (Paris), Suppl. no. 6, **49**, C2-409 (1988).
62. F. Argoul, A. Arneodo, and P. Richetti, Phys. Lett., **120A**, 269 (1987); F. Argoul, A. Arneodo, P. Richetti and J.C. Roux, J. Chem. Phys., **86**, 3325 (1987); P. Richetti, J.C. Roux, F. Argoul, and A. Arneodo, J. Chem. Phys., **86**, 3339 (1987).
63. L.P. Shil'nikov, Dokl. Akad. Nauk SSSR, **160**, 558 (1965); Mat. Sbornik, **77**, (119), 461 (1968); **81** (123), 92 (1970); [Eng. trans.: Sov. Math. Doklady, **3**, 394 (1962); **6**, 163 (1965), **8**, 54 (1967); Math USSR Sbornik **6**, 427 (1968); **10**, 91 (1970)].
64. J. Guckenheimer and P. Holmes, Nonlinear Oscillations, Dynamical Systems and Bifurcations of Vector Fields (Springer, Berlin, 1984).
65. C. Sparrow, in Chaos, A. Holden, ed. (Adam Hilger, Bristol, 1986) p. 111.
66. D.S. Broomhead and G.P. King, in Nonlinear Phenomena and Chaos, S. Sarkar, ed. (Adam Hilger, Bristol, 1986) p. 136.
67. Among various references relevant to the question of polarization dynamics in "two-mode lasers" several recent ones include: G.P. Puccioni, M.V. Tratnik, J.E. Sipe and G.L. Oppo, Opt. Lett., **12**, 242 (1987); G. Ropars, A. Le Floch, G. Jezequel, R. Le Naour, Y.C. Chen and J.-M. Liu, IEEE J. Quantum Electron., **QE-23**, 1027 (1987); V.G. Gudelev, A.Ch. Izmailov, and V.M. Yasinskii, Sov. J. Quantum Electron., **18**, 166 (1988); D.S. Bakaev, V.M. Ermachenko, V.Yu. Kurochkin, V.N. Petrovskii, E.D. Protsenko, A.N. Rurukin, and R.A. Shananin, Sov. J. Quantum Electron., **18**, 20 (1988); and M.L. Anishchenko, V.M. Ermachenko, N.V. Naumov, V.N. Petrovskii and E.D. Protsenko, Sov. J. Quantum Electron., **18**, 33 (1988).
68. U. Hübner, N.B. Abraham, and C.O. Weiss, Phys. Rev. A, to be published (1989); U. Hübner, W. Klische, N.B. Abraham and C.O. Weiss, Applied Physics B, to be published (1989); and in Coherence and Quantum Optics VI, L. Mandel, E. Wolf and J. Eberly, eds., (Plenum, NY, 1990).

69. E. Roldán, G.J. De Valcárcel, R. Vilaseca, F. Silva, J. Pujol, R. Corbalán, and F. Laguarta, Opt. Commun., to be published (1989) and in Coherence and Quantum Optics VI, L. Mandel, E. Wolf and J. Eberly, eds., (Plenum, NY, 1990); R. Corbalán, F. Laguarta, J. Pujol and R. Vilaseca, Opt. Commun. **71**, 290 (1989); F. Laguarta, J. Pujol, R. Vilaseca and R. Corbalán, in Dynamics of Nonlinear Optical Systems, eds., L. Pesquera and J. Bermejo (World Scientific, Singapore, 1989); J. Pujol, F. Laguarta, R. Vilaseca and R. Corbalán, J. Opt. Soc. Am. B **5**, 1004 (1988).

70. D. Hennequin, M. Lefranc, A. Bekkali, D. Dangoisse and P. Glorieux, in Measures of Complexity and Chaos, N.B. Abraham, A.M. Albano, A. Passamante, and P.E. Rapp, eds., (Plenum, NY to be published); D. Dangoisse, A. Bekkali, F. Papoff and P. Glorieux, Europhys. Lett. 6, 335 (1988); F. Papoff, A. Fioretti, E. Arimondo and N.B. Abraham, in Measures of Complexity and Chaos, N.B. Abraham, A.M. Albano, A. Passamante, and P.E. Rapp, eds., (Plenum, NY to be published); D. Hennequin, F. DeTomasi, B. Zambon and E. Arimondo, Phys. Rev. A, **37**, 2233 (1988); F. de Tomasi, D. Hennequin, B. Zambon and E. Arimondo, J. Opt. Soc. Am. B **6**, 45 (1989).

EXPERIMENTAL MEASUREMENTS TO IDENTIFY AND/OR CHARACTERIZE CHAOTIC SIGNALS

J.R. Tredicce
Department of Physics and Atmospheric Sciences
Drexel University
Philadelphia, PA 19104, USA
and
N.B. Abraham
Department of Physics
Bryn Mawr College
Bryn Mawr, PA 19010-2899, USA

Outline

I- Introduction.
- a) Chaos - A practical definition.
- b) Properties of chaos.
- c) Classification of Lasers (*based on chaotic possibilities*).

II- Qualitative Experimental Techniques to Measure and Characterize Chaotic Signals.
- a) Time Series, Power Spectra, Phase-space Portraits, and Poincaré Sections.
- b) Identifying Chaos.
- c) Return Maps.
- d) Bifurcation Diagrams.

III- Illustrations of some of these methods in the study of a Chaotic Modulated Laser.
- a) Time Series.
- b) Power Spectra.
- c) Phase-space Portrait.
- d) Poincaré Sections.
- e) Bifurcation Diagrams.

IV- Quantitative Methods to Characterize or Identify Chaos: Measurement of dimensions
- a) Reconstruction of an attractor in phase space.
- b) Measuring the dimension

I- Introduction

I. a) CHAOS - A practical definition -

Since 1963, interest in irregular behavior of nonlinear dynamical systems has grown exponentially. The existence of "stable" aperiodic solutions in apparently simple deterministic systems sharply conflicts with the expectations we have developed from studies of linear physical systems. Our conventional training suggests to us that irregularity appears only in complicated transients which die out. The persistence of irregularity is not only anomalous but also conficts with many of our concepts about the order and regularity of deterministic processes in nature. Even more remarkable has been the discovery that deterministic systems are not necessarily predictable ones.

We must now come to the more sophisticated recognition that irregularity and unpredictibility are not just connected with random processes, but they may also arise from purely deterministic situations. In other words, *the teological concept of a determined but not predictable future recently has found simple examples in nature* . Indeed it is just that unpredictable behavior of a deterministic system which we now take as the definition of the scientific and technical term "chaos" [1].

For those who are unfamiliar with chaos, it may be important to stress the differences between determinism and predictibility.

A deterministic system is one for which there are equations of motion which we can use together with a precise description of the present conditions to know exactly its evolution in time, forward and backward. A simple example is the forced damped harmonic oscillator:

$$m(d^2x/dt^2) + c(dx/dt) + kx = F_0 \cos \omega t \qquad (1)$$

For a given initial condition we can write down the exact solution. In presenting this solution we might give the variable x as a function of time or we might choose a number of variables corresponding to the number of degrees of freedom and represent the solution as a trajectory in this "phase space" with time as a parameter.

If $F_0 = 0$, a phase space for describing this system is uniquely specified by two variables x and (dx/dt) because the differential equation is one second-order equation.

(In general, if our system of equations is n first-order differential equations, we will need an n-dimensional phase space.)

In the absence of the forcing term, the solution is

$$x(t) = A \exp(-\gamma t/2) \cos(\omega_0 t + \phi) \quad (2)$$

where $\omega_0 = [(k/m)-(\gamma/2)^2]^{1/2}$, $\gamma = c/m$

and A and ϕ are constants determined by the initial conditions x(0) and (dx/dt)(0).

We see that as time extends to infinity, the solution comes exponentially close to x=0. Furthermore, if x(0)=0 and (dx/dt)(0)=0 then the solution remains the same for all time. To represent this solution we can plot x as a function of time, or we can plot (dx/dt) versus x using time as a parameter. In this phase space the trajectory of the solution beginning at x=(dx/dt)=0 is a point. A point solution in the phase space is called a "fixed point" and it is called a "stable fixed point" if perturbations of this solution relax back to this value.

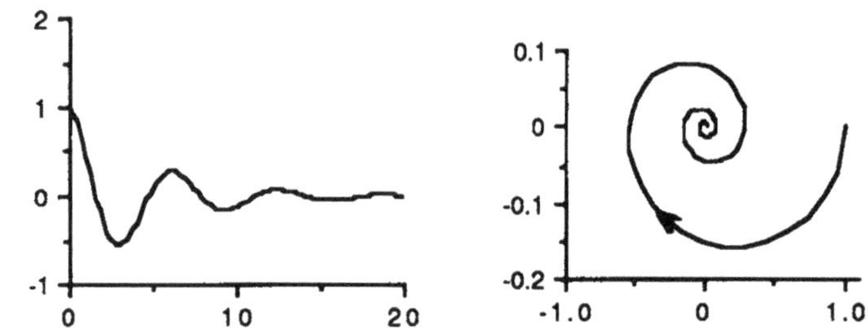

a) b)
Fig.1 : a) **x** as a function of time **t**, and b) **dx/dt** as a function of **x** for the damped harmonic oscillator.

In the presence of the forcing term, the asymptotic solution is

$$x(t) = B \cos(\omega t + \phi), \quad (3)$$

where B and ϕ are **not** determined by the initial conditions x(0) and (dx/dt)(0),.but by all the parameters of the system (F_0, γ, ω_0, and ω). In this case the solution is oscillatory and the plot of the solution in phase space gives an ellipse. Changing the initial conditions will not change the values of B and ϕ and therefore the asymptotic solution in the phase space after a transient is the same ellipse.

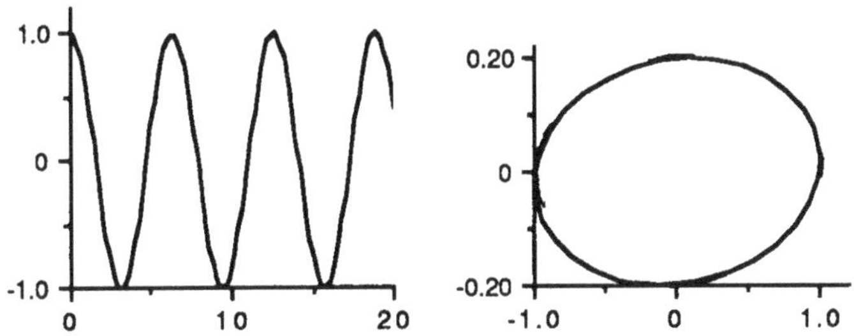

a) **b)**

Fig.2 : a) **x** as a function of time ; and b) **dx/dt** as a function of **x** for a forced harmonic oscillator

Two kinds of systems can be considered. One is conservative, in which the energy remains constant. This would be the case in an unforced harmonic oscillator system with c=0. The other is dissipative, c≠0. For a bounded conservative system, there is a solution which originates at each initial condition and which then returns to that condition in a finite time.

In contrast, the decay of transients is a property of dissipative systems. The set of possible points in phase space for asymptotic solutions is much smaller than the set of possible initial conditions. Hence one says that the phase space contracts for a dissipative system and one refers to the asymptotic set of points forming an asymptotic solution as an "**attractor**".

The main property of a deterministic system (regardless of whether it is conservative or dissipative) is the **uniqueness** of the solution: if **x(t)** is a solution beginning from some initial conditions, then that is the only solution having those initial conditions.

If the system one is looking at is constantly disturbed by noise, a degree of unpredictability in its behavior is introduced. This can be simulated in models by adding random noise in the equations. Thereby the precisely deterministic aspect of the system is removed. However, if the system is strongly dissipative, then the influence of the noise may be relatively small and the solutions in phase space may remain very similar to those of the noise-free case. Very often we can continue to model these systems by noise-free equations, an important consideration if we are modelling experiments where noise is unavoidable.

The solution $\mathbf{x}(t)$ of (1) has another important property -- it is predictable. If we observe $\mathbf{x}(t)$ for a time longer than $4\pi/\omega$ (two periods), we discover that it is periodic and therefore it is not necessary to know the initial conditions and the equations of motion to predict what $\mathbf{x}$ will be at any time in the future or what it was in the past. The amplitude, frequency and phase of $\mathbf{x}$ carry all the necessary information about the system. In a general way, any periodic evolution is a predictable one.

In much of the mathematical physics we have studied previously, we have reduced, approximated and simplified the physics in order to arrive at linear problems. When a restoring force is linear, the solutions are periodic, arising from simple harmonic oscillations. Other simple linear problems have solutions in the form of polynomials or exponentials. From these examples we were conditioned to believe that determinism and predictibility were synonymous.

However, this is not always so. In the "new" world of nonlinear dynamics we have chaotic solutions of a deterministic system which are aperiodic and therefore unpredictable. To be deterministic yet aperiodic in a bounded region of phase space requires that trajectories never cross. (By definition of a phase space and a deterministic (and unique) solution, if two solutions meet at some point, then they share the same future. Therefore by uniqueness they share the same past and are identical.) Hence chaotic solutions can only occur in systems with at least three degrees of freedom (that is, in phase spaces with at least three variables) where the trajectories can stay in a bounded region without intersecting. The unpredictability of chaotic solutions lies in the surprising behavior of the evolution of chaotic systems starting from neighboring points in the phase space. The trajectories that make up the asymptotic solution are dense -- near every loop of a trajectory is another trajectory. Within any finite neighborhood of a point on a trajectory there is an infinite set of other trajectories. For chaotic solutions, the set of points which make up the asymptotic solution forms a fractal object with a dimension greater than two.

Despite their closeness, neighboring trajectories do not asymptotically collapse together, rather when they are close enough they separate exponentially. The exponential divergence is much faster than the diffusive separation of trajectories in the presence of noise. And yet, with all this expansiveness, the full set of possible initial conditions contracts to a subset of values which are visited by the asymptotic trajectory.

Such a complicated chaotic trajectory is called a "strange attractor".

Because of the exponential divergence, the more distant future can be determined only with a more precise specification of the initial condition. Alternatively the future from a starting point of a certain approximate condition involves the selection from among various possible future paths, requiring a steady rate of new information. This information rate is called the "entropy" of the solution.

A formal mathematical definition of chaos is that it is a solution which has at least one positive Lyapunov exponent. This positive exponent gives the exponential divergence of closely neighboring trajectories. Hence the future of a chaotic trajectory is very sensitive to the starting point. However this definition does not offer much help to the experimentalist who is seeking to recognize chaotic dynamics underlying the irregularities in a quantity that is being measured: Lyapunov exponents (with only a few exceptions) are not the quantities we are able to calculate easily from experimental data. [In some cases, the positive Lyapunov exponent (if there is only one) which is also the rate of entropy production of the system, can be determined because it gives the inverse of the autocorrelation time of the signal or the bandwidth of the spectrum (see e.g., Ref. 15 and 16)]. However, usually this is difficult and we must rely on other properties of chaos when we are trying to recognize it experimentally.

I.b) Properties of Chaos

Based on its definition some properties of a chaotic solution are obvious:

1) the behavior in time of any variable of the system is aperiodic;

2) a power spectrum of any variable has some broadband parts;

3) the dimension of a chaotic attractor must be greater than or equal to two;

4) the entropy of the system (defined as the sum of all positive Lyapunov exponents) is bigger than zero but finite.

and some other properties are not so obvious but still true:

5) a Poincaré section shows an infinite number of points;

6) the dimension of the attractor for a chaotic solution is usually a fractal number.

The first three properties are, from the experimental point of view, those easiest to check, but their existence does not unambigously confirm that the signal was generated by chaotic dynamics. In fact, noisy systems can present the same properties.

Quantitative measures such as those of determining dimensions and entropies are required in order to distinguish chaos from noise. While it is rarely difficult to determine these features from numerical solutions of model equations, it is more difficult when one is dealing with experimental systems. Nevertheless, in some cases it is feasible to calculate these characteristics from laboratory data without excessive demands on computation time, data acquisition rates, or data precision.

However, before trying sophisticated experimental techniques to observe and characterize complex chaotic dynamics, we have to know when and where we can expect to observe chaos. This question is still unresolved but we know two necessary (though not sufficient) conditions: 1) the existence of nonlinearities and 2) at least three degrees of freedom. Unfortunately (or maybe fortunately) almost every real experimental system satisfies these conditions. Hydrodynamical [2], chemical [3], biological [4], mechanical [5], electrical [6],....and.....**LASER** [7] systems are some examples where chaos has been recognized.

The relevance of lasers in the general study of nonlinear dynamical systems lies in two important factors: a) the existence of relatively simple, yet highly accurate, theoretical models allowing a direct comparison between numerical and experimental results and b) time scales for the pulsations ranging from μsec to msec which are fast enough to eliminate long time stabilization requirements and yet slow enough to make quantitative observations feasible.

From here on we will concentrate much of our discussion on laser systems and draw many of our examples from the literature of laser physics. This is because lasers are the subject of this summer school and also because we have direct experience with them. However many of the methods we describe in this report are valid for other nonlinear systems.

I.c) Classification of Lasers (based on their possible types of dynamical behavior)

It is well known that most lasers can be described in terms of a closed set of differential equations called the Maxwell-Bloch equations (see the chapter by L. M. Narducci *et al.* in this volume) that in their simplest form are:

$$\frac{dE}{dt} = -\kappa E + 2C\kappa P$$

$$\frac{dP}{dt} = -\gamma_{\perp} (P - E D) \qquad (3)$$

$$\frac{dD}{dt} = -\gamma_{\parallel} (D - D_o + \frac{1}{2} (E^* P + E P^*)$$

where E, P and D are the field amplitude, the polarization and the population inversion respectively, κ, $\gamma_{\perp}$, and $\gamma_{\parallel}$ are their respective loss rates and D_o is the pump rate.

This model has five coupled variables (since E and P are complex) with nonlinear terms and therefore chaotic solutions are a possibility. However, the dynamics of most lasers are usually limited by the values of the parameters. When several decay rates are much faster than others, the dissipation is so great that the dynamical evolution in the phase space quickly moves to a subspace that can be described by a smaller number of variables. The asymptotic behavior and hence the attractors are then limited by the dimension of this subspace. The contraction reduces the number of relevant degrees of freedom from the dimension of the phase space to the dimension of the subspace. If the subspace is one dimensional, then the only posibility is a fixed point. If the behavior is limited to two dimensions, then the solutions might be constant (fixed points) or periodic. If the dynamical behavior occurs in three or more dimensions, then the behavior could be constant, periodic, quasiperiodic (helical motion on the surface of a torus that does not repeat) or chaotic. Thus by classifying lasers in terms of the number of variables required to describe their asymptotic subspaces, as was done in [8], we can anticipate their possible dynamical behavior.

i) Class A Lasers

In the case e.g. of He-Ne, dye, and Ar^+ lasers the relation among the decay rates is $\kappa << \gamma_{\parallel}, \gamma_{\perp}$. In such cases, the polarization and the population inversion will adiabatically follow the behavior of the field and the dynamical behavior of the system can be described in terms of a single variable

$$\frac{dE}{dt} = -\kappa E + 2 C \kappa E [D_o / (1 + |E|^2)]. \qquad (4)$$

The phase space is unidimensional and therefore periodic and chaotic solutions are impossible. More complex behavior can be observed only if more variables are needed to describe the system. This would be true, for example, in the case of multimode operation (see, e.g., refs. 9-10), where one might need a variable to describe the amplitude of each mode of the electromagnetic field. Alternatively, if not all atoms or molecules of the material are the same, one might need additional variables to describe the material. For example, in the case of inhomogeneously broadened lasers [11],or lasers with saturable absorbers [12], or lasers with a modulated injected signal [13], periodic and chaotic pulsations have been observed or theoretically predicted.

ii) Class B Lasers

When $\kappa, \gamma_{\parallel} << \gamma_{\perp}$ the dynamical behavior of a laser is well described in terms of two coupled variables [14]

$$\frac{dE}{dt} = -\kappa E + 2 C \kappa E D$$

$$\frac{dD}{dt} = -\gamma_{\parallel} (D - D_o + D |E|^2) \qquad (5)$$

Semiconductor, Nd:YAG, and CO_2 lasers belong to this category. In principle both constant and periodic solutions are possible, but with a little analysis we can show that the structure of the laser equations prevents there from being periodic solutions.

However, if there are different nonlinearities (as is the case for more complicated models of semiconductor lasers) periodic solutions can be found. Only one more variable is needed to make chaos feasible in these systems. This might come from other equations for other modes or from other equations to describe the inhomogeneities of the material, but the easiest way to do it is to make one of the two equations non-autonomous by making a parameter time dependent:

e.g: $\kappa(t) = \kappa_0 \,(1 + m \cos\omega t),$ (6)

where m is the modulation depth, and ω is the modulation frequency.

Loss [15-18], frequency [19-20] and pump [21] modulation cause qualitatively similar dynamical behavior. A system described in these terms is one of the simplest displaying complex time structures because it is limited to three dimensions with a well-known driving force.

Other ways to observe chaos in Class B lasers while remaining inside the validity of a model based on a set of first order differential equations is obtained when we allow the interaction between two counterpropagating fields (two modes) inside a ring resonator [22-24], or we inject an external signal [25]. In such cases the slow decay of the grating formed by the population inversion couples the two beams, even if one of them is much stronger than the other, producing a deterministic switching between the strengths of the intensities of the two beams. Others have successfully added a saturable absorber in the laser cavity in order to see pulsations and chaos in Class B lasers [26-27].

iii) Class C Lasers

If $\kappa \sim \gamma_{\parallel} \sim \gamma_{\perp}$ then none of the variables E, P or D can be adiabatically eliminated by simple arguments. In the limits of validity of such models, the steady states of the laser will be unstable if:

$$\kappa > \gamma_{\parallel} + \gamma_{\perp}$$

and

$$(2\,C\,D_0/\kappa) > (\kappa + \gamma_{\parallel} + \gamma_{\perp})(\gamma_{\parallel} + \gamma_{\perp})/(\kappa - \gamma_{\parallel} - \gamma_{\perp}).$$

The resulting pulsations are typically chaotic if $\gamma_{\|}/\gamma_{\perp} > 0.2$. Some optically pumped infrared lasers belong to this class and chaotic dynamics of such lasers have been experimentally studied [28-29, see also N. B. Abraham elsewhere in this volume].

One of the first experiments in which chaos was identified in laser systems was performed in a Class B laser with a modulated parameter [15]. In the next section, we will use it to illustrate how to apply different experimental techniques to study nonlinear dynamics.

II- Qualitative Experimental Techniques to Characterize Dynamical Behavior

II.a) Time series, Power spectra, Phase space portraits, and Poincaré sections.

By "qualitative experimental techniques" we mean those that do not involve the use of digital oscilloscopes and/or computational analysis of the data. In other words we refer here to analog measurements.

Four relatively simple measurement techniques can be discussed first.

i) Time series

ii) Power spectra

iii) Partial construction of a phase space portrait

iv) A Poincaré section.

For a "time series" one uses an oscilloscope to view the signal in time. If the signal is periodic, usually the triggering setting on the oscilloscope can be set so that the displayed pattern is stable. If the behavior is irregular then usually the pattern blurs as similar starting conditions that satisfy the trigger circuit evolve differently.

For a "power spectrum" one uses a spectrum analyzer to Fourier analyze the fluctuations. Periodic signals have spectra made up of sharp spectral lines with a dominant fundamental frequency and some harmonics if the signal is not sinusoidal. Irregularities in the temporal evolution of a signal usually tend to blur the narrow lines of a periodic spectrum or to introduce many different frequencies, a broadband portion of the spectrum.

For time series and power spectra one must buy the appropriate analog instrument, but for a partial view of a phase space, one needs only an oscilloscope and some cleverness. In principle, a phase space (or a projection thereof) would require that you measure more than one variable of a system. Then you could use the x-y display mode of the oscilloscope to plot one variable against another with time as a parameter. The advantage of such displays is that periodic solutions are immediately seen as simple closed curves, while irregular signals reveal themselves as bands. The structure of a chaotic attractor is often seen in the pattern of dark and bright regions which suggest that there are definite rules for behavior despite the irregularities. Such plots give a preliminary idea of the appearance of the attractor. In most experiments only a single physical quantity is measureable but it is still possible to have two signals for the x-y display. Equally useful for such plots are (x,y), (x(t),(dx/dt)(t)), and $(x(t),x(t+\tau))$. Simple circuits that give differentiation or delay (or a combination of both) are easy to build. For a system externally driven, like a laser with modulated losses, it is also possible to display the external driving force versus the laser intensity. All of these graphs will give exactly the same amount of information about the dynamical behavior of the system.

A Poincaré section is a subset of the information in a phase portrait. Formally, we might define a Poincaré section as the set of points where a trajectory crosses a preselected plane of the phase space. However, there are many equivalent, and more practical definitions such as:

i) The values of the other variables when one variable has a maximum;

ii) The values of the variables when a periodic driving force has a selected value;

These can be generated by using the phase portrait plots on the x-y oscilloscope display described above with an addtional condition. For example, using the z-axis "blanking" one can suppress the phase portrait except when some condition is reached such as the conditions mentioned above. It is relatively easy to construct these "sections" when there is a periodic forcing of the system because then it is easy to generate the trigger for the display. When a system is spontaneously pulsing, the frequency of the pulsation is determined by the strength of the nonlinearities of the system which in turn depend on the parameters. As the parameters change the pulsing frequency changes, and one then needs much more sophisticated electronics to detect and trigger the oscilloscope appropriately in order to get a useful display.

II.b) Identifying chaos.

All of these techniques are useful for characterizing a particular signal, but how can they be helpful in identifying chaotic behavior? One important way is that they can be used to follow changes in the behavior of a system as parameters are changed. In the last ten years it has been found that there is a small number of relatively simple sequences of behaviors by which systems go from stable or periodic pulsations to chaotic pulsations [30]. Examples of these "routes to chaos" are described in the pictures below:

i) Period Doubling

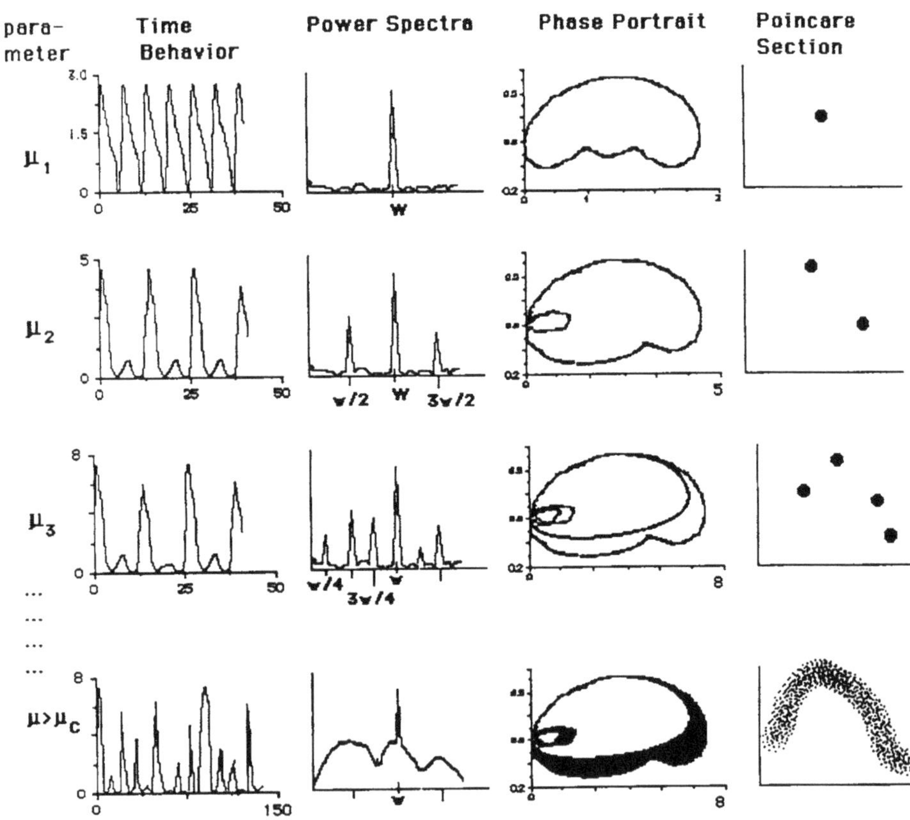

The universality of this period-doubling lies in the fact that the sequences occur in many different systems and that they occur with the following similar features. The succession of appearance of subharmonics converges at a finite value of the parameter. Defining μ_n as the value of the parameter where the nth subharmonic appears, the geometric convergence of the sequence of μ_n's to an accumulation point at which chaos begins is governed by the relation [31]

$$\delta = \lim \mu_n > 0 \; (\mu_n - \mu_{n-1})/(\mu_{n+1} - \mu_n) = 4.669\ldots$$

ii) Quasiperiodicity

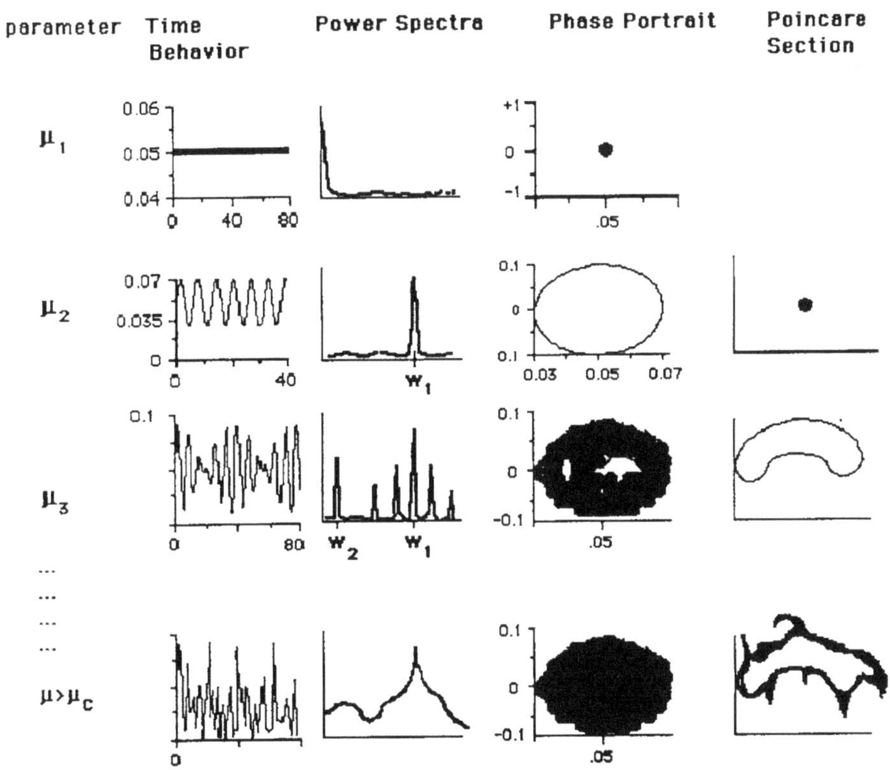

These sequences involve first periodic behavior followed by the onset of a new frequency in addition to the old one and its harmonics. Although the new frequency may initially be irrationally related to the first, usually the nonlinear coupling first generates many combination tones $pf_1 + qf_2$, where p and q are integers and then the frequencies f_1 and f_2 are nonlinear "pulled" into some rational ratio. After this there may be period-doubling to chaos or there may be an abrupt rupture to chaotic behavior.

It must be noticed that the transition from a periodic state to a quasiperiodic one and then to chaos is another generic route well studied theoretically [30,32] and also observed experimentally in lasers [33].

iii) Intermittency

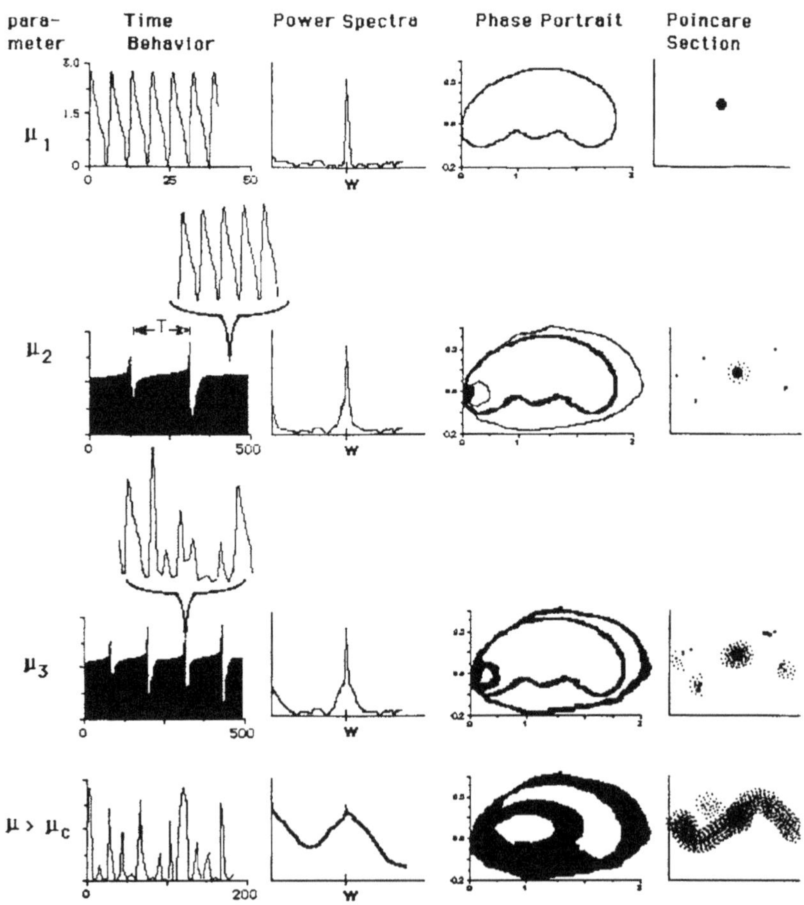

Made up of periodic "laminar" regions and bursts of irregularity, an intermittent system changes as the parameter changes with the average length of the laminar periods scaling as $\langle l \rangle \sim l_o / (\mu - \mu_c)$ [34], where μ_c is the critical value of the parameter where the intermittency begins.

II.c) Return Maps

A return map is somewhat harder to generate without specialized electronics, but because the concept is a nice one we will describe it anyway. By a return map we indicate that we think the important behavior of the system may be described in the relation of successive characteristic values such as

i) The sequence of heights of the peaks;

ii) The sequence of time intervals between the peaks;

iii) The sequence of time intervals between zero-crossings;

Assuming a deterministic relation and assuming that a short prior history is sufficient to determine the future, by plotting successive pairs of these values, one attempts to find a simple curve which represents the rule for how the present controls the future. The simpler the shape of the curve, the simpler the rule. The more blurred the curve the more complex the rule. The absence of an obvious "rule" indicates that the irregularity is more complicated (requiring multiple conditions for specifying the future) or that its origin lies in random noise. Samples of the "rules" that would be found for the different routes to chaos are shown in the figure below

Sample sequences of Return maps

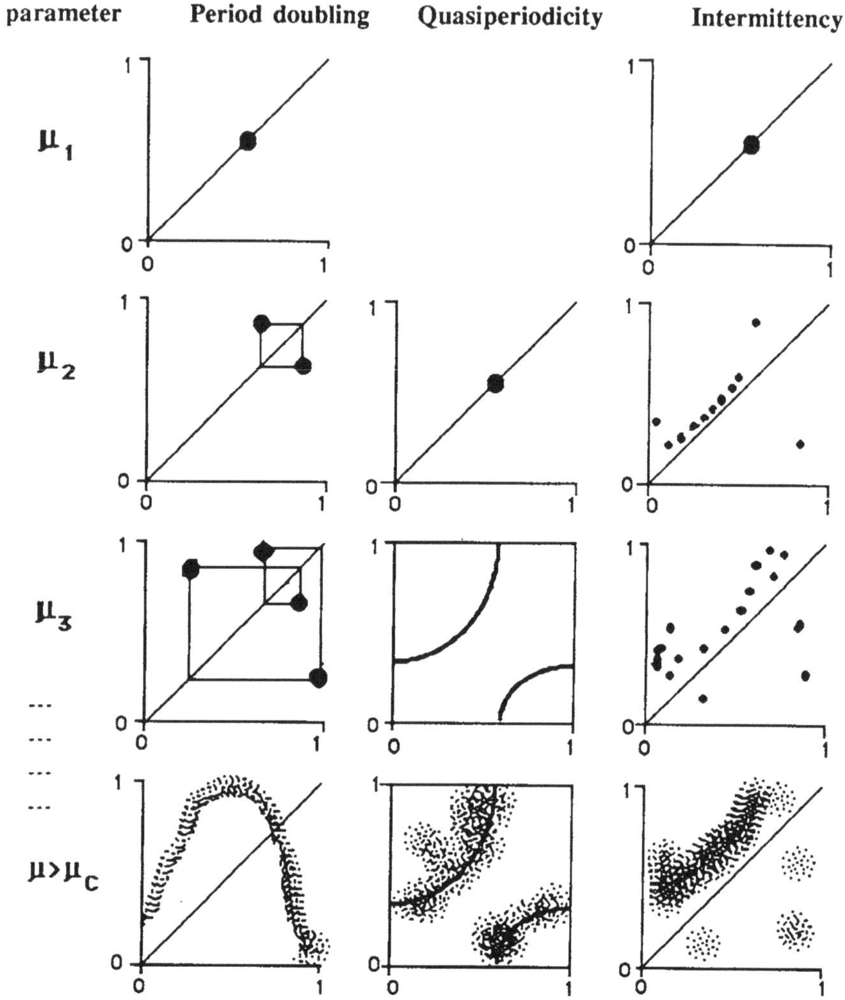

II.d) Bifurcation Diagrams

One can also take the Poincaré section information and use it to generate bifurcation diagrams. By slowing varying the parameter and plotting it on the horizontal axis and by plotting the output of one variable in the Poincaré section on the vertical axis, one can record a pattern of behavior. Schematic bifucation diagrams from the three cases are shown below

Diagrams

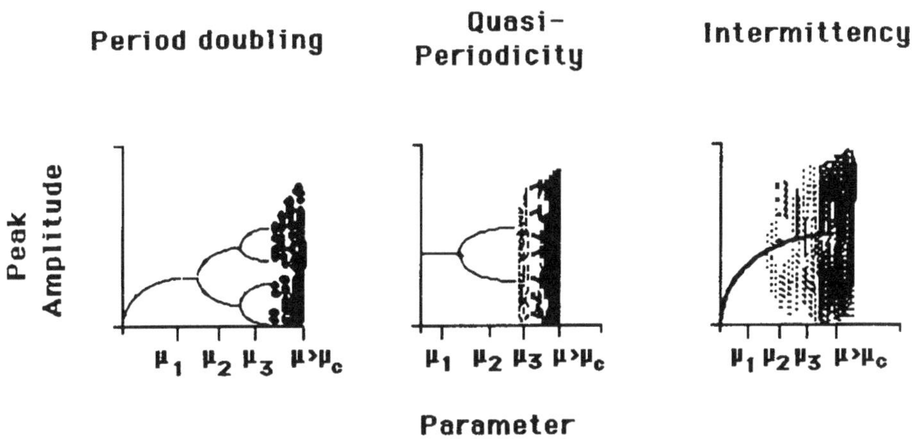

In real experiments, one can not vary the parameter arbitrarily slowly. This leads to some anomalies which are illustrated for the period doubling case below. For example, if the parameter is swept at a constant rate first increasing and then decreasing, the behavior does not seem to be single valued.

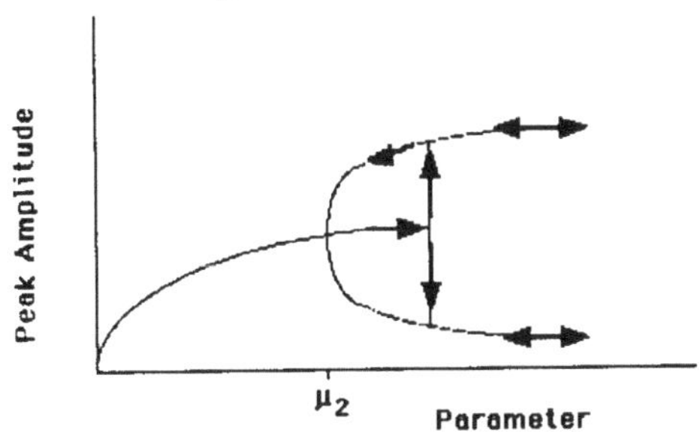

However, there are times when the system suddenly and abruptly changes from one type of behavior to another due to a process called "crises", and the bifurcation diagram then has discontinuities [35-36].

III- Illustrations of some of these techniques in characterizing a modulated CO_2 laser.

In this section we will describe different experimental arrangements used to analyze the dynamics of a CO_2 laser with modulated losses [15]. Similar set-ups can be used for many other systems, though some methods are uniquely appropriate to systems which have an externally imposed periodic clock. We also explain practical problems and how to resolve them depending on the particular situation and we discuss what kind of information is obtained by different tests.

The CO_2 experimental set-up is shown in Fig. 3.

CO_2 lasers are known for having relaxation oscillations, that is, upon a sudden switching of the parameters the response is a damped train of pulses at a frequency which we will denote by ω_R. The pulsation frequency typically lies in the range of 100-300 kHz. It depends somewhat on the current, reflectivity of the mirrors, and gas pressure. The damping of these oscillations is relatively weak meaning that the damped resonance has a relatively high Q.

A sinusoidal modulation of the losses, induced by the electro-optic modulator (EOM) inside the cavity leads to sinusoidal modulation of the laser output. But if $\omega_o \sim \omega_R$, the amplitude of the response can excite nonlinear interactions. This is called "resonant destabilization". Because of the high Q of the resonance, the modulation can destabilize the system at relatively very small values of the modulation amplitude if the modulation frequency is sufficiently near the relaxation oscillation frequency.

The EOM is made of CdTe antireflected coated for a wavelength $\lambda = 10.6\ \mu m$ and usually requires a driver voltage of several kilovolts. This is a problem if we have to modulate this voltage at a rate of hundreds of kilohertz. However, using a crystal with a small cross section (~ 4 mm^2) and relatively long length (60 mm) and by biasing the driver voltage, an amplitude modulation of about 500 volts is enough to produce an 45% amplitude modulation of the loss parameter. However, because of the small cross section of the crystal we found it necessary to introduce a beam expander inside the cavity to match the transverse mode size to the crystal without excessive losses that would have made laser action impossible.

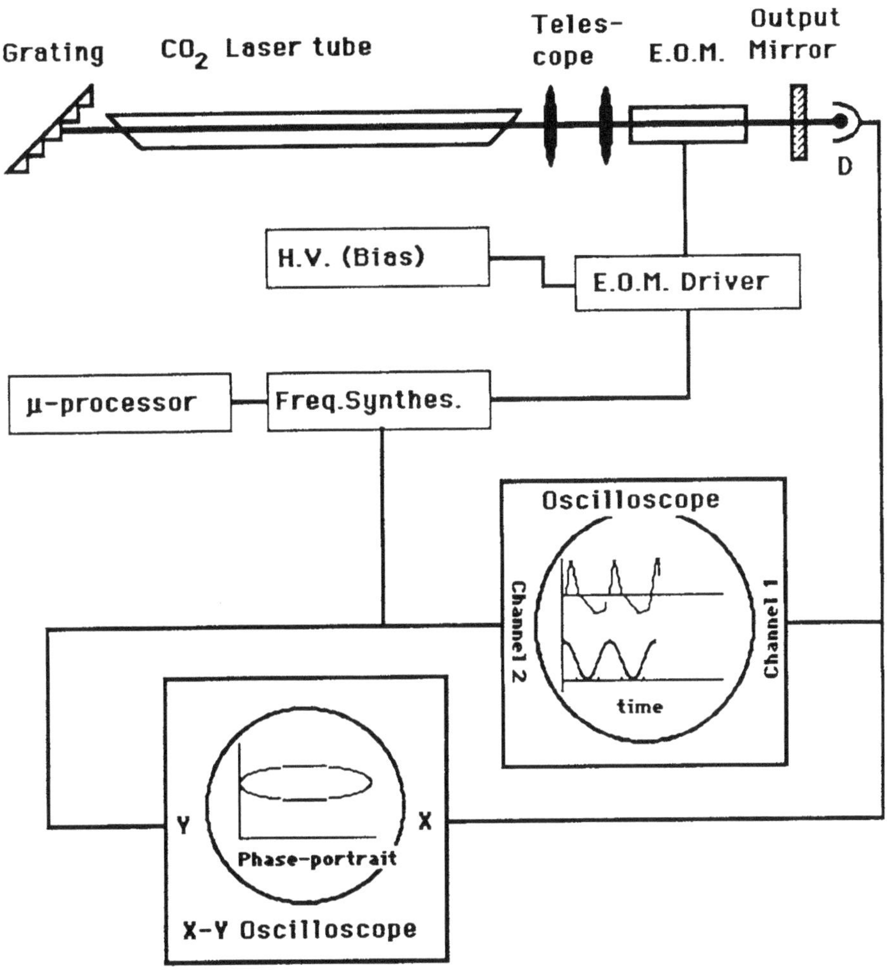

Fig. 3 : Experimental set-up : CO_2 laser with loss modulation.
D: HgCdTe detector ; EOM: CdTe electro-optic modulator

The driver of the EOM is an amplifier with a voltage gain of 20 and a bandwidth of 300 kHz with maximum output of 500 volts. We can apply also a bias of 1 kilovolt maximum. Both frequency and amplitude of modulation are controlled by a programmed frequency synthesizer, but almost any signal generator is probably stable enough to reproduce our results at least qualitatively.

The voltage applied to the EOM induces a change in the linear polarization of the field travelling through it and then the Brewster-angle windows act as a polarizer changing accordingly the amount of the field amplitude coupled out of the cavity. Special care must be taken in checking the orientation of the crystal axis with respect to the plane of polarization of the input field, defined by the Brewster window in order to avoid phase modulation instead of amplitude modulation. If the orientation of the extraordinary axis of the crystal is not a 45º with respect to the polarization of the input field, then a change in the input voltage induces a change in the effective longitudinal index of refraction corresponding to a change on the speed of light and therefore on the effective length of the cavity.

The EOM could be replaced by an acousto-optic modulator (AOM) which would avoid the problems of drive voltage and phase modulation mentioned above but in this case one must carefully shield the detection system to avoid pick-up noise from the RF driver of the AOM which then interferes with the measurements.

The detector is made of HgCdTe and is operated at liquid nitrogen temperature. It is attached to a pre-amplifier which has a bandwidth of 100 MHz. For all the analog measurements we will describe here a fast pyroelectric detector with a bandwidth of 1 MHz would have been sufficient.

III.a) Temporal behavior of the laser intensity

For a modulation frequency near the relaxation oscillation frequency ($\omega \sim 200$ kHz in our laser), a modulation amplitude of $m \sim 10\%$ produces an intensity output signal on the oscilloscope of the type shown in Fig. 4.

Various interpretations of Fig. 4 are possible. In fact, an experimentalist who observed it in the 1960's would try to stabilize the laser against sources of noise causing this "unfortunate irregularity". Instead in the 1980's, the experimentalist would say

"This result is very interesting, we are observing deterministic chaos", and she would work to enhance the instability.

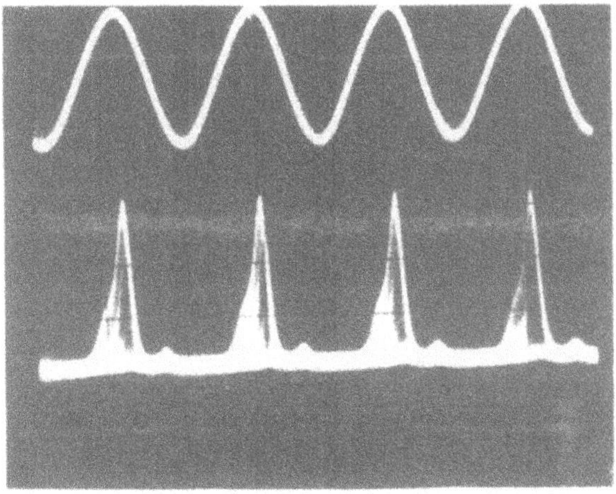

Fig. 4 : The upper trace is the voltage , V, applied to the EOM as a function of time; the frequency of this sinusoidal signal is 191.5 kHz. The lower trace is the laser intensity, I, as a function of time. The aperiodic behavior is demonstrated by the fact that successive traces are not superimposed.

It is clear that the aperiodic behavior observed in Fig. 4 does not provide much help to distinguish noise from chaos. Such a time signal taken alone seems insufficient for the study of chaotic behavior even if it carries all the necessary information about the dynamics. Note that it is not the irregularity alone but the regularities of the correlations hidden within it that are the indicators of chaos.

However, if we monitored the evolution of the time signal as we change one of the parameters, this simple technique is able to give useful information. One example is given in Fig. 5a-d. The upper trace shows the voltage applied to the EOM as a function of time, while the lower one is the output intensity of the laser, I, as a function of time. For small values of m < 4%, the period of the intensity equals that of the modulation as in any forced damped linear oscillator. When the modulation amplitude was increased, the period of the laser intensity pulsations doubled (Fig. 5b). A further increase in m caused the period to double again (Fig. 5c-5d); and for even larger m we observed the trace shown in Fig. 4.

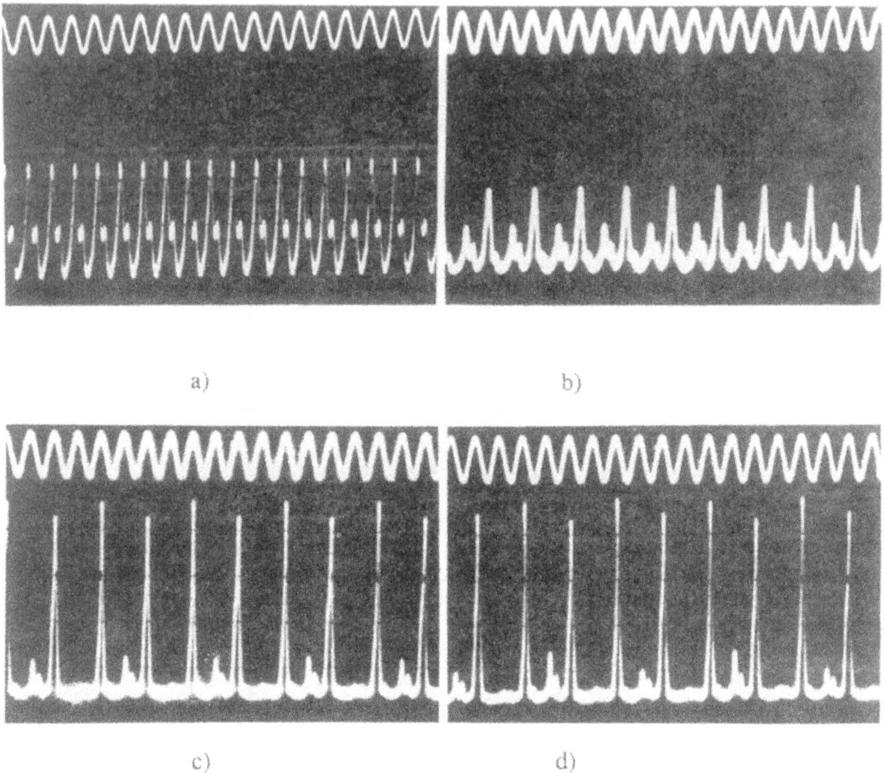

Fig.5: In each picture the upper traces are the modulation voltage **V** as a function of time while the lower traces are **I** as a function of time: a) period one (P1) ; b) period two (P2); c) period four (P4); d) period eight (P8). The frequency of modulation is 191.5 kHz and the modulation amplitude increases from a) to d). The intensity I is measured in arbitrary units and the vertical scale in a) is different from b)-d).

This is an example of a period doubling cascade. Even though the observation of this sequence does not insure that the final state is a purely deterministic one, it is a good indicator that the system reaches Fig. 4 by a standard route to chaos and from this we can infer that Fig. 4 is itself an example of chaotic behavior.

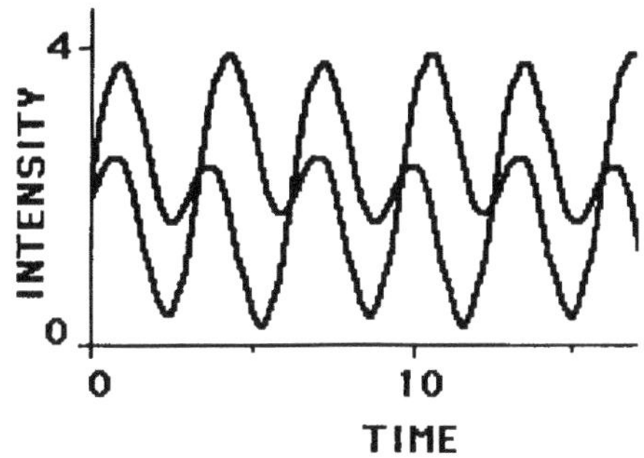

Fig. 6: Time signal for period two pulsing under badly triggered condition.

It must be noted that the use of a simple analog oscilloscope becomes less useful as the periodicity of the signal increases because it becomes harder to trigger the oscilloscope appropriately. As the triggering process becomes unable to distinguish among different peaks, one sees confusion as in Fig.6. If the amplitudes of the successive peaks within one period are very different, the oscilloscope is a useful tool, but if the shapes of the pulses or the times between pulses are different while the amplitudes are very similar, one may fail to get a clear indication of the presence of the period-doubling cascade. There is less difficulty when there is an external forcing signal, triggering on that together with adjustment of the length of the trace may restore some order to the displayed pattern. With an external forcing, one can also select the phase of the external signal that gives the optimally clear display.

In some cases, one can use the time series display on an oscilloscope to give an indication of the nature of the chaos.

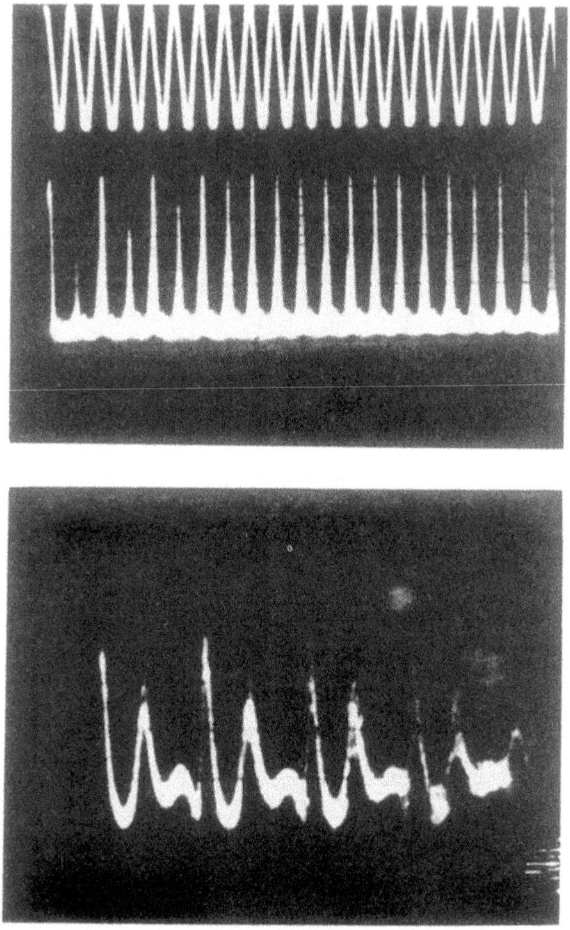

Fig.7: Two examples of a chaotic signal well triggered in different time scales. Both of them indicate that the laser intensity remains correlated for 4 to 5 periods of the modulation frequency.

If the system is a deterministic one, the evolution of the trajectories away from a point in phase space retain correlation, losing it exponentially at a rate that gives the Lyapunov exponent. If one can be sure to start at (or near) the same place in the phase space on successive traces, one can use the exponential blurring of the oscilloscope trace as an indicator of chaos. Noisy signals will show a trace which blurs linearly with time characteristics of diffussion with a finite bandwidth. The major difficulty is in triggering the oscilloscope in a sufficiently controlled way so that the traces all begin near the same point in phase space. This can be done by triggering on very extreme traces which are sufficiently unique. Or one might use not only the trigger but also some other condition to gate the z-axis modulation of the display. If such a display can be achieved, in the case of chaos we observe a very well-defined trace on the screen of the oscilloscope (Fig. 7) for some time until it broadens exponentially.

Quantitative analysis of the breadth of the trace was used by Derstine *et al.* [37] to identify and measure a Lyapunov exponent.

III.b) Power Spectrum

Examples of the power spectra of the modulated laser are shown in Fig. 8a-d where the spectra correspond to the time signals of Fig. 5b-d. The period doubling process is characterized by the appearance of peaks at the subharmonic frequencies (and their harmonics) of the fundamental one, until it becomes broadband. Analysis of the power spectrum does not have the problems we mentioned in section 2a for the time signals observed with an oscilloscope. Using spectra one can easily distinguish period doublings arising from changes in phase or shape of the pulses, however the limited bandwidth of the available spectrum analyzers and their relatively high cost are some of the draw-backs of this test.

This test is also powerful for distinguishing period doubling from the presence of two nearly locked sinusoidal frequencies. In fact, period doubling is a consequence of the increasing relevance of the nonlinearities in the system and therefore is characterized by the appearance of sub-harmonics and harmonics which are not observable in the presence of two sinusoidal frequencies.

On the other hand, it is not recommended to use time series to try to interpret the case of two or more incommensurate frequencies (quasiperiodicity). In contrast, power spectra give clear indications by showing bunches of peaks equally spaced with respect to the center frequency. The spacing between peaks is the beating frequency (Fig. 9).

a)

b)

c)

d)

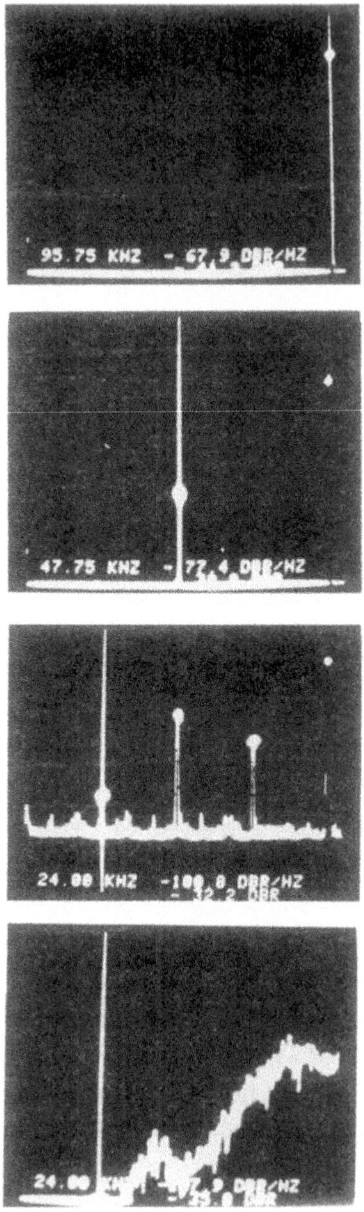

Fig.8 a)-d): Power spectra S(f) vs. f corresponding to Fig. 5b) to 5d) and to Fig. 4 respectively. The marker indicates the lowest frequency at which a well distinguishable peak appears. The fundamental frequency is 191.5 kHz.

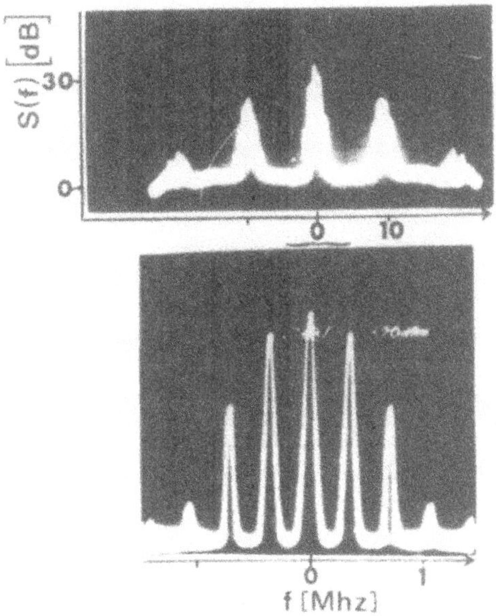

Fig. 9: Power spectra for quasiperiodic behavior [from Ref. 49]

Power spectra are also useful in the observation of the so-called "Inverse Lorenz cascade" [38] that follows a period doubling sequence after the accumulation point is reached and chaos is observed. At first the broadband portion of the spectrum is weak, but as the parameter is increased the broadband noise "floor" rises to swamp out each subharmonic in turn, with peaks disappearing in the inverse order from the the order in which they appeared. Details on the observation of this process is found in Ref. 36.

c) Phase-Space Portraits

A simple set-up for generating a phase space portrait is shown in Fig. 10; an analog differentiator takes part of the output signal as its input and the original signal and the output of the differentiator are sent to an oscilloscope set in the x-y mode.

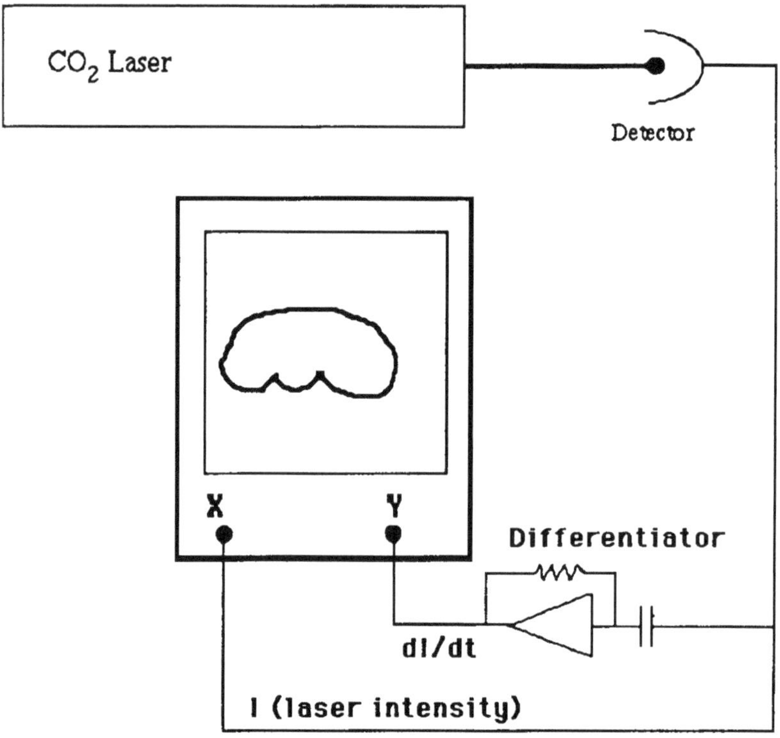

Fig.10 Experimental diagram for observing a phase space portrait.

In Fig. 11 we show two different phase space portraits corresponding to the signals of Fig. 5b)-d) and Fig. 4. The period doubling sequence is characterized by the appearance of different loops in this projection of the phase space.

a)
b)
c)
d)

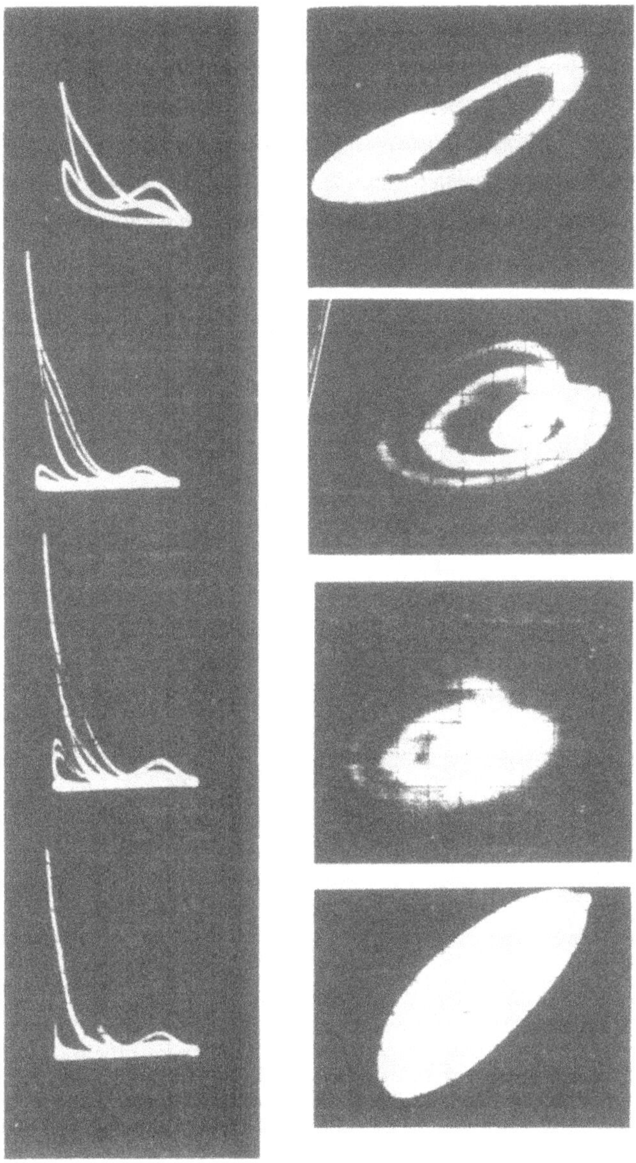

Fig.11: Different phase-space portraits corresponding to Fig. 5b) to 5d) and Fig.4). On the left side we plot V as a function of I; on the right, (dI/dt) as a function of I. In the former case we filter the signal to obtain a better representation on the oscilloscope.

This test does not indicate the periodicity of the system if it is not complemented by some additional data. In fact, the existence of more than one loop in the dx/dt vs. x plane may be a consequence of multiple peaks in the same period of time (as is shown in Fig. 12a). Also, if a period doubling involves the near disappearance of one peak for a full cycle of the modulation, the phase space portrait shows one fewer loop than the periodicity of the system (Fig. 12b). Adjustment of the gain of the x and y displays of the oscilloscope may be needed to detect peaks that are nearly missing.

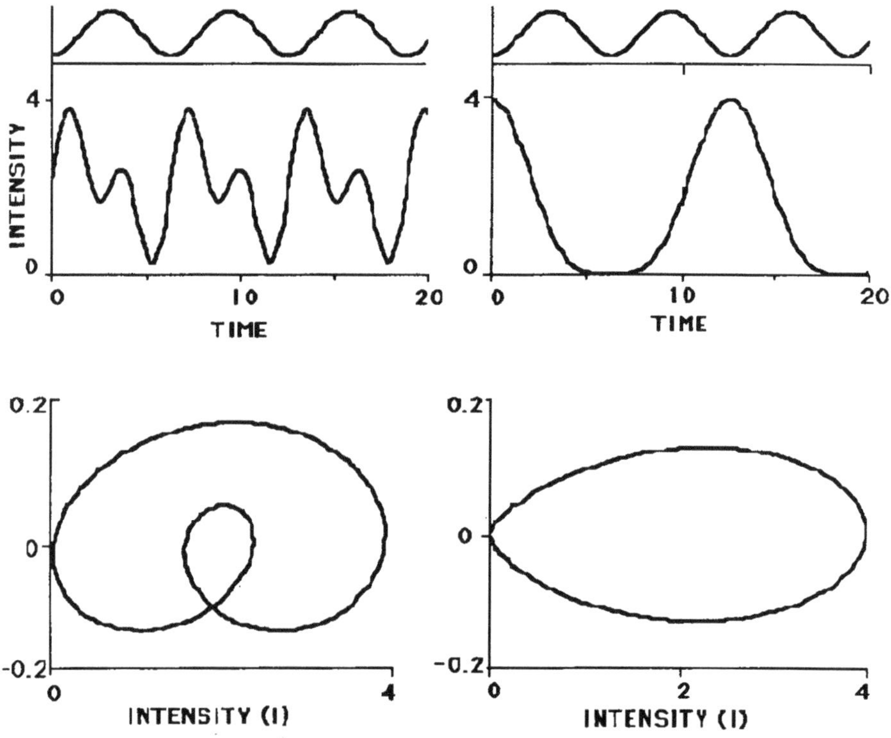

Fig. 12 : Anomalous phase space portraits : a) a period one attractor that appears with two loops in the phase space portrait.; b) a period two attractor with just one loop in the phase space portrait. From top to the bottom we observed the forcing as a function of time, the intensity as a function of time , and the derivative of the intensity with respect to time (dI/dt) as a function of intensity I. Other examples of anomalous phase-space portraits are in Fig. 11.

As a complement, when the signal is chaotic the phase space portrait can gives an indication of the "size" in phase space of the attractor, showing the regions of at least two variables visited by the attractor and also how frequently those regions are visited. In this way, this test could have a relevant role in the observation of structural changes of the strange attractor due to a process called crisis [35-36] (Fig. 13).

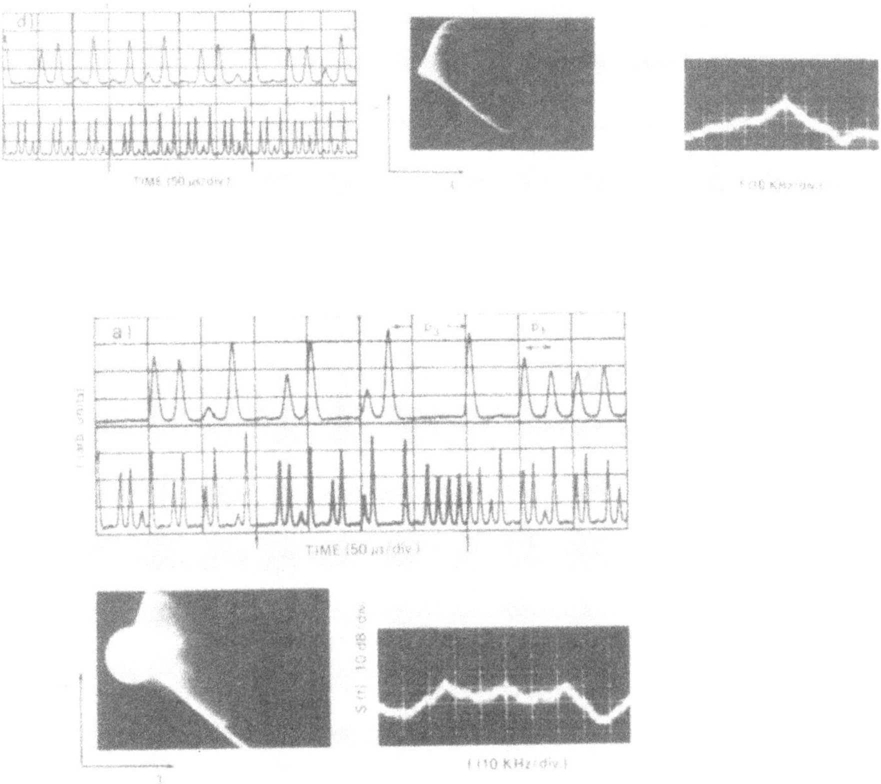

Fig.13: Intensity as a function of time, phase-space portraits [(dI/dt) as a function of I] and power spectra before and after an external crises. All scales in the different figures are identical. The upper traces of the time signals are the expansion of the lower ones corresponding to the region between arrows. The expansion of the attractor is clearly noticed in the phase space portraits

III.d) Poincaré sections

A Poincaré section is a sub-space of the phase-space defined by a given value of one of the variables of the system. Experimentally we are able to construct a two dimensional projection of a Poincaré section, which in a three dimensional system will coincide with the Poincaré section itself.

Taking as an example the laser with modulated losses, we have two different ways to construct a Poincaré section:

i) Plot the modulation voltage (V) as a function of the intensity (I) when I is maximum (dI/dt = 0)

ii) Plot (dI/dt) as a function of I for $V(t) = V_0$ (Stroboscopic signal).

The first one requires an experimental set-up such as that shown schematically in Fig. 14.

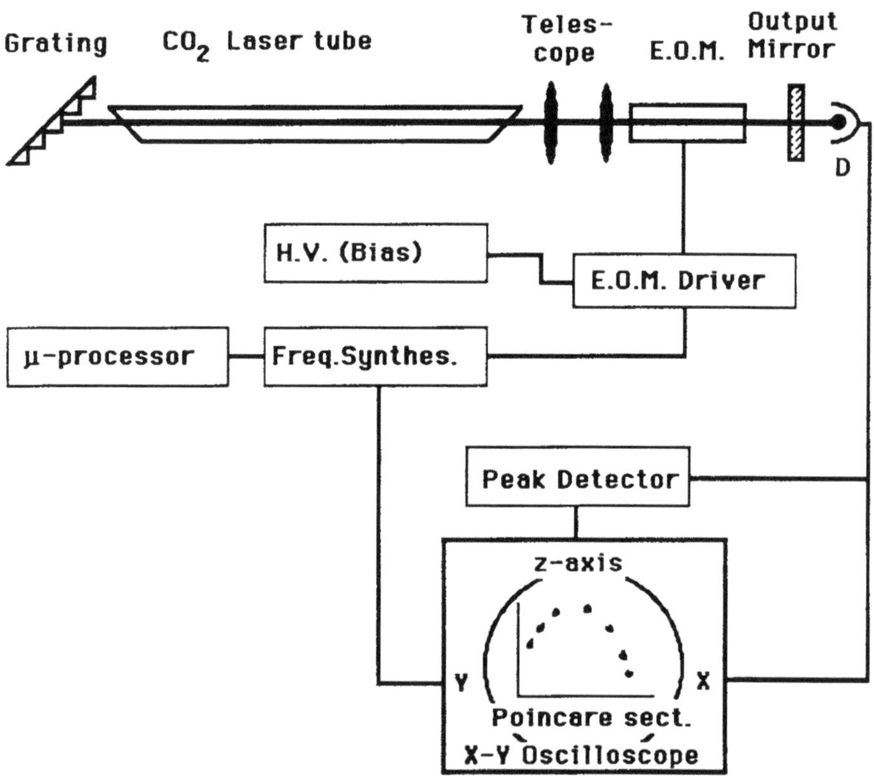

Fig. 14 : Experimental set-up to construct a Poincaré section.

The peak detector generates a signal that triggers a short pulse (much shorter than the period of the modulation frequency) acting on the z-axis of an oscilloscope (the z-axis controls the brightness of the spot on the screen). Thus a spot on the oscilloscope will appear in coincidence with the pulse indicating the value of the modulation voltage and the laser intensity at that instant.

The stroboscopic section is easily constructed by using a synchronous output of the signal generator that commands the EOM driver. In some cases signal processing is necessary to convert this synchronous output into a pulse of short duration and the right polarity to achieve fine control of the z-axis of the oscilloscope.

Typical projections of the Poincaré section are shown in Fig. 15. Usually the number of points in the Poincaré section indicates the periodicity of the orbit in phase space; however it can give a false indication due to the appearance or disappearance of peaks inside the same periodicity. This ambiguity is the result of the missing information about the evolution of the system during the time between points in the section (the so-called flying time).

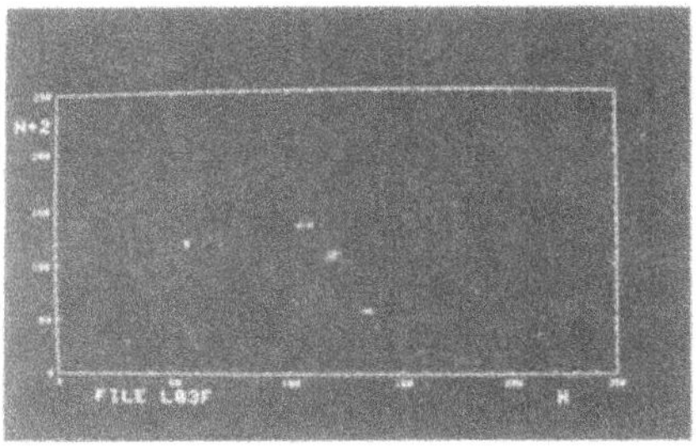

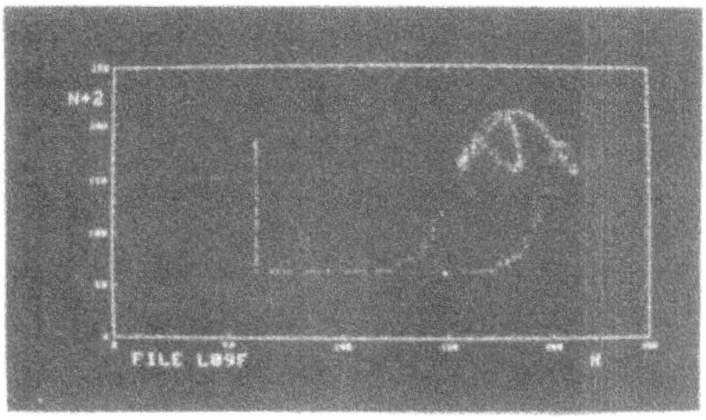

a) b)

Fig.15 Poincaré sections for a) a period eight and b) a chaotic signal

In the presence of quasiperiodicity we observe a continuous line in the Poincaré section that becomes a discrete number when the two frequencies lock at a given rational ratio. The number of points in the section and their succession in time indicates the value of the locking ratio.

Although this test can not be used to discriminate between chaos and noise, it proved to be very useful when complemented with digital analysis of the data obtained as described in the next section [39].

III e) Bifurcation Diagrams

The experimental construction of a bifurcation diagram is based on a one dimensional projection of a Poincaré section plotted as a function of one control parameter of the system.

An "analog" bifurcation diagram is achieved by the experimental set-up of Fig. 16. The signal generator provides a ramp and the modulation amplitude m (our control parameter) is proportional to the voltage output of the signal generator. The synchronous output of the frequency synthesizer controls the brightness of the spot on the oscilloscope given one point per period of the modulation. Eventually we plot, on the screen, the intensity I as a function of m.

It is worth noting that the rise time of the ramp must be much larger than the periodicity of the system to avoid, as much as possible, transients and delays, generating transient bistability, when the parameter is swept accross the bifurcation points[40-41].

A typical bifurcation diagram for a laser with modulated losses is shown in Fig. 17. At very low values of m the intensity is single valued for each m; when m is increased the first bifurcation takes place, and a period two orbit is represented by two different values of the intensity. Note that the slope of I(m) changes drastically at the bifurcation. In the chaotic region the intensity covers practically a continous range of values for each m. A further increase in m shows the existence of new periodic solutions and/or other aperiodic situations following different bifurcation cascades or the appearance of higher peaks in the intensity.

Other examples of bifurcation diagrams can be found in Ref. 42.

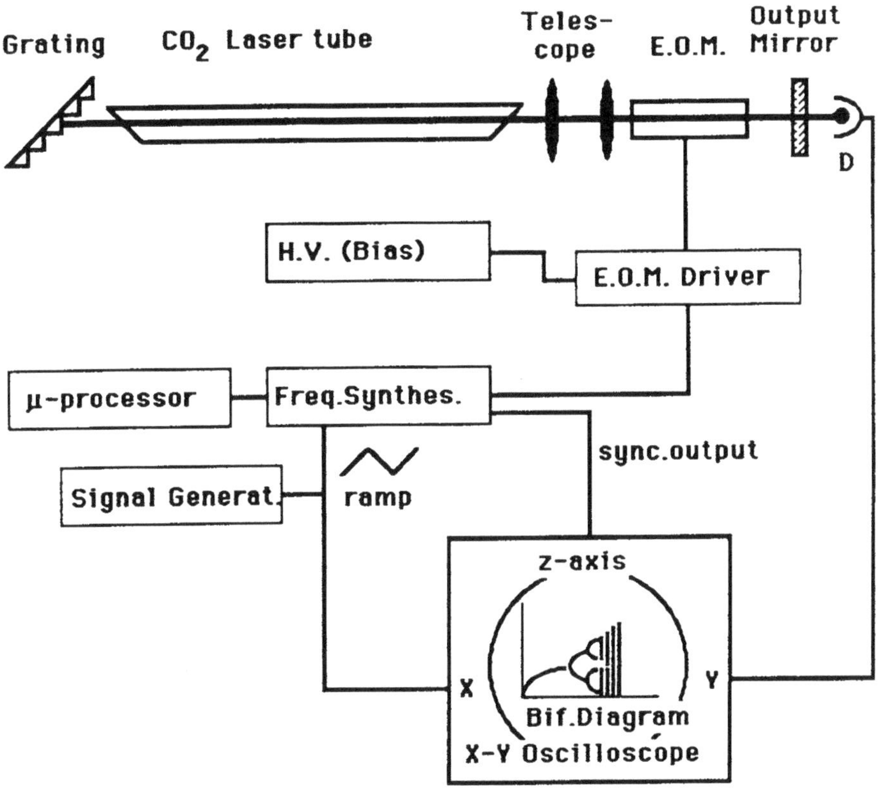

Fig. 16: Experimental set-up for the observation of Bifurcation Diagrams.

Bifurcation diagrams are very useful in the detection of a different phenomena involved in nonlinear dynamical systems: the so-called generalized multistability [15, 36, 43]. Generalized multistability is defined as the existence of different orbits in phase space for identical points in the parameter space. An infinite number of orbits coexist in a conservative system, each one defined by a given initial condition. In a dissipative system the coexistence of different attractors is related to the amount of dissipation. In the limit of infinite dissipation there exists only one attractor for each point in the parameter space independent of the initial condition choosen. In a real system the amount of dissipation usually lies somewhere between the two limits described above,

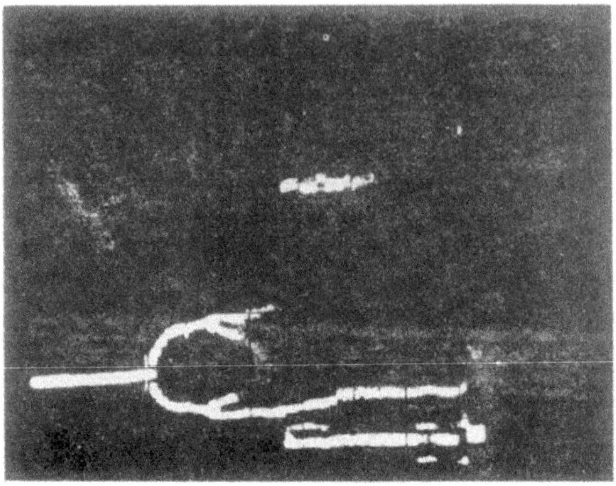

Fig.17: Bifurcation diagram. Peak intensity as a function of modulation amplitude, m. The period doubling process is clearly noticed as well as a discontinous jump in the "size" of the attractor for high values of m.

thus it is possible to observe two or more coexisting attractors. At the same time the amount of overlapping of this solutions in the parameter space can be used to estimate the dissipation in the phase space [36].

In some cases the overlapping becomes evident in a bifurcation diagram obtained by sweeping the control parameter forward and backward (Fig. 18b) and in some other cases a hard excitation is necessary in order to observe coexisting attractors (Fig.18a) [36].

The localization of orbits in parameter space can influence the whole dynamical behavior of the system by forcing collisions among them (orbits not attractors) which may change the structural and topological characteristics of the attractors, especially the chaotic ones, or which may even sweep them off the entire phase space by destroying their stability. For this reason bifurcation diagrams are a major tool in the analysis of nonlinear dynamical systems and are one of the few unambigous ways to determine the equivalence between two apparently different systems [43].

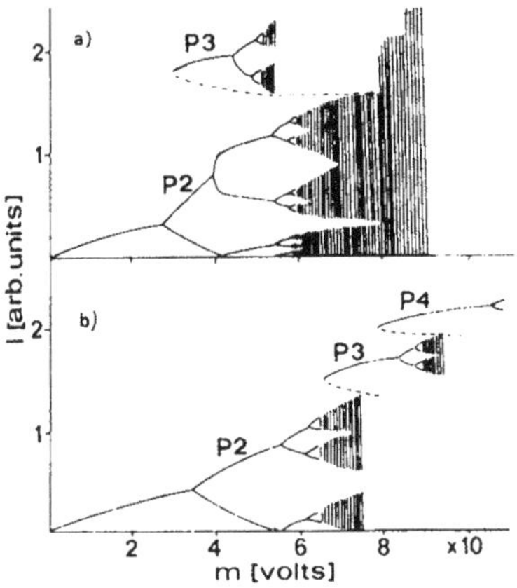

Fig. 18: Schematic bifurcation diagrams for a) high, and b) low pumping rates. The dashed zones indicate the chaotic regions, and the broken lines are the assumed location of unstable orbits. The collision of an unstable orbit with a chaotic attractor is called crisis and it can result in a large expansion of the chaotic attractor (a) or on its dissappearance (b)

Summary

The relevance of these analog experimental tests is that they can be performed in real time and they usually do not involve expensive and/or sophisticated instrumentation. However none of these tests or other known tests can unambigously distinguish chaos from noise when the system does not show any of the generic transitions from periodic states to chaotic ones. This problem is treated from an exclusively experimental point of view in the next section.

IV- Quantitative Methods to Characterize or Identify Chaos: Measurement of dimensions

In recent years there has been a general dissatisfaction with the identification of chaos by qualitative measures when there are distinct measurable properties that can be used unequivocally. In this section we will review the notions of the measurement of the dimension of a strange attractor and show that the requirements in computation and in data recording are not necessarily too severe.

What do we mean by the dimension of an attractor? Intuitively we must think of the topological character of the asymptotic trajectory of the solution in the phase space of the system. If the solution is asymptotically a constant, then the asymptotic trajectory is a point which we would define as having a dimension of zero. If the solution is asymptotically periodic, then the trajectory is a curve that closes on itself in a finite time. Locally such a curve is just a line and so we would like that to have dimension one. If the solution is quasiperiodic and made up of two incommensurate frequencies, then the attractor is a wrapping on the surface of a doughnut (a torus) and if the whole surface is covered asymptotically, then we would say the attractor has dimension two. The trouble comes when we have higher dimensional objects and those of noninterger (fractal) dimension.

A simple way to define a dimension is to consider our attractor in a phase space of say N dimensions and to ask, how many N-dimensional hypercubes of side ε it would take to completely cover the attractor. In the limit as ε goes to zero, the number of hypercubes that it takes to cover the attractor should grow as ε^D, where D is the dimension. This "box counting" method gives a dimension known as the Hausdorff dimension. For a digitized attractor, where we record points on equal time intervals, we can do either coverage of this subset of points on the attractor, or we can ask about the neighbors of each point. Let $N(\varepsilon)$ be the number of neighbors of the ith point that are within ε of the ith point. In the limit as ε tends to zero and as the total number of sampled points tends to infinity, N(e) will scale as ε^D where D is known as the correlation dimension. It is this dimension which Grassberger and Procaccia have popularized for calculation.

But first we must reconstruct the attractor and we begin with the fact that we usually can only detect one of the variables of the system. What should we record given that we can measure a single variable x(t)? Among the possibilities are the following

$\{x(t + n\tau)\}$, $\{x_{peak}\}$, and $\{t_{peak}\}$. The first is a sequence of data recorded at a steady rate, with data values recorded after every time interval τ. The second is a sequence of data which are the values of the peaks, or the peaks above some preset discriminator level. The third is the time of occurrence of the peaks. This last data set might be best transformed by calculating successive differences to form a new set.

With the data set in hand, we would like then to determine the underlying attractor and its characteristics.

IV. a) Reconstruction of the attractor

No matter which algorithm is going to be used to evaluate the dimension of an attractor, it is necessary to reconstruct the orbit in phase space or its topological equivalent. Unlike when you are dealing with a numerical calculation from a model, where the number of variables and their values are known, experiments usually give only the values of one variable from a system of an unknown number of degrees of freedom. Fortunately it is possible to reconstruct the topology of an attractor from that information. An embedding theorem [44] tells us that a topologically equivalent n-dimensional phase space can be constructed by using vectors whose components are the values of the single variable delayed successively by equal time intervals. Without entering into a formal demonstration of the logic behind it, this theorem is based on the fact that a vector $X^{(j)} = (x_1(t), x_1(t+\tau),\ldots\ldots, x_1(t+(2n+1)\tau))$ carries information sufficient to obtain a vector $X'^{(j)} = (x_1, (dx_1/dt),\ldots\ldots\ldots, (d^n x_1/dt^n))$ which is a point in the real phase space, if τ is infinitesimally small.

In making the reconstruction we have to deal with several choices:

i) what is an appropriate value for n (called the "embedding dimension");

ii) what is an appropriate value for the digitizing time (how often should the data be recorded?);

iii) what should be the "embedding time ", τ

iv) what should be the time separation of the first components of successvie vectors.

The first problem is the easiest to resolve. It if the topology is properly recovered for a sufficiently large n and for all larger values of n, then we can apply our dimension calculation algorithm in different embedding dimensions. The result should converge for sufficiently large n, reaching a saturation at some value of the embedding dimension. If convergence does not occur then we probably are analyzing a signal whose reconstruction is not an object of a well defined dimension. This is probably caused by a high dimensional dynamical signal or stochastic noise (and at this level the two are indistinguishable).

It would appear that a high value of n will insure the best reconstruction of the attractor, however that is not always the case because if n becomes much larger than the real dimension of the phase space then any attractor looks like a zero-measure point in phase space and distances between the points are increasingly dominated by the noise in the irrelevant variables. Usually by beginning with a low value of n and successively increasing it we are able to choose the right value.

The troubles introduced by the choice of τ are much more complex. A very small τ brings no new information, while if measurements are of only finite precision, a very large τ results in values that are uncorrelated with each other except as limited by the probability distribution of the variable. As τ increases, information about the high frequencies of the spectrum of the variable are increasingly ignored as indicated in Fig. 18. τ must be selected to give information on the time scale of the nonzero part of the autocorrelation function and the period (or quasiperiod, if the system is noisy).

The power spectrum of the signal can be used as an indicator of the maximum sensible value of τ. In fact, even if the spectrum is broadband it will decrease for high frequencies and the 3dB cut-off point (f_m) gives the $\tau_m \approx 1/f_m$. Typically a value 10 times smaller makes more sense.

Usually there is no reason to have the digitizing time different from the embedding time, but as we do not apriori know the best embedding time, a good experimental rule of thumb is to digitize with at least 10 points within a period or at least 10 points within the autocorrelation time.

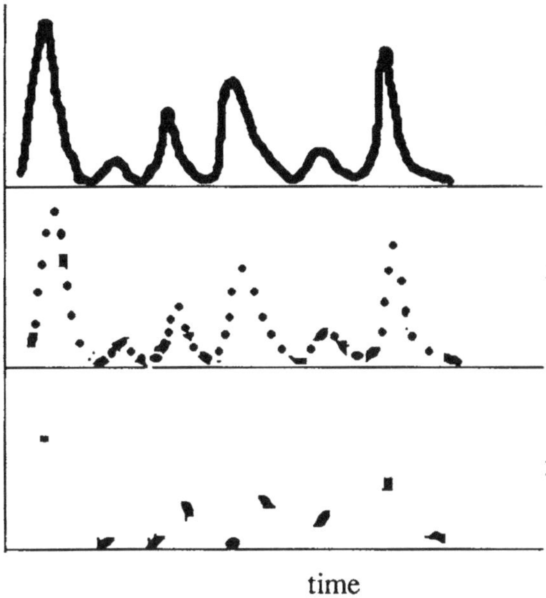

Fig. 18: Reconstruction of a signal as a function of time using a digitized series with 10 points per period and 1 point per period.

IV.b) Measuring the dimension

Having selected the data set one can now plot a topological equivalent of the attractor. What then is its dimension? In evaluating the exponents defined earlier, it is easy to see that we must have a good statistical visitation of the attractor [45]. If we take points close together in time, we will not have sufficent information about neighbors transverse to each trajectory. By emphasizing the recent past and future of each point, we will make the attractor locally appear to be a one-dimensional object. It is easy to estimate how to avoid this. Assume that we have N as the total number of data values and that the average number of points per period or correlation time is p. Suppose that the data are spread in a bounded region of phase space. Suppose the span of the data is given by Δx. The distance between neighbors along a typical portion of the trajectory is given by $\Delta x/2p$. But if the attractor is a fractal object of dimension D, then the hypervolume per data point is roughly given by $[(\Delta x)^D/N]$. For a good statistical representation of the attractor, we want the occurence of statistically important neighbors to be greater than of the short time deterministically important neighbors.

Hence we want $\Delta x/2p >> [(\Delta x)^D/N]^{1/D}$ or $N >> (2p)^D$, this can be viewed as a constraint on the relationship between p and N.

To give an example of how this problem arises, it is necessary to describe qualitatively the usual algorithm to calculate dimension. Assume a point $X^{(j)}$ in the embedding space belonging to the trajectory to be analyzed is taken as a starting point (see Fig. 19). We define an hypersphere of radius ε around that point (some algorithms use a hyper-ellipsoid, but this does not affect our discussion here); then we measure the number of data points, L(ε) lying inside that hypersphere,

$$\lim_{\varepsilon \to 0} L(\varepsilon) \sim \varepsilon^D$$

where D is a characteristic exponent called the correlation dimension and it is the number we want to measure. Usually by plotting the log L(ε) versus log ε one can find a straight region which has a slope which gives D. More accurate methods include fitting a region of the plot of L(ε) versus ε to the power law, to avoid overemphasis of the data at small ε where the statistical accuracy may be bad.

Now, if τ is too small so that the number of points per period is much larger than the number of loops around the attractor, only consecutive data points lie inside the hypersphere and we overemphasize the information about the "line" of the immediately past and future trajectory, which clearly has a dimension of 1. A limited number of data points does not allow to our partial reconstruction of the total trajectory to come close again to $X^{(j)}$ as would be necessary if we were to determine the "fatness" of the strange attractor. Hence the apparent value of D is reduced toward unity (see Fig. 20).

The restrictions stated above make the choice of τ a crucial problem in the experimental measurements of quantities like dimensions and entropies.

It was proposed to use τ determined by the first minimum of the autocorrelation function [46] which clearly will correspond to 1/4 the period of a sinusoidal signal. However this choice is completely arbitrary and usually not appropriate. More recently, [46] the use of the first minimum of the mutual information was suggested as a right choice for τ. In our experiments we have used a time delay of 1/4th. to 1/8th. of the period or autocorrelation time.

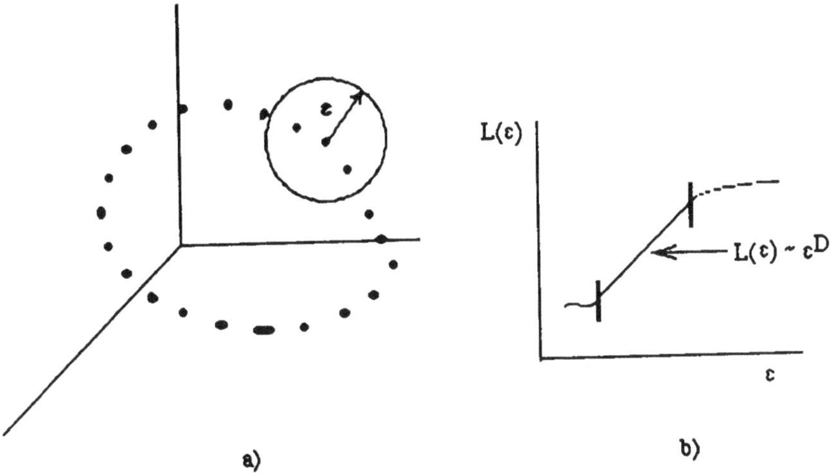

Fig. 19: a) Reconstruction of an attractor in the embedding phase space; b) typical plot of L(ε) as a function of ε showing the scaling region.

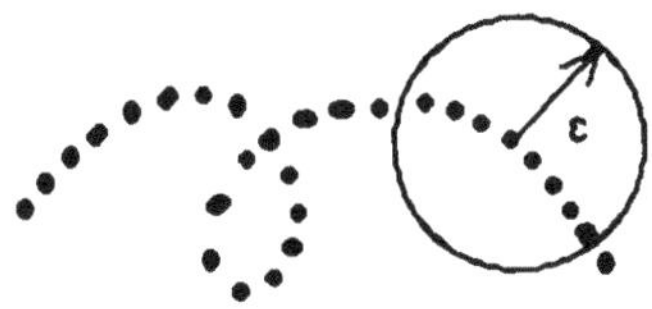

Fig. 20: Part of a trajectory in phase space showing how a small time delay τ overemphasizes the information about the near future and past around the point taken as origin for the calculation of the dimension.

Many of the problems arising from the choice of τ can be avoided by measuring the dimension of a projection of a stroboscopic or Poincaré section which will have a dimension one less than that of the full attractor. In such a case τ is taken to be equal to the "flying time" between points in the Poincaré section and therefore all the information about the recent past and future of the trajectory is automatically destroyed.

Experimental results obtained from a time series and from the corresponding stroboscopic section on the laser with modulated losses are shown in Fig. 21.

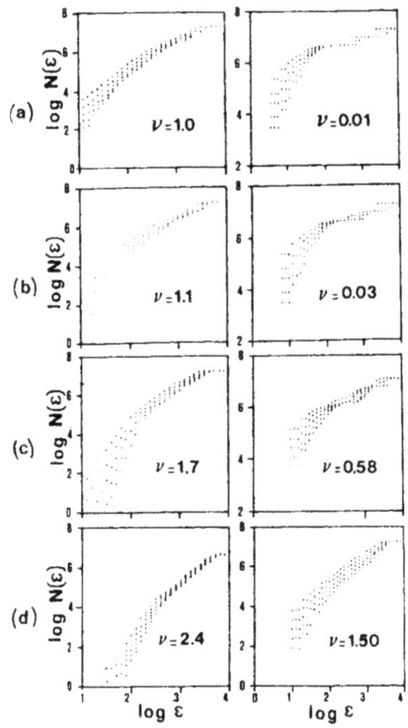

Fig. 21: Plots of L(ε) as a function of ε for different values of the embedding dimension n calculated from the time series (left) and from a Poincaré section (right) for different subharmonic bifurcations: (a) P2, (b) P4, (c) P8, (d) chaos (from Ref.15)

In practical cases, there is not enough data to measure the neighbors of only one point on the reconstructed attractor. The common step then is to measure the neighbors of all points and to average L(ε) before finding the power law. Technically, since the attractor may be stretched and folded differently at different points, it is better to find the local power law around each point and average the exponents (this is called the point-wise dimension). Depending on the uniformity of the attractor the averaging of all interpoint distance distributions may or may not be valid. So caution must be exercised even when this method seems to give a valid region of a power law dependence (as evidenced by a straight line region on the log-log plot). One quick check is to record the distribution of interpoint distances (rather than the cumulative integral) and to see if this has a power law in the same length scale region of epsilon (the scaling region) and if so, to verify that the exponent of this power law is one less than for the correlation dimension.

Other possible sources of error, are the effects of noise and the effects of finite accuracy of the measurement of the data. As demonstrated in Ref. 47, noise disturbs the correct positions of the points on the reconstructed attractor, and in an oversize embedding space this tends to "diffuse" the points into a larger dimension on the length scale of the standard deviation of the noise. This error causes the apparent power law to be greater than the true value and to steadily increase with increasing embedding dimension. Surprisingly the errors introduced by the finite precision of the data, which disturbs the true positions of the points on the attractor to lattice points in our space of finite precision, does not have the same effect. Because more points are moved closer to each other than are moved apart, there is a systematic error that causes an underestimate of the exponent in the power law. This underestimate is corrected by plotting the $L(\varepsilon)$ versus $\varepsilon+d/2$ where d is the precision of the measurements of the data.

The calculation of any of these algorithms increases as more data points are included. On small laboratory computers, it is generally prohibitive to try to calculate dimensions using more than about 10,000 points. However, Smith [48] has shown that a dimension that is estimated with fewer than 42^D points may be in error. Others use the rule of thumb that there must be at least 10^D points to have a reliable calculation of a dimension. Generally calculation of dimensions bigger than 3 or 4 is a tricky business and reports of such results should be viewed with extreme caution.

As a general procedure we recommend the following steps in measuring a dimension.

1) Check the numerical alogrithm you have written against a numerically generated digitized periodic signal (which should have dimension one) and a digitized two-torus (sum of two incommensurate frequencies) and against digitized random electronic noise. Be careful of testing random numbers generated by computer

algorithms, as many of them are really chaos of a low dimension rather than truly random.

2) Check the algorithm against a periodic signal generated by your system to see the effects of intrinsic experimental noise.

3) Calculate the dimension of the aperiodic signals you think may be chaotic (fractal) strange attractors and look for saturation of the dimension as a function of the dimension of the embedding. If there is no saturation, that may be an indication that you are measuring noise.

4) Calculate the dimension for different values of the delay time τ and look for a region where the dimension is independent of τ. If it is possible, make measurements independent of τ by measuring the dimension in a Poincaré section (most easily done if the system is being modulated by an external "clock" -- a sinusoidal signal generator).

5) If the result is in the range of 2-4, then you may have found chaos.

What does the value of the dimension tell you? First, if it is fractal, it says the aperiodic signal is generated by low dimensional dynamics rather than by noise. The specific value, rounded up to the nearest higher integer, gives the minimum number of collective variables in the physical system that are necessary for a complete description of the dynamics. This also gives the minimum number of equations needed in a model for the system. (It will not always be easy to reduce physical models to such subsets, but this may give some indications that the physical variables can be grouped or slaved in some interesting ways worth exploring).

Summary

While we have not discussed the different definitions of dimensions and the different possible algorithms to calculate them in detail, we have reviewed the experimental methods and problems in trying to apply one particular algorithm to calculate the dimension from experimental data. We have also given a recipe for checking the validity of the results and rules of thumb for the many choices that must be made.

Acknowledgements

We are grateful to our collaborators on many of the issues covered in this review, specially to A. M. Albano, F. T. Arecchi, W. Gadomski, C. Green, G. L. Lippi, R. Meucci, L. M. Narducci, G. L. Oppo, A. Poggi, G. P. Puccioni, and W. Scharpf.

References

[1] See e.g. " Universality in Chaos", Ed. P. Cvitanovic, (Hilger, Bristol U.K., 1984); "Chaos", Ed. Hao Bai-Lin, (World Scientific, Singapore, 1984) ; "Order within Chaos", P. Berge, Y. Pomeau, and C. Vidal. (Hermann, Paris, 1984). ; "Directions in Chaos", Ed. Hao Bai-Lin, (World Scientific, Singapore, 1988).

[2] See e.g., H. L. Swinney, and J. P. Gollub, Physics Today **31**, 41 (1978), and references therein; A. Libchaber, S. Fauve, C. Laroche, Physica **7D**, 73 (1983).

[3] See e.g., F. Argoul, A. Arneodo, P. Richetti, and J. C. Roux, J. Chem. Phys. **86**, 3325 (1987), and references therein.

[4] See e.g., G. J. Mpitsos, H. C. Creech, C. S. Cohan, and M. Mendelson, in: "Dynamic Patterns in Complex Systems", J. A. S. Kelso, *et al.,* eds. (World Scientific, Singapore, 1988); P. E. Rapp, I. D. Zimmerman, A. M. Albano, G. C. Deguzman, and N. N. Greenbaun, Phys. Lett. **110A**, 335 (1985).

[5] See e.g., V. Croquette, Pour la Science **62**, 62 (1982); V. Croquette, and C. Poitou, Comptes Rendus de l'Academie des Sciences de Paris, **C292**, 1353 (1981).

[6] T. Braun, J. A. Lisboa, R. E. Francke, and J. A. C. Gallas, Phys. Rev. Lett. **59**, 613 (1987).

[7] See e.g. Special Issues on "Instabilities in Active Optical Media", N. B. Abraham, L. A. Lugiato, and L. M. Narducci, eds. J. Opt. Soc. Am. **B2**, January 1985; and "Nonlinear Dynamics of Lasers". D. K. Bandy, A. N. Oraevsky, and J. R. Tredicce, eds. J. Opt. Soc. Am. **B5**, May 1988.

[8] J. R. Tredicce, F. T. Arecchi, G. L. Lippi, and G. P. Puccioni, J. Opt. Soc. Am. **B2**, 173 (1985).

[9] For theoretical work on instabilities in Class A lasers, see e.g.: W. Brunner, and H. Paul, Opt. Quant. Electron. **15**, 87 (1983) ; W. Brunner, R. Fisher, and H. Paul, J. Opt. Soc. Am. **B2**, 202 (1985).; H. Fu, and H. Haken, Phys. Rev.A **36**, 4802 (1987) , and J. Opt. Soc. Am. **B5**, 899 (1988).

[10] For experiments on Class A Lasers see e.g.: N. J. Halas, S. N. Liu, and N. B. Abraham, Phys. Rev.A **28**, 2915 (1983).; L. W. Hillman, R. W. Boyd, J. Krasinsky, and C. Stroud,Jr. , Phys. Rev. Lett. **52**, 1605 (1984).

[11] V. S. Idiatulin, and A. V. Uspenskii, Radio Eng. Electron. Phys. **18**, 422 (1973); L. W. Casperson, IEEE J. of Quant. Electron. **QE14**, 756 (1978); R. S. Gioggia, and N. B. Abraham, Phys. Rev. Lett. **51**, 650 (1983); M. F. H. Tarroja, N. B. Abraham, D. K. Bandy, and L. M. Narducci, Phys. Rev. A **34**, 3148 (1986).

[12] J. C. Englund, J. Opt. Soc. Am. **B5**, 1083 (1988). J. C. Antoranz and M. A. Rubio, J. Opt. Soc. Am. **B5**, 1070 (1988).

[13] T. Yamada, and R. Graham, Phys. Rev. Lett. **45**, 1322 (1980); H. J. Sholz, T. Yamada, H. Brand, and R. Graham, Phys. Lett.**82A**, 321 (1982).

[14] H. Haken, "Synergetics: An Introduction" 3rd. ed. (Springer-Verlag, Berlin 1983); and "Advanced Synergetics (Springer-Verlag, Berlin 1983).

[15] F. T. Arecchi, R. Meucci, G. P. Puccioni, and J. R. Tredicce, Phys. Rev. Lett. **49**, 1217 (1982).; J. R. Tredicce, F. T. Arecchi, G. P. Puccioni, A. Poggi, and W. Gadomski, Phys. Rev. A **34**, 2073 (1986).

[16] E. Brun,B. Derighetti, R. Holzner, and D. Meier, Helv. Phys. Acta **56**, 825 (1983); E. Brun, B. Derighetti, D. Meier, R. Holzner, and M. Ravani. J. Opt. Soc. Am. **B2**, 156 (1985).

[17] I. I. Matorin, A. S. Pikovskii, and Ya. I. Khanin, Sov. J. of Quant. Electron. **14**, 1401 (1984).

[18] W. Lauterborn, R. Steinhoff, J. Opt. Soc. Am. **B5**, 1097 (1988).

[19] T. Midavaine, D Dangoisse, and P. Glorieux, Phys. Rev. Lett. **55**, 1989 (1986).

[20] D. Dangoisse, P. Glorieux, and D. Hennequin, Phys. Rev. Lett. **57**, 2657, (1986).

[21] J. R. Tredicce, N. B. Abraham, G. P. Puccioni, and F. T. Arecchi, Opt. Commun. **55**, 131 (1985).

[22] P. A. Khandokhin, and Ya. I. Khanin, J. Opt. Soc. Am. B**2**, 226 (1985), and references therein; G. V. Perevedentseva, P. A. Khandokhin, and Ya. I. Khanin, Sov. J. of Quant. Electron. **10**, 71 (1980).

[23] Ye. L. Klochan, L. S. Kornienko, N. V. Kravtsov, Ye. G. Lariontsev, and A. N. Shelayev, Radio Eng. Electron. Phys. (USSR) **19**, 58 (1974).

[24] G. L. Lippi, J. R. Tredicce, N. B. Abraham, and F. T. Arecchi, Opt. Commun. **53**, 123 (1985); H. Zeghlache, P. Mandel, N. B. Abraham, L. M. Hoffer, G. L. Lippi, and T. Mello; Phys. Rev. A **37**, 470 (1988); and references therein.

[25] J. L. Boulnois, A. Van Lerberghe, P. Cottin, F. T. Arecchi, and G. P. Puccioni, Opt. Commun. **58**, 124 (1986).

[26] E. Arimondo, and E. Menchi, Appl. Phys. B **37**, 55 (1985); E. Arimondo, P. Bootz, P. Glorieux and E. Menchi J. Opt. Soc. Am. B**2**, 193 (1985); E. Arimondo, F. Casagrande, L. A. Lugiato, and P. Glorieux, Appl. Phys.B **30**, 57 (1983).

[27] D. Hennequin, F de Tomasi, B. Zambon, and E. Arimondo, Phys. Rev. A **37**, 2243 (1988) ; D. Dangoisse, A. Bekhali, F. Papoff, and P. Glorieux, Europhys. Lett. **6**, 335 (1988).

[28] C. O. Weiss, and J. Brock, Phys. Rev. Lett. **57**, 2804 (1986).

[29] C. O. Weiss, N. B. Abraham, and U. Hübner, Phys. Rev. Lett. **61**, 1587 (1988).

[30] J. Guckenheimer, and P. Holmes, "Nonlinear Oscillations, Dynamical Systems and Bifurcations of Vector Fields" (Springer-Verlag, New York, 1986); J. P. Eckmann, Rev. Mod. Phys. **53**, 643 (1981).

[31] M. J. Feigenbaum, J. Stat. Phys. **19**, 25 (1978).

[32] D. Ruelle, and F. Takens, Commun. Math. Phys. **20**, 167 (1971); S. Ostlund, D. Rand, J. Sethna, and E. Siggia, Physica **8D**, 303 (1983).

[33] L.A. Melnikov, E. M. Rabinovich, and V. V. Tuchin, J. Opt. Soc. Am. B**5**, 1134 (1988); and references therein.

[34] Y. Pomeau, and P. Manneville, Commun. Math. Phys. **74**, 189 (1980); P. Berge, M. Dubois, and M. A. Rubio, Phys. Rev. Lett.

[35] C. Grebogi, E. Ott, and J. Yorke, Physica **7D** , 181 (1983); H. Solari, and R. Gilmore, Phys. Rev. A **38**, 1566 (1988).

[36] R. Meucci, A. Poggi, F. T. Arecchi, and J. R. Tredicce, Opt. Commun. **65**, 151 (1988).

[37] M. W. Derstine, H. M. Gibbs, F. A. Hopf, and L. D. Sanders, IEEE J. of Quant. Electron. **QE21**, 1419 (1985).

[38] E. N. Lorenz in "Nonlinear Dynamics", ed. R. H. G. Hellerman, Annual N. Y. Acad. Sci. **357**, 282 (1980).

[39] G. P. Puccioni, A. Poggi, W. Gadomski, J. R. Tredicce, and F. T. Arecchi, Phys. Rev. Lett. **55**, 339 (1985).

[40] P. Mandel, and T. Erneux, Phys. Rev. Lett. **53**, 1818 (1984); W. Scharpf, M. Squicciarini, D. Bromley, C. Green, J. R. Tredicce, and L. M. Narducci, Opt. Commun. **63**, 344 (1987).

[41] R. Kapral, and P. Mandel, Phys. Rev. A **32**, 1076 (1985).

[42 M. Ravani, B. Derighetti, G. Broggi, E. Brun, and R. Badii, J. Opt. Soc. Am. **B5**, 1029 (1988).

[43] H. Solari, E. Eschenazi,R. Gilmore, and J. R. Tredicce, Opt. Commun. **64**, 49 (1987); H. Solari, and R. Gilmore, Phys. Rev. A **38**, 1566 (1988).

[44] H. Whitney, Ann. Math. **37**, 645 (1936).

[45] R. Badii, and A. Politi, J. Stat. Phys. **40**, 725 (1985).

[46] A. M. Fraser, and H. L. Swinney, Phys. Rev. A **33**, 1134 (1986).

[47] M. Möller, W. Lange, F. Mitschke, N. B. Abraham, and U. Hübner, Phys. Lett. **138A**, 176 (1989).

[48] L. A. Smith, Phys. Lett. **133A**, 283 (1988).

[49] J. R. Tredicce, E. Quel, A. Ghazzawi, C. Green, M. Pernigo, L. M. Narducci, and L. A. Lugiato, Phys. Rev. Lett. **62**, 1274 (1989).

SEMICONDUCTOR LASER PHYSICS

Hartmut Haug
Institut für Theoretische Physik
Universität Frankfurt
Robert-Mayer Strasse 8
D-6000 Frankfurt
Federal Republic of Germany

Abstract:

An introduction to the description of the optical properties of semiconductors is given. The dependence of the real and imaginary part of the gain function on the plasma density is discussed. The laser equations for the complex field and the plasma density are derived. An introduction to the calculation of noise spectra and the description of optical instabilities is presented.

I. Basic Optical Properties of Semiconductors

For the description of semiconductor lasers like for other lasers one has to write down the equation for the *electric field* E of the laser light. Most often a classical treatment of the light field by the classical Maxwell equations, if necessary with stochastic noise sources, is sufficient. The field is driven by the *dielectric polarization* P which in term is determined by the *inversion* of the electrons in the bands of the semiconductor. Thus the general structure of semiconductor laser theory is the same as that of conventional lasers, which are described by Maxwell-Bloch equations. However, the most important difference between semiconductor and conventional lasers is that the optically active electrons are not localized at one atom, but can move through the whole crystal.

The energy levels in a crystal consist of allowed bands seperated by gaps. In an unexcited crystal the last completely filled band is called valence band. The first empty band is called conduction band, it is seperated from the valence band by an energy gap

E_g which corresponds to visible and infrared light frequencies. The states in these bands can be described by the momentum **k**, because the momentum is a good quantum number due to the translational invariance present in a perfect crystal. The maximum of the valence band and the minimum of the conduction band are in the semiconductor laser materials at $\mathbf{k} = 0$, which is the center of the Brillouin zone. The optical transitions in a laser take place between states close to the extrema of the conduction and valence band, where one can expand the energy function E_{ik} quadratically (i = c, v)

$$E_{ik} = E_i^0 + \frac{\hbar^2 k^2}{2m_i} \quad . \tag{1.1}$$

We take $E_v^0 = 0$ and $E_c^0 = E_g$. Because E_c has a minimum, $m_c > 0$, but $m_v < 0$. The "masses" m_i describe just the curvature of the bands around the extrema, and are called effective masses. One can get rid of the negative effective mass m_v by considering holes, i.e. missing electrons, in the valence band, rather than all the electrons in this band:

$$E_{hk} = - E_{vk} = - \frac{\hbar^2 k^2}{2m_v} = \frac{\hbar^2 k^2}{2m_h} \quad , \tag{1.2}$$

with $m_n = - m_v$. Introducing the hole concept, one has to change also the sign of the momentum $\mathbf{k}_h = - \mathbf{k}_v$, in fact the transition to the hole picture is nothing but a time inversion transformation. An optical transition of an electron from the valence to the conduction band caused by the absorption of a photon generates thus a hole in the valence and an electron in the conduction band. Vice versa the emission of a photon is accompanied by the recombination of an electron-hole pair.

The electrons and holes from a neutral plasma. Due to the Coloumb interaction one has rapid collisions between these carriers, which establish in very short times τ ($\tau = 100$ ns is a typical relaxation time) thermal distributions of the electrons in the conduction band and of the holes in the valence band.

For the calculation of the optical gain in a semiconductor laser, one has to integrate over the contributions of all **k**-states, in this respect it has an inhomogeneously broadened gain spectrum; because of the short relaxation time the electron-hole plasma reacts collectively, in this respect it resembles a laser with a homogeneously broadened gain spectrum. The rapid carrier-carrier collisions cause a fast relaxation of the dielectric polarization **P**, so that the polarization always follows adiabatically the values of the laser field and of the population. Thus, **P** is no longer a dynamical variable and can be eliminated adiabatically. The electrons and holes are always in a local equilibium, described by Fermi distributions. The parameters in this distributions can however still very slowly in space and time according to the hydrodynamic equations. Only if these slow variations can be disregarded, the electron-hole population is determined by a simple rate equation for the total number of electron-hole pairs.

In the following we will discuss the equation for the dielectric polarization in a semiconductor and incorporate sucessively the effects of the interactions between the charged carriers [1,2].

The momentum of a photon $\hbar k = \hbar\omega/c$ is small compared to a typical electron momentum in a crystal $\hbar \mathbf{k} \simeq \hbar 2\pi/a$, where a is the lattice constant. The absorption (or emission) of a photon is thus accompanied by the creation (or annihilation) of an electron with momentum $\hbar \mathbf{k}$ and of a hole with momentum $-\hbar \mathbf{k}$. The optical polarization is the expectation value of the dipole operator:

$$P = \frac{1}{V} \int d^3x < \psi_c^\dagger(\mathbf{x})\ ex\ \psi_v(\mathbf{x}) > \quad . \tag{1.3}$$

Here $\psi_c^\dagger(\mathbf{x})$ and $\psi_v(\mathbf{x})$ are the creation and annihilation operators of an electron in the conduction and valence band, respectively. Expanding $\psi_i(\mathbf{x}) = \sum a_{i\mathbf{k}} \phi_{i\mathbf{k}}(\mathbf{x})$ into Bloch functions $\phi_{i\mathbf{k}}(x)$, the polarization is

$$P = \sum_{\mathbf{k}} d_{\mathbf{k}} < a_{c\mathbf{k}}^\dagger a_{v\mathbf{k}} > \ , \tag{1.4}$$

where

$$d_{\mathbf{k}} = \int d^3x\ \phi_{c\mathbf{k}}^*\ ex\ \phi_{v\mathbf{k}}$$

is the optical dipole matrix element. The annihilation of an electron with momentum $\mathbf{k}$ in the valence band corresponds to the creation of a hole with $-\mathbf{k}$:

$$a_{v\mathbf{k}} = a_{h-\mathbf{k}}^\dagger$$

The polarization

$$P = \sum_{\mathbf{k}} d_{\mathbf{k}} P_{\mathbf{k}} \quad \text{with} \quad P_{\mathbf{k}} = < a_{e\mathbf{k}}^\dagger a_{h-\mathbf{k}}^\dagger > \ , \tag{1.5}$$

is thus determined by the pair amplitude $P_{\mathbf{k}}$. $P_{\mathbf{k}}$ is just the off-diagonal element of a density matrix.

The stationary equation of motion for $P_{\mathbf{k}}$ for a single noninteracting e-h pair driven by a coherent classical light field is

$$\hbar(\omega - e_{ek} - e_{hk} + i\delta)\, P_k = -d_k E \quad . \tag{1.6}$$

Here E and ω are the amplitude and frequency of the light field and

$$\hbar e_{ik} = E_{ik} \text{ for } i = e, h.$$

The infinitesimal damping δ arises if one switches on the field adiabatically at $t = -\infty$. We get the polarization

$$P = -\sum_k \frac{d_k^2\, E}{\hbar(\omega - e_{ek} - e_{hk} + i\delta)} \quad . \tag{1.7}$$

The polarization has a resonance denominator, the poles are in the lower half of the complex frequency plane. Using Dirac's identity

$$\frac{1}{x + i\delta} = \mathcal{P}\,\frac{1}{x} - i\,\pi\,\delta(x) \quad , \tag{1.8}$$

where $\mathcal{P}$ denotes the principal value and $\delta(x)$ is the delta-function, one obtains the real and imaginary part of the polarization:

$$\chi(\omega) = \frac{P}{E} = -\,\mathcal{P}\sum_k \frac{d_k^2}{\hbar(\omega - e_{ek} - e_{hk})} + i\pi \sum_k d_k^2\, \delta(\hbar(\omega - e_{ek} - e_{hk})) \quad . \tag{1.9}$$

Here, $\chi(\omega)$ is the complex optical succeptibility which is linked with the complex gain per unit time $g(\omega)$ by

$$g(\omega) = 4\pi\omega i\chi(\omega) \tag{1.10}$$

The real part of the gain, i.e. g', is positive for negative absorption χ''. In a semiconductor without pumping we have naturally only absorption, so that $g'<0$.

As a next step we take the attractive Coulomb interaction V_q into account, where $V_q = 4\pi e^2/\epsilon_0 q^2$ is the Fourier transform of $V(r) = e^2/\epsilon_0 r$. Here ϵ_0 is the dielectric constant of the crystal in the ground state. The equation of motion for one electron-hole pair is then

$$\hbar(\omega - e_{ek} - e_{hk} + i\delta)\, P_k = -d_k E - \sum_{k'} V_{k-k'}\, P_{k'} \ . \qquad (1.11)$$

The homogeneous part of this equation

$$\sum H_{k,k'}\, \phi_{nk'} = E_n \phi_{nk} \ . \qquad (1.12)$$

with

$$H_{k,k'} = \hbar(e_{ek} + e_{hk})\, \delta_{k,k'} - V_{k,k'} \qquad (1.13)$$

is the Schrödinger equation of the hydrogen atom (the hole plays the role of the proton). In real space representation, the k^2 terms become the operator of the kinetic energies of the electron and hole and the convolution integral $\sum V_{k,k'}\, \phi_{nk'}$ simply becomes the product of the potential and wave function. The solutions of the hydrogen atom are known for the bound states as well as for the ionization continuum. We expand the pair function P_k with respect to these eigenfunctions

$$P_k = \sum_n a_n \phi_{nk} \qquad (1.14)$$

and find from the inhomogeneous equation

$$P_k = \sum_n \frac{\sum_{k'} \phi^*_{nk'}\, \phi_{nk}}{\hbar(\omega - \epsilon_n + i\delta)}\, d_k\, E \ . \qquad (1.15)$$

The susceptibility is thus

$$\chi(\omega) = d^2 \sum_n \frac{|\phi_n(x = 0)|^2}{\hbar(\omega - \epsilon_n + i\delta)} \ , \qquad (1.16)$$

where $\phi_n(x{=}0) = \sum_k \phi_{nk}$. The dipole matrix element d_k has been assumed to be constant. The resonances are now at the pair energies ϵ_n. Each transition has an oscillator strength $|\phi_n(x{=}0)|^2$ which is the probability to find the electron and hole at the same place, because only if they are at the same place they can make an optical transition.

As a next step, we assume that a thermal plasma is present in the crystal. This plasma causes various modifications of the polarization equation.

1) Pauli principle effects

The probability that in a thermal plasma with the temperature T the state i, **k** is populated is given by the Fermi distribution function

$$f_{ik} = \frac{1}{e^{\beta(e_{ik} - \mu_i)} + 1} \quad , \qquad (1.17)$$

where $\beta = 1/k_B T$ and $\mu_i(n,T)$ is the chemical potential.

The optical transition from the valence to the conduction band is only allowed if the initial state is populated and if the final state is empty

$$(1 - f_{ck})\, f_{vk} = (1 - f_{ek})\,(1 - f_{hk}) \quad . \qquad (1.18)$$

The reverse process can only take place if both the electron and hole state are populated

The net absorption probability is thus proportional to $f_{ek}\, f_{hk}$

$$(1 - f_{ek})(1 - f_{hk}) - f_{ek}\, f_{hk} = 1 - f_{ek} - f_{hk} \quad , \qquad (1.19)$$

which is a measure of the inversion and ranges from 1 for the unexcited crystal to -1 for the crystal which is completely inverted at the momentum **k**.

2) Screening

The Coulomb potential is screened by a plasma

$$V_s(q,\omega) = \frac{V_q}{\epsilon(q,\omega)} \quad , \qquad (1.20)$$

where $\epsilon(q,\omega)$ is the longitudinal dielectric function. A qualitively good approximation is obtained by the so-called static plasmon-pole approximation in which

$$\epsilon(q,\omega) \simeq \epsilon(q) = \epsilon_0 \left(1 - \frac{\omega_{pl}^2}{\omega_q^2}\right) , \qquad (1.21)$$

where ω_{pl} is the plasma frequency

$$\omega_{pl}^2 = \frac{4\pi ne^2}{\epsilon_0 m} . \qquad (1.22)$$

m is the reduced electron-hole mass $m = m_e m_n/(m_e + m_h)$, n is the plasma density, ω is the frequency of the effective plasma mode:

$$\omega_q^2 = \omega_{pl}^2 \left(1 + \frac{q^2}{\kappa^2}\right) + \frac{(e_{eq} + e_{hq} - E_g/\hbar)^2}{2} . \qquad (1.23)$$

Here κ is the inverse screening length defined by

$$\kappa^2 = \frac{4\pi e^2}{\epsilon_0} \sum_{i=e,h} \frac{\partial n}{\partial \mu_i} , \qquad (1.24)$$

where $n = \sum f_{ek} = \sum f_{hk}$. If we disregard the last term in ω_q^2 which varies as q^4, we find the well-known statically screened potential

$$V_s(q) = \frac{4\pi e^2}{\epsilon_0} \frac{1}{q^2 + \kappa^2} , \qquad (1.25)$$

whoes Fourier transform is

$$V_s(r) = \frac{e^2}{\epsilon_0 r} e^{-\kappa r} . \qquad (1.26)$$

3) Energy renormalization

The energies of the electron and hole change in the presence of a plasma. These changes lead to a reduction of the land gap with inscreening plasma density

$$\Delta E_g(n) = V_s(r{=}0) - V(r{=}0) - \frac{1}{2} \sum V_s(q) (f_{eq} + f_{hq}) . \qquad (1.27)$$

The first term is just the difference of the self-interaction with and without a plasma. Using (1.26) one gets

$$V_s(r=0) - V(r=0) = \frac{e^2}{\epsilon_0 r}\left(e^{-\kappa r} - 1\right)\Big|_{r=0} = -\frac{e^2\kappa}{\epsilon_0} \quad , \qquad (1.28)$$

The second term in (1.27) is the screened exchange self-energy. It can be estimated by assuming that the momentum transfer is in the order of the Fermi wave number $q \simeq k_F$, so that

$$-\frac{1}{2}\sum_q V_s\,(k_F)\;(f_{eq} + f_{hq}) = -V_s\,(k_F)\,n \quad . \qquad (1.29)$$

Fig. 1 gives an example of the variation of the energy gap with the plasma density n.

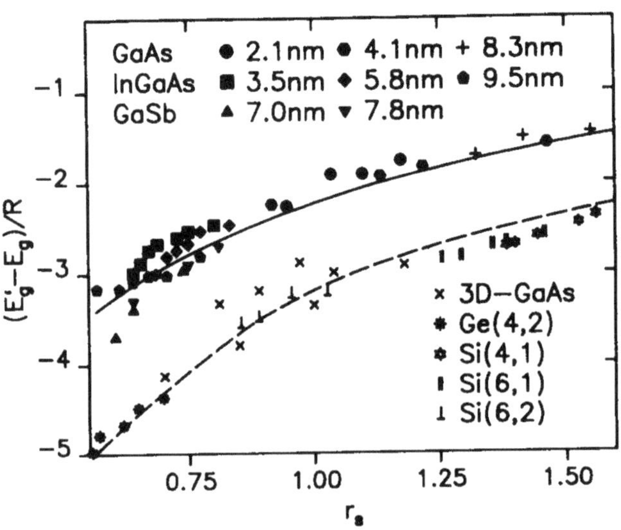

Fig. 1 : Calculated and measured band gap shifts (in units of excitonic rydbergs R) against the dimensionless interparticle distance r_s for 2 and 3 dimensional semiconductors (according to ref. [5])

Now we can put all 3 many-body effects -the Pauli principle effect, the screening and the energy renormalization- into the polarization equation and get in the general nonstationary case

$$\hbar\left(\frac{i\partial}{\partial t}+\omega-\epsilon_{ek}-\epsilon_{hk}+i\gamma_k\right)P_k = -(1-f_{ek}-f_{hk})\left[d_k E+\sum_{k'} V_s(k-k')P_{(k)'}\right] . \quad (1.30)$$

Here $\hbar(\epsilon_{ek}+\epsilon_{hk}) = \hbar(e_{ek}+e_{hk}) + \Delta E_g(n)$ are the renormalized energies, γ_k is the energy broadening due to phonon scattering, carrier-carrier scattering etc.. Eq. (1.30) connects the polarization function P_k with the complex field amplitude E(t) in the laser diode and the plasma density n(t). In order to complete the laser model we have to formulate the equations of motion for these variables.

Before doing this we will discuss how one can construct an approximate solution of the integral equation (1.30) in stationary situation. Introducing a susceptibility function $\chi(k,\omega)$ by

$$P_k = \chi(k,\omega)E \quad (1.31)$$

we get from (1.30)

$$\chi(k,\omega) = \chi^0(k,\omega)\left[1-\frac{1}{d_k}\sum_{k'} V_s(k-k')\,\chi(k',\omega)\right] , \quad (1.32)$$

where

$$\chi^0(k,\omega) = -\frac{d_k\,(1-f_{ek}-f_{hk})}{\hbar(\omega-\epsilon_{ek}-\epsilon_{hk}+i\gamma_k)} . \quad (1.33)$$

is the free-carrier susceptibility. The integral equation (1.32) can be simplified further by introducing a vertex function $Q(k,\omega)$ which describes the deviations between $\chi(k,\omega)$ and the free-carrier susceptibility function $\chi^0(k,\omega)$:

$$\chi(k,\omega) = \chi^0(k,\omega)\,Q(k,\omega) . \quad (1.34)$$

Then it follows from (1.32) that

$$Q(k,\omega) = 1 + \frac{1}{d_k} \sum_{k'} \lambda V_s(k - k')\, \chi^0(k',\omega)\, Q(k',\omega) \quad , \tag{1.35}$$

where we have introduced a formal interaction parameter λ . A power expansion of $Q(k,\omega)$ in terms of λ yields

$$Q(k,\omega) = \sum_{n} \lambda^n\, q_n(k,\omega) \quad , \tag{1.36}$$

where

$$q_0 = 1 \quad \text{and} \quad q_1 = \frac{1}{d_k} \sum_{k'} V_s(k-k')\, \chi^0(k',\omega) \quad . \tag{1.37}$$

Approximating q_n by q_1^n and summing up the resulting geometrical series we get [3,4]

$$Q(k,\omega) = \frac{1}{1 - q_1(k,\omega)} \quad . \tag{1.38}$$

Eq. (1.38) is the simplest nontrivial Padé approximation for $Q(k,\omega)$. Physically $Q(k,\omega)$ describes the complex excitonic enhancement due to the attractive screened Coulomb potential so strongly, that bound electron and hole states do no longer exist, the Padé approximation (1.38) yield good results. In Fig. (2) the spectra of absorption and refraction calculated with the Padé approximation are shown for 300 K GaAs for three plasma densities with and without excitonic enhancement.

II. The Laser Equations [3]

The rapid carrier-carrier scattering processes thermalize an electron-hole plasma within ca. 100 fs, i.e. 10^{-13} s. For variations which are slow on this time scale, we can describe the electron and hole distribution by a local equilibrium Fermi distribution in which the parameters can still be slowly varying functions in space and time. For simplicity we will assume that the system is in a spatially homogeneous state and that no temperature variations occur. Than only the total plasma density n(t) and the related chemical potential $\mu_e(t)$ and $\mu_h(t)$ can vary with time:

$$n(t) = \frac{1}{V} \sum_{k} \frac{1}{e^{\beta(e_{ik} - \mu_i(n(t)))} + 1} = n_i(t) \quad . \tag{2.1}$$

Unfortunately, (2.1) can not be evaluated analytically. Aguillera-Navarro et al. [6] have

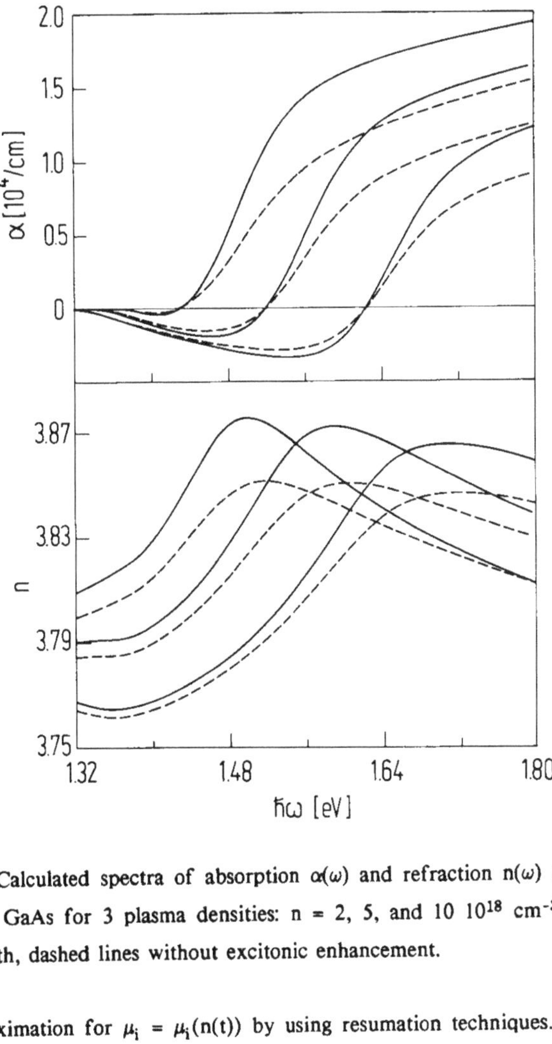

Fig. 2: Calculated spectra of absorption $\alpha(\omega)$ and refraction $n(\omega)$ for 300 K bulk GaAs for 3 plasma densities: n = 2, 5, and 10 10^{18} cm^{-3} . Full lines with, dashed lines without excitonic enhancement.

given an approximation for $\mu_i = \mu_i(n(t))$ by using resumation techniques. Their result is valid for a very wide range of densities and temperatures. The relation is

$$\beta\mu_i = \ln \nu_i + K_1 \ln (K_2\nu_i + K_3) + K_4 \nu_i + K_5 \qquad (2.2)$$

with

$$\nu_i = \frac{n}{n_{0i}} \quad \text{and} \quad n_{0i} = \frac{1}{4}\left(\frac{2 m_i}{\hbar^2\pi^2\beta}\right)^{\frac{3}{2}}$$

and

$K_1 = 4.8966851$ $K_2 = 3.3105795$

$K_3 = 73.6264033$ $K_4 = 0.133376$

$K_5 = -21.0508644$

valid for $-\infty < \mu_i\beta \leq \dot{3}0$.

The rate equation for the plasma density is

$$\frac{dn}{dt} = r_p - r_{st} - r_{sp} - r_{nr} \quad , \tag{2.3}$$

where the rates on the rhs of eq. (2.3) are due to pumping (r_p), due to stimulated emission (r_{st}), due to spontaneous emission (r_{sp}) and due to the nonradiative transitions (r_{nr}), respectively.

The pump rate due to an injection current density j is given by

$$r_p = \frac{j\eta}{ed} \quad ,$$

where η is the quantum efficiency and d the transverse dimension of the active region. The loss rate due to stimulated emission is

$$r_{st} = \text{Im} \sum_k d_k \, P_k \, (t) \, E^*(t) \quad . \tag{2.4}$$

Because the carrier-carrier relaxation time is so short and the related damping γ_k is so large, P_k can be eliminated adiabetically, which means we can use the stationary solution of eq. (1.30) and get

$$r_{st} = - \sum_k \chi''(k,\omega) \, |E(t)|^2 = \frac{1}{4\pi\omega} g'(\omega) \, |E(t)|^2 \quad . \tag{2.5}$$

The rate of stimulated emission can be put into the form

$$\chi''(\omega) = (1 - e^{\beta(\hbar\omega-\mu)}) \, I(n,\omega) \quad , \tag{2.6}$$

where

$$I(n,\omega) = \text{Im} \sum \frac{d_k^2 \, f_{ek} \, f_{nk}}{1 - q_1} \, \frac{1}{\hbar(\omega - \epsilon_{ek} - \epsilon_{hk} + i\gamma_k)} \quad . \tag{2.7}$$

The relation (2.6) shows that the cross-over between gain ($\hbar\omega < \mu$) and absorption ($\hbar\omega > \mu$) occurs exactly at

$$\hbar\omega = \mu = \mu_e + \mu_n + E_g + \Delta E_g(n) \quad . \tag{2.8}$$

The rate of spontaneous emission into the continuum of all modes is

$$r_{sp} = \int_0^{\infty} \frac{d\omega}{2\pi} \left(\frac{\omega n(\omega)}{\pi c^2}\right)^2 I(n,\omega) \quad , \tag{2.9}$$

where $n(\omega)$ is the index of refraction. The nonradiative recombination rate can due to multi-phonon emission processes via deep-trap levels or due to Auger recombination processes

$$r_{nr} = Bn + Cn^3 \quad . \tag{2.10}$$

The electric field of the laser will be treated classically. From Maxwell's equation one gets

$$\left(\Delta - \frac{\epsilon_0}{c^2}\frac{\partial^2}{\partial t^2} - \frac{4\pi\sigma}{c^2}\frac{\partial}{\partial t}\right) E(r,t) = \frac{4\pi\epsilon_0}{c^2}\frac{\partial^2 P(r,t)}{\partial t^2} \quad , \tag{2.11}$$

where ϵ_0 is the dielectric constant of all not explicitely treated polarizabilities. σ is the conductivity, which is used to introduce the lasers of the cavity. Eq. (2.11) is simplified as usual by expanding E into a complete set of resonator eigenfunctions $u_m(r)$ and by making a slowly-varying amplitude approximation. The resulting equations for the amplitude and phase of a single mode are:

$$\dot{E}(t) + \frac{\kappa}{2} E = \frac{g'(\Omega)}{2} E \quad , \tag{2.12}$$

$$\Omega + \dot{\phi} = \omega - \frac{\Delta g''(\Omega)}{2} \quad . \tag{2.13}$$

Here E(t) is the real amplitude of the active laser mode, ϕ is the phase of the laser field. The complex gain function has been introduced in eq. (1.9). It depends on the plasma density n(t). $\Delta g''(\Omega)$ is the imaginary part of $g(\Omega)$ (see (1.10)) which describes the changes of the refraction due to a finite plasma density n

$$\Delta g'' = 4\pi\omega i \,(\chi' - \chi_0') \quad , \tag{2.14}$$

where χ_0' is the real part of the susceptibility of the unpumped crystal. Ω is the frequency of the single laser mode, ω is the frequency of the resonator filled with the unpumped semiconductor. κ are the resonator losses per unit time. The laser threshold is given by

$$\kappa = g'(\Omega) \quad , \tag{2.15}$$

i.e. the laser mode closest to the gain maximum, which fulfills eq. (2.15) first, will start to lase. In our simple spatially homogeneous model in which spectral hole burning is impossible due to the rapid carrier-carrier scattering, the plasma can support only one stable laser mode. The equations for the plasma density n(t) and the light amplitude E(t) are coupled nonlinear equations, but eq. (2.13) for the phase ϕ does not couple back to eqs. (2.12) and (2.3). The dynamics of a solitary semiconductor laser is thus gouverned by two coupled differential equations and cannot exhibit chaotic behaviour. In order to get chaotic behaviour one needs at least three nonlinear coupled differential equations.
In the next chapter we will add stochastic noise sources to the equations for n(t), E(t) and $\phi(t)$ and calculate the noise properties of semiconductor lasers. In the final chapter we will show how modulation of the pump current or the feedback in an external cavity leads to chaotic instabilities.

III. Noise Spectra of a Semiconductor Laser [7,8,9,10]

All dissipative processes give rise to related fluctuations F(t). Often the fluctuations can be approximately described as Markoffian noise sources, which are uncorrelated at different times:

$$\langle F(t)\, F(s)\rangle = 2\, \Gamma\delta(t - s) \tag{3.1}$$

Eq. (3.1) is called dissipation-fluctuation relation because it links the correlation of the fluctuations with the function Γ which describes the dissipation rate. The resulting noise sources in the rate equation which has been obtained after adiabatic elimination of the polarization equation can be shown to have all shot noise character [8]:

$$\frac{dn}{dt} = r_p - r_{st} - r_{sp} - r_{nr} + F_p + F_{st} + F_{sp} + F_{nr} \quad , \tag{3.2}$$

where the fluctuation have the following correlations:

a) pump noise [10]

Diode lasers are pumped by an injection current, which is driven by a voltage source via a serial resistor R_s. If the serial resistor is larger than the differential resitance of the diode, R_s suppresses the current fluctuations in the diode. The resulting pump noise is as Yamamota has shown simply the Nyquist noise in the serial resistor

$$\langle F_p(t)\, F_p(s)\rangle = \frac{4\, kT}{e^2 V R_s}\ \delta(t - s) \qquad (3.3)$$

Obviously by increasing R_s one can essentially subpress the pump noise.

b) Stimulated emission

$$\langle F_{st}(t)\, F_{st}(s)\rangle = 2\pi\Omega\ (E_{cv} + E_{vc})\ \frac{E^2(t)}{V\hbar}\ , \qquad (3.4)$$

where

$$g(\Omega) = 2\pi\ \Omega\ (E_{cv} - E_{vc}) \quad .$$

Eq. (3.4) gives just the sum of the interband transition rates due to the laser action.

c) Spontaneous emission

$$\langle F_{sp}(t)\, F_{sp}(s)\rangle = \frac{r_{sp}}{V}\ \delta(t - s) \qquad (3.5)$$

d) Nonradiative transition

$$\langle F_{nr}(t)\, F_{nr}(s)\rangle = \frac{r_{nr}}{V}\ \delta(t - s) \qquad (3.6)$$

For the light field we formulate the stochastic equations for the intensity $I = E^2 = 2\pi\hbar\Omega N$ (N is the number of photons in the laser mode) and the phase ϕ.

$$\left(\frac{\partial}{\partial t} + \kappa - g'\right) I = 2EF'' \qquad (3.7)$$

$$\left(\frac{\partial}{\partial t}\phi + \Omega - \omega - \frac{1}{2}g''\right) I = EF' \qquad (3.8)$$

The second moments of the fluctuation are

$$2 < F'(t)\ F'(s)> = 2 < F''(t)\ F''(s)> = 2\pi\hbar\Omega\ R\ \delta(t - s) \qquad (3.9)$$

with

$$R = \kappa\ (2\ n_{th} + 1) + 2\pi\ \Omega\ (E_{cv} + E_{vc}) \qquad (3.10)$$

and

$$n_{th} = \frac{1}{e^{\beta\hbar\Omega} - 1} \quad . \tag{3.11}$$

Real and imaginary part of F are not correlated.

Next we linearize our 3 stochastic equations around their stationary values:

$$I = I_0 + j(t);\; n = n_0 + \rho(t);\; \phi = 0 + \phi(t) \tag{3.12}$$

and find for the mean intensity from (3.7)

$$(\kappa - g'_0)\, I_0 = 2\langle EF''\rangle \; . \tag{3.13}$$

The correlation between the fluctuating field E and F″ gives just the light intensity in the laser mode due to spontaneous emission:

$$2\,\langle EF''\rangle = 2\pi\hbar\Omega\,(2\pi\Omega\, E_{cv}) \quad . \tag{3.14}$$

In terms of the photon number (3.13) yields with $g_0 = g(n_0)$

$$2\pi\,\Omega\, E_{cv}(n_0) + g_0'\, N_0 = \kappa\, N_0 \; , \tag{3.15}$$

which describes just the balance between the generation rate of photons due to stimulated and spontaneous emission and the cavity loss rate.
The fluctuating part of the intensity obeys after a Fourrier transform the equation

$$(-i\omega + \kappa - g_0')\, j_\omega + \frac{\partial g'}{\partial n_0}\, I_0\, \rho_\omega = 2(EF'' - \langle EF''\rangle)_\omega \; . \tag{3.16}$$

From eq. (3.8) we get an equation for the laser frequency

$$\left(\Omega - \omega - \frac{1}{2}g_0''\right) I_0 = \langle EF'\rangle_\omega \quad , \tag{3.17}$$

where $\langle EF'\rangle$ describes the frequency shift due to the spontaneous emission. The fluctuating phase is

$$(-i\omega\, I_0\, \phi_\omega) = \frac{1}{2}\,\frac{\partial g''}{\partial n_0}\, I_0\, \rho_\omega - \left[\Omega - \omega - \frac{1}{2}\, g_0''\right] j_\omega + (EF' - \langle EF'\rangle)_\omega \tag{3.18}$$

Above threshold, where I_0 is large, we can disregard the terms not proportional to I_0 in (3.17) and (3.18). Thus

$$\rho_\omega \simeq \frac{2}{\frac{\partial g'}{\partial n_0} I_0} (EF'' - \langle EF''\rangle)_\omega \tag{3.19}$$

and

$$-i\omega\, \phi_\omega I_0 = \frac{1}{2}\,\frac{\partial g''}{\partial n_0}\, I_0\, \rho_\omega + (EF' - \langle EF'\rangle)_\omega$$

$$= \alpha(EF'' - \langle EF''\rangle)_\omega + (EF' - \langle EF'\rangle)_\omega \quad , \tag{3.20}$$

where

$$\alpha = \frac{\frac{\partial g''}{\partial n_0}}{\frac{\partial g'}{\partial n_0}} \tag{3.21}$$

is the linewidth enhancement factor. The laser linewidth is obtained from

$$\langle E^*(t+\tau)\, E(t)\rangle \simeq E^2\, e^{i\Omega\tau} < e^{i(\phi(t+\tau)-\phi(t))} >$$

$$\simeq E^2\, e^{i\Omega\tau} \exp\left[-\frac{1}{2}\, \frac{\langle(\phi(t+\tau)-\phi(t))^2\rangle}{\tau}\, \tau\right]$$

$$\simeq E^2\, e^{i\Omega\tau} e^{-\frac{1}{2}\langle\frac{d\phi^2}{dt}\rangle\,\tau} \quad . \tag{3.22}$$

We see that the laser linewidth is given by

$$\Delta\omega = \omega\, \langle|\phi_\omega|^2\rangle = \frac{1}{I_0\Omega}\, (1+\alpha^2)\, \langle|(EF' - \langle EF'\rangle)_\omega|^2\rangle \tag{3.23}$$

or

$$\Delta\omega = \frac{2\pi\, \hbar\, \Omega}{I_0} (1+\alpha^2)\, 2\pi\, E_{cv} \quad , \tag{3.24}$$

which is the enhanced Schalow-Townes linewidth formula. The result (3.24) has been derived already in 1967 (see Ref. [7]). Only later it has been recognized that the linewidth enhancement factor α in semiconductors is very large, typically around 5, because the refractive changes of the complex gain function with the plasma density are quite large in semiconductors as compared to atomic systems.

The general structure of (3.18) and the corresponding equation for the fluctuating part of the plasma density ρ_ω which one gets from (3.2) is [9]

$$(-i\omega + \Omega_{jj}\ ) \ j_\omega + \Omega_{j\rho}\ \rho_\omega = F_{j\omega} \tag{3.25}$$

$$\Omega_{\rho j}\ j_\omega + (-i\omega + \Omega_{\rho\rho})\ \rho_\omega = F_{j\omega} \quad , \tag{3.26}$$

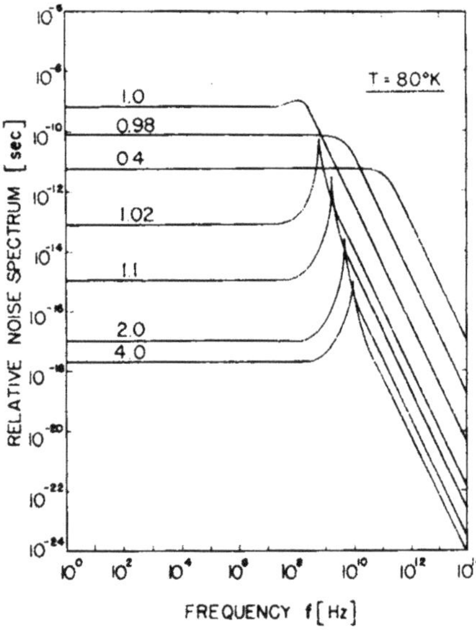

Fig. 3: Intensity noise spectra for 80K bulk GaAs at various pump levels according to ref. [8].

from which the intensity and plasma density fluctuation $\langle|j_\omega|^2\rangle$ and $\langle|\rho_\omega|^2\rangle$, respectively can easily be calculated . In Fig. 3 we show a calculated intensity noise spectrum due to Ref. [8]. A relaxation oscillation in the MHz region is clearly seen. In these damped oscillation the light intensity and the plasma density oscillate out of phase around their stationary values. These oscillations are activated by the noise sources. In Fig. 4 the complex field amplitude spectra are shown due to Ref. [9]. They are obtained by calculation j_ω and ρ_ω from (3.25-26) and inserting the results in the phase eq. (3.18). The relaxation oscillation are seen as side modes to the laser line.

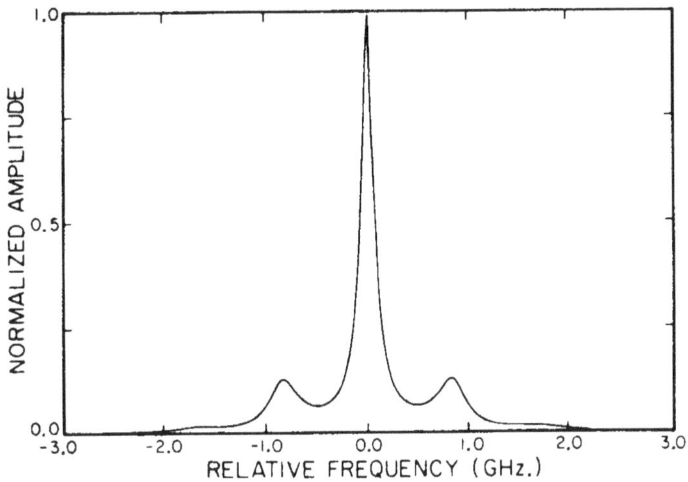

Fig. 4: Complex field amplitude spectrum for a GaAs laser according to ref. [9].

IV. Semiconductor Laser Instabilities

We will now study the nonlinear dynamics of a semiconductor laser, and omit for simplicity the noise terms. As mentioned already the dynamical properties of a semiconductor laser are essentially described by the coupled equations for the light intensity and the plasma density. In order to find chaotic instabilities, we need at least one more coupled differential equation. This can be adchieved in various ways. The situations in which chaos can arise will be seen to be arrangement which arise very naturally in the application of diode lasers in optical communication. The two situations which we will discuss are a) the modulation of the pump current and b) the delayed feed-back in an external cavity.

a) Modulation of the pump current [11,12].

We will discuss first a periodically modulated pump current

$$j(t) = j_0 + j_1 \cos\psi(t) \tag{4.1}$$

with

$$\frac{\partial\psi(t)}{\partial t} = \omega_0 \quad . \tag{4.2}$$

Eq. (4.2) is the third autonomous differential equation which is a necessary prerequirement for chaotic dynamics. The coupled equation of the modulated laser diode are

$$\frac{\partial I}{\partial t} + (\kappa - g'(n))I - 2\pi\hbar\Omega\,(2\pi\Omega\, E_{cv}(n)) = 0 \tag{4.3}$$

$$\frac{\partial n}{\partial t} - \frac{j\eta}{ed} + g'(n)\,\frac{I}{2\pi\,\hbar\,\Omega} + r_{sp}(n) + r_{nr}(n) = 0 \tag{4.4}$$

$$j(t) = j_0 + j_1\cos(\omega_0 t) \quad . \tag{4.5}$$

Chen et. al. [11] have analyzed these equations and found in agreement with experiment subharmonic bifurcations and a Feigenbaum route to chaos. The modulation frequency ω_0 has to be chosen to be in the vicinity of the frequency of the relaxation oscillations.

Winful et. al. [12] modified eq. (4.3) by adding an absorption rate of a saturable absorber to the l.h.s. of (4.3)

$$\frac{hI}{1 + \frac{I}{I_s}} \ , \tag{4.6}$$

where I_s is the saturation intensity. Laser diodes with a saturable absorber exhibit self-pulsing with a frequency ω_s. If the modulation frequency ω_0 is chosen in the vicinity of ω_s, the ratio of these frequencies tend to lock into simple ratios. In other words, the winding number $\rho = \omega_s/\omega_0$ locks to rational numbers $\rho = p/q$ (where p and q are integers), if one varies e.g. the modulation amplitude j_1. The simpler the ratio p/q is, the larger is the range of stability of this state. If an oscillation with the ratio p_1/q_1 exists and another one with p_2/q_2, then one finds also an oscillation with $(p_1+ p_2)/(q_1+ q_2)$. Such a construction is called a Farey tree. For larger modulation amplitudes one finds also regions where quasi-periodic and aperiodic solutions exist. This scenario is called the *quasi-periodicity route to chaos* introduced by Ruelle and Takens.

b) Delayed feedback in an external cavity [13]

Suppose a laser diode is placed in front of an optical fiber, the light coming from the diode is reflected partly at the fiber surface and coupled back into the diode with a time delay of $2\tau = 2L/c$, where L is the distance between the diode mirror and the fiber end.

The equation for the complex field amplitude $\tilde{E}$ is

$$\frac{\partial \tilde{E}}{\partial t} = -i\Omega' \tilde{E}(t) + \frac{1}{2}(g(n) - \kappa)\tilde{E}(t) + k\tilde{E}(t - 2\tau) \tag{4.7}$$

supplemented by the rate equation (4.7) for the plasma density. Introducing again

$$\tilde{E} = Ee^{-i(\Omega t + \phi(t))} \text{ and } n = n_0 + \rho(t) \tag{4.8}$$

and linearizing all transition rates in $\rho(t)$, e.g.

$$g(n) = g_0 + \frac{\partial g}{\partial n_0}\rho(t) \tag{4.9}$$

we find

$$\frac{\partial E}{\partial t} = \frac{1}{2}\left(g_0' + \frac{\partial g'}{\partial n_0}\rho - \kappa\right)E + kE(t + 2\tau)\cos(\phi(t+2\tau) - \phi(t) - 2\Omega\tau) \tag{4.10}$$

$$\frac{\partial \phi}{\partial t} = \frac{1}{2}\frac{\partial g''}{\partial n_0}\rho - \frac{kE(t-2\tau)}{E(t)}\sin(\phi(t+2\tau) - \phi(t) - 2\Omega\tau) \tag{4.11}$$

$$\frac{\partial \rho}{\partial t} = \frac{i\eta}{ed} - \left(g_0' + \frac{\partial g'}{\partial n_0}\rho\right)\frac{E^2}{2\pi\hbar\Omega} - r_{sp_0} - r_{nr_0} - \frac{\partial}{\partial n_0}(r_{sp} + r_{nr})\rho \ . \tag{4.12}$$

Due to the retarded feedback the phase becomes an active variable. Again two competing frequencies are present, namely the relaxation oscillation frequency ω_r and the roundtrip frequency $\omega' = 2\pi c/L$. If one varies the feedback constant k, one finds again frequency locking between ω_r and ω' and a quasiperiodic route to chaos. Chaos can easily be seen as a spectral broadening of the laser line. Thus chaos in laser diodes leads to a coherence collapse.

References

[1] H. Haug and S. Schmitt-Rink, Progr. Quantum Electronics 9, 3 (1984)

[2] H. Haug (ed.), Optical Nonlinearities and Instabilities in Semiconductors, Academic Press, N. Y. 1988

[3] H. Haug and S. W. Koch, Phys. Rev. A to be published

[4] C. Ell, H. Haug and S. W. Koch, Optics Letters, to be published

[5] G. Tränkle, E. Lach, A. Forchel, F. Scholz, C. Ell, H. Haug, G. Weinmann, G. Griffith, H. Kroemer, ans S. Subbanna, Phys. Rev. B36, 6712 (1987)

[6] V. C. Aguillera-Navarro, G. A. Estévez, and A. Kostecki J. Appl. Phys. 63, 2850 (1988)

[7] H. Haug and H. Haken, Z. Physik 204, 262 (1967)

[8] H. Haug, Phys. Rev. 184, 338 (1969)

[9] K. Vahala and A. Yariv, IEEE-QE 19, 1102 (1983)

[10] Y. Yamamoto and S. Machida, Phys. Rev. A. 35, 5114 (1987)

[11] Y. C. Chen, H. G. Winful and I. M. Lin, Appl. Phys. Letters 47, 208 (1985)

[12] H. G. Winful, Y. C. Chen and I. M. Lin, Appl. Phys. Letters 48, 616 (1986)

[13] G. C. Dente, P. S. Durkhin, K. A. Wilson and C. E. Moeller, preprint.

DYE LASERS PHYSICS (IN ITS 23rd BIRTHDAY)

Hector F. Ranea Sandoval
Departamento de Fisica
Facultad de Ciencias Exactas
Tandil, Prov. de Buenos Aires, Argentina.

Outline:

I.- Introduction
II.- Historical Landmarks
III.- Topics on Laser Dyes
IV.- Topics on Dye Lasers
V.- Other topics
VI.- Applications
VII.- Outlook

I. Introduction

After 23 years of intense research in the field of Dye Lasers by very many scientist and companies, to elaborate notes sufficiently comprehensive and at the same time introductory to such a field, might seem like writing many times the very same already written book.

In fact, only five years after its discovery, the first book on Dye Lasers [1], was published. Almost 400 contributions are cited in five review chapters. In its last edition (1977), the book was enlarged with more than 450 references. And then was 1977, more than a decade ago!!

It is not in the aim of these notes but to introduce the reader to some basic aspects of the field, and to update some data of the last years. At the end, I shall present a personal view of the trends apparent in recent publications (which continue at a pace consistent with the expectations of 23 years ago).

The field of applications of dye lasers is large. For reasons of space, I shall only mention but a very few of them, selecting from the newest ones.

II. Historical landmarks

Organic compounds were proposed as laser media, early in 1961 by Rautian and Sobel'man, and by Brock et al. [1]. Although several attempts were made to prove those assertions, it was only in 1966 that Sorokin and Lankard at IBM labs., and Schafer-Schmidt and Volze - by the time at Marburg University- independently found laser emission in liquid solutions of certain organic compounds termed dyes [1].

The IBM group found laser action in a Phtalocyanine, a molecule studied because of its structural similarity with hemoglobin and chlorophyll, i.e. in connection with biological research.

The studies of the Marburg group were oriented towards the Carbocyanines, a class of molecules that proved to be more efficient as IR laser active media. In fact several Carbocyanines (DODCI, DTTC, for example) are still used today in connection with ultra short light pulses generation, and as laser media.

The next years were full of very important theoretical and experimental progresses. In particular, I consider necessary to mention the ones displayed in Table I, together with a brief comment about Laser Dyes and of Dye Lasers which will complete the notes.

The external tuning of the output, proved with time to be one of the most prominent features of these lasers. By tunability is intended not only the variation of the output "color" but most important, the capability of concentrate the energy of the untuned to the tuned output, which means a many folded-multiplied spectral brillance.

This feature is responsible for most of the avalanche in publications in many different fields of science and technology. The reason for this behavior should be traced in the homogeneous broadening of the levels that take part in emission and absorption processes in the couple Dye-Dye host (i.e., the solvent).

The same broadening mechanism, helps the Dye lasers to achieve the shortest light pulses technologically available, which in today's state of the art represents a few cycles of the carrier wave.

The liquid solution that constitutes tha active medium, also provides the clues for attaining CW output, high average power in pulsed operation, and even high power per pulse.

These subjects are condensed in Table I, in which several of the most relevant steps are just mentioned.

TABLE I. HISTORICAL LANDMARKS

Organic Compounds as Laser Media	RAUTIAN-SOBEL'MAN	1961
Stimulated Emission from an Org. Dye in Liquid Solution 1966	SOROKIN-LANKARD	
	SCHAFER-SCHMIDT-VOLZE	1967
Visible Dye Laser Emission	SCHAFER et al	1967
	SOROKIN-LANKARD	1967
External Tuning of Emission	SOFFER-Mc FARLAND	1967
Flash-Lamp Pumped Dye Laser	SOROKIN-LANKARD	1967
	SCHMIDT-SCHAFER	1967
CW Dye Laser Feasibility	SNAVELY-ACHAFER	1969
(Triplet Quenchers)	PAPPALARDO et al.	1970
	MARLING et al.	1970
Cw Dye Laser	PETERSON, TUCCIO, SNAVELY	1970
Three-Element Caveties	HERCHER, PIKE	1971
	TUCCIO, STROME	1972
Anastigmatic Folded Cavity	KOGELNIK et al.	1971
	KOHN et al.	1971
Ultraviolet Dye Lasers	BROIDA, HAIDON	1970
Synthesis-Characterization of New Dyes	DREXHAGE	1972
First Book on Dye Lasers (2 Edition in 1977)	F. P. SCHAFER, Ed	1973
N2 Pumped Dye Lasers	MYERS et al.	1971
	HANSCH	1972

TABLE I (cont)

Planar and Cilindrical Waveguide Dye Lasers	BURLAMACCHI, PRATESI	1973
Frequency Stabilization Single Mode	BARGER, SOREM, HALL	1973
Velocity Fluctuations of Dye Jet	SCRODER, WELLING, WLLEGE-HAUSEN	1973
Injection Locking[2]	BURLAMACCHI, SALIMBENI	1976
	GANIEL, HARDY, TREVES	1976
(Two Stage I. L.)	OKADA, MAEDA, MIYAZOE	1979
Excime Laser Pump of Dye Lasers	SUTTON, CAPELLE	1976
[3]	BUCHER, CHOW	1977
	TISONE et al . (5J)	1975
	RULLIERE et al.	1977
High Peak Power	ALEKSEEV (10 MW)	1972
	BUNKENBURG (11 MW)	1972
	BALTAKOV (30 MW)	1974
High Average Power	MOREY, GLENN (114 W)	1976
[4]	JETHWA, SCHAFER (100 W)	1978
[5]	MASSINGHI et al . (200 W)	1980
Mode Locking i n Pulsed Operation	GLENN et al.	1968
	BRADLEY, DURRANT	1968
	SOFFER, LIN	1968
CW Mode Locking of Dye Lasers	DIENES et al.	1971
	DESBOIS et al.	1971
	SHANK, IPPEN , DIENES	1972
	CHAN, SHARI	1974
[6]	BRITO CRUZ et al.	1987
Distributed Feedback Dye Lasers	SHAND,BJORKHOLM, KOGELNIK	1972
[7]	Zs. BOR	1979
ASE in Dye Lasers	GANIEL et al.	1975
[8]	HNILO et al.	1980
[9]	NAIR-DASGUPTA	1985
Optical Bistability in the Operation of Dye Lasers [10]	HEUDORFER, MAROWSKY	1978
	SCHAEFER, WILLIS	1986
	HAAS, MUNZ, MAROWSKY	1981
Chaos in Dye Lasers [11]	STROUD et al.	1986
	ATMANSPACHER, SCHEINGRABE	1986
	CHYBA et.	1987
Review on Pulsed Dye Lasers[12]	WALLENSTEIN	1979
Semiclassical theory of Dye Lasers [13]	FU-HAKEN	1987/8

III. Topics in Laser dyes

Of course, there is not such a chemical called lasere dye. Instead, it is just an abreviated way to name organic compounds of the class of dyes, which have been suitable synthesized, purified AND IDENTIFIED, so you know exactly what label to put in your flask at the laser lab.

There are several excelent books on these substances, how to know them better, and so on. However, a comprehensive and at the same time indispensable knowledge about them is found in Ref.1, Chapter 4, so I will review here the most important features of the Physics of Laser Dyes that can be useful for the understanding of the basic features of the Dye Lasers.

From the structural point of view, Organic Dyes are (organic) compounds composed by carbon chains linked through unsaturated bonds (i.e. alternating double and single valence bonds). The cycles or chains formed by this distribution of electrons are prominently governed by the Π-bond structure. This bond is responsible for most of the physical behavior of the molecule in its interaction with light, and mostly responsible also for their chemical stability.

Since Π-bonds have maximum bonding energy, at maximum overlap the resulting structure is a normally plane molecule (since at maximum overlap the simmetry axis of the molecules are parallel). This result in a rigid, planar skeleton which as can be shown, is of extreme importance in the operation of these substances as laser media.

Examples of this kind of bonds is found in very many carbonchain-compounds. Benzene ring is probably one of the better known examples.

One of the important consequences of the molecule to be linked through this special bonding, is the light absorption band, whose principal peak lies in the blue visible region of the spectrum ensuring good photostability, against compounds bounded by σ-bonds which absorbs mostly in the region of 160 nm, corresponding to photons of energy capable to photolize the molecule.

The other characteristic of conjugated bonds is that they providde the molecule with high dipole moments, which is responsible for high levels of light absorption. On the other hand, this rigidity helps the energy absorbed to be re-emitted as photon as well, which means that a substance in contact with a Dye will be brightly dyed.

Wide absorption and emission bands in Dye molecules are also consequences not only of its structure but of the environment (A Dye named Rhodamine 6G exhibits a 100nm wide absorptyion spectrum (which corresponds roughly to 3000 cm^{-1}) and a similarly wide emission spectrum). In fact, width of the spectrum and spectral emission range will also be dependent on solvent properties.

This apparently messy physics is nevertheless simplified when one is faced to the energy level diagram of a simingly complicated group of atoms (even hundreds of them) conforming molecules, with several radicals, giving by its own another peculiarities to it.

Before getting into more details, let me introduce the level manyfold in order to interpret the discussions that follows. The promotion of an electron, according to Pauli's principle can only be in such a way that the electron should have the spin parallel or antiparallel to the electrons in the ground state, thus giving rise to singlets (total spin null) or triplets manifolds.

The singlets are responsible for most of the good things of Dye Lasers we are going to talk about. The triplets are the bad guys in this story and are responsible for most of the precautions in design and limitations in this kind of lasers.

However, there are other sources of troubles, indeed. All these having to do with the energy path (and degradation) through the manifolds before the molecule cann re-emit the radiation -absorbed as a photon- as another photon.

The main channels for this degradation are :

1) Loss of energy stored in vibro-rotational levels, through collisions with the solvent. This path, populates the lowest level of each state with time conmstants faster than 10^{12};

2) Intersystem crossing between Singlet and Triplet. This is a non-radiative energy transfer, its time constant is in the order of 100ns;

3) Very fast intrasystem crossing, i.e., conversion of energy from the lowest excited Singlet state, to the Ground state.

Up to now I mentioned the non radiative channels only.

4) Fluorescence, the emission of a photon between the lowest level of the first excited Singlet state and the Ground state. Meantime on the order of 1 -5 ns. This is the one that builds up laser oscillation;

5) radiation re-absorbed from ground state to excited Singlets, and from excited Singlets to higher lying ones;

6) Fluorescence absorption between excited Triplets;

7) emission from the lowest Triplet to the Singlet Ground state. This is a forbidden transition (phosphorescence), and its velocity constant is thus very low. It is responsible for Ground state population depletion.

I shall show next, a way people found to circumvent this drawback by adding some Triplet quencher agents to the solution. These details are taken into account in the modelling of a Dye Laser in the next section. However, I hope that this introduction will serve also to present several classes (the most famous ones) of Laser dyes used in today's and yesterday's technologies.

Of the most striking features of Dye lasers is its possibility not only of tuning the radiation of a single dye output , but , with different Dyes, to cover larger spectral regions or even mix Dyes to have multiple-color output. The first is due to the molecule and its environment; the second is to the great many classes of Dyes there are; the third is due to the fact that several dyes of different classes are compatible with each other and can be mixed without any harm [14].

The next two tables, summarize the most significant aspects of eqach of the most important classes of dyes today used in dye laser research. The latter will introduce to some aspects of newly introduced ones. The reader should know where to refer in case he needs some more details about the data presented therein..

TABLE II

SUMMARY OF IMPORTANT PROPERTIES OF DYES FROM SEVERAL FAMILIES

	Scintillators	Coumarines	Xanthenes	Oxazines	Phenoxazones	Cyanines
Structural mobility	high	v.low	v low	v.low	v.low	high
Mesomeric form A						
Φ_F		high (70%)	v.high (95%)	v.high (90%)	high	30%
Absorption	UV	UV	Green	Red	Green	Red
Emission	UV-violet	Blue-Green	Orange-Red	Deep red	Red	Near IR
Region Stoke's shift (λ_{em}-λ_{abs}) (nm)	60-70	80-100	50-60	30	80-100	50

Solvent Polarity shift (solvato-chromatic effect (SV))		Absorption maxima - shift tow.longer λ's with increasing solvent pol. (Mesomeric Form B in cage) ≅ 30 nm	Similar to Coumarines		Very high SV effect ≅100 nm	Low SV effect ≅ 10nm
pH shift		At low pH Absorption shift to l's to the red A decrease in Fluorescence is aparent	PROTONATION occurs as polar solvents increa.			
Triplet formation		low	v.low	v. low	low (higher than xanthenes)	v.low
Photochemical stability	good (polyphenil 1 operate C.W. 15hrs. with 10% decrease in energy)	good	v. good	v. good	v. good (better than xanthenes)	poor (fresh preparation are often needed)
Examples	POPOP	Coumarin 120	Rhodamine 110	Oxazine 4	Nile Blue-A-Oxazine	DODCI

	BiBuO	Coumarin 102	Rh6G	Cresyl violet		IR-Dye
	Polyphenil 1	Coumarin 6	Rh B	Nile blue		DTDCI

AND...

TABLE III

VERY FEW EXAMPLES OF NEWLY DEVELOPED DYES

DYES (* denotes old ones)	λ_{abs}(max)	λ laser	Tuning range	Operation Mode	Solvent	(%) En. Conv.	Concentration (M/l)
* POPOP	358	381		Vapor (e-beam pumping	Ar^+-N_2		
* BiBuQ	305	365-382	359-405	XeCl . Pulsed	CH	17	$3.6\ 10^{-4}$
[15]	305	365-382	350-395	XeCl. Pulsed	CH	16	
Polyphenyl1	305	385	362-412	XeCl. Pulsed	EG	19	$6\ 10^{-4}$
[15]	305	385	365-412	C.W. Ar^+-Kr	EG	8	$2.5\ 10^{-3}$
*p-Terphenyl	276	337	322-365	KrF.Pulsed	CH	13	$8\ 10^{-5}$
[3-15]	276	341	335-355	Flash-Lamp	DMF		$1\ 10^{-4}$
* Rh B	544	587	579-601	SH. Nd-YAG	MeOH		$3\ 10^{-4}$
	544	609	594-643	N_2	EtOH		$5\ 10^{-3}$
	544	620	596-647	Flash-Lamp	EtOH		$8\ 10^{-5}$

	544	637	608-682	C.W. Ar^+	EG		$2\ 10^{-3}$
Nile Blue-A-Oxazone	540	615	600-740	N_2. Pulsed	EtOH		$1\ 10^{-3}$
	540	615	600-740	Flash-lamp 15/s	EtOH		$1.5\ 10^{-4}$
	540			C.W.	Eg/Et		
Hexacyanine 3 (HITC)	751	868	837-905	XeCl.Pulsed	DMSO	5.8	$2.2\ 10^{-3}$
"IR-Dye"[25]	1090	1300	1.2-1.3 μm	Nd-YAG. C.W.	BzOH	0.3	$1\ 10^{-3}$
* IR-140	779	893	882-913	Nd-YAG	DMSO		$1\ 10^{-3}$
	779	950		Flash Lamp	DMSO		
[3]	779	964	906-1018	N_2	DMSO		$1\ 10^{-3}$

Caption to tables. Numbers in brackets refer to numbered references. When no number is present, Ref.1 should be undertood. More data can be found in [15]. Unless otherwise specified, wavelengths and wavelengths ranges are in nm. Solvents are: EtOH, Ethyl alcohol; MeOH, Methyl alcohol; BzOH, Benzyl alcohol; DMSO, DiMethyl Sulfoxide; CH, CycloHexane; DMF, DiMethyl Fluoride.
Details on excitation procedures should be found in the original references.

IV.- TOPICS ON DYE LASERS

Dye Lasers are very useful; instruments in an extensive fieldof applicatioins, that range from ultrasensitive spectroscopic techinques, to applications in Biology and Medicine. Non - linear Optics, single atom - detection are among them also. The reasons are to be traced in the distinctive feature of Dye Lasers I already mentioned.

Due to space limitations, the discussions on modelling a Dye Laser and on different types of Pump;i;ng an Tuning arrangements will be breief and only infromative.

Very recently, a general semiclassical theory of Dye Lasers has been developed[13]. Nevertheless, an approach using the rate equation approximations, is still correct for most of the facts,Dye Lasers offer to the experementalist.

For each particular pumpi;ng arrangment, a different set of rate equations modelsd the situation. However, all of them are based on a four-level quantum description of the molecule. This approximation was partly justified in the first section of these notes..

The following Figure displays the energy conversion pathway in a general dye molecule (particular cases as excimer formation, [photoisomerization or aggregation excluded). Also shown are the symbols used for the rates of each process.

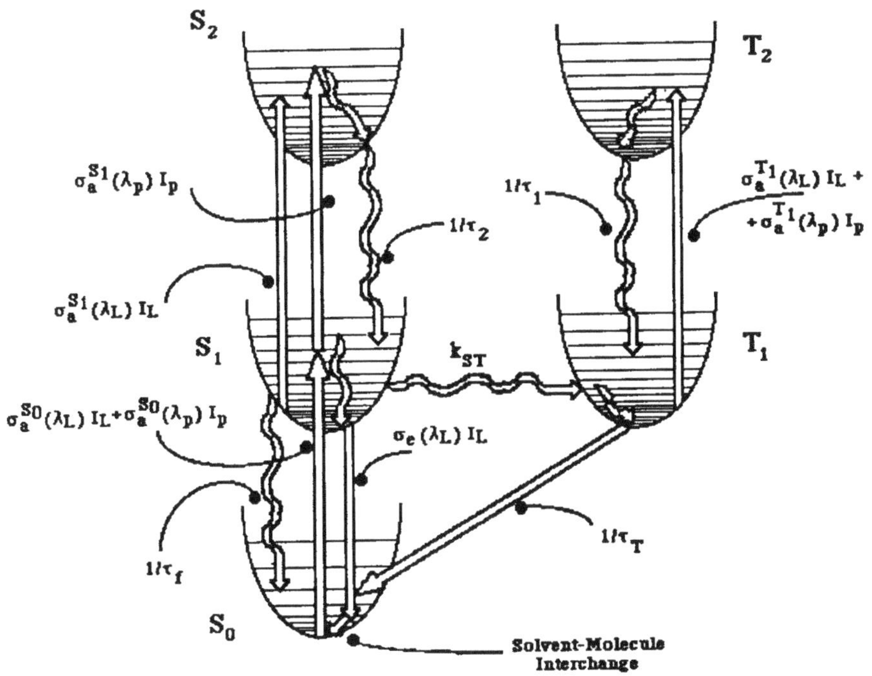

As it was mentioned, only two of the processes are radiative, of which only the fluorescence es useful for laser action [1]. Thus, a good figure of merit is the ratio of fluorescent events,to all other possible ones, i.e. the fluorescent quantum yield.

$$\Phi_f = \frac{(1/\tau_{rf})}{(1/\tau_{rf}) + k_{st} + k_{sg} + k_Q . [Q]} \qquad (1)$$

[Q] is the concentration of a triple quencher agent that can be added to the couple dye-solvent to partly inhibit the deletereous effects that appear when the triplet levels population became important, depleting the Ground state,as it was mentioned before. All other symbols are defined in Figure 1. The radiativelifetime τ_{rf} is related to the fluorescence ljifetime through $\Phi_f = \tau_f / \tau_{rf}$. In Ref. [1], the reader shall find expressions of τ_{rf} in terms of molecular parameters.

A list of Φ_f for several Laser Dyes is shown in the preceeding Tables.

In what follows, the molecules that constitute the medium, are suposed to absorb a number of photons, out of the photons emmited by a certain pump (sometimes a laser (Nitrogen Lasers, Excimer Lasers, Cw ion Lasers, mainly), a Flash-Lamp, or even.another pump)

The energy stored in the first singlet state, can undertake its degradation through the many channels already sketched above. In Figure 1, those marked by a straight line are the radiative ones. Those marked with a wiggled one are the non- radiative pathways.

These energy transfer mechanisms must be introduced in the mentioned rate equations. Figure 1 also presents the probabilities for each event to occur in terms of cross sections, used in writing such equations.

For the sake of clarity I shall suppose a Flash-Lamp-Pumped-Dye Laser. A schematic diagram of such an apparatus in shown in Figure 2.

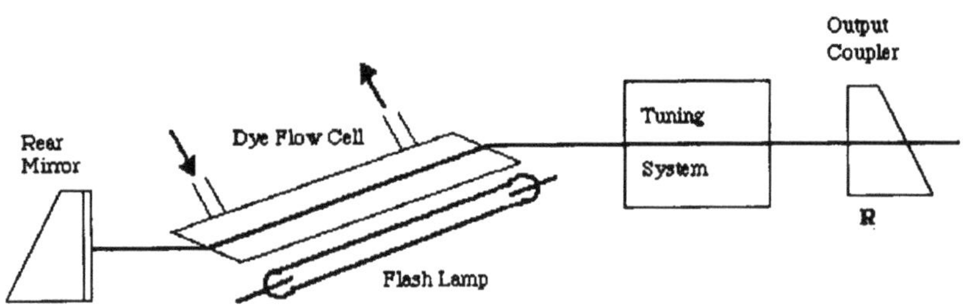

Fig.2 : Schematic Diagram showing the fundamental parts constituent of a Flash-lamp pumped Dye Laser

Specializing the problem a little more, let us suppose that the pump pulse last about 1 μs . According to the discussed energy kinetics, the rate- equations are

$$\frac{dN_1}{dt} = \left[\sigma_a^{So}(\lambda_p)\ I_p + \sigma_a^{So}(\lambda_L)\ I_L \right] N_o - \sigma_E(\lambda_L)\, I_L\, N_1 -$$

$$- \left[\sigma_a^{S1}(\lambda_p)\ I_p + \sigma_a^{S1}(\lambda L)\ I_L \right] N_1 - N_1\ (k_{st} + (1/\tau_f)) + N_2 \quad (2)$$

for the first excited singlet,

$$\frac{dN_T}{dt} = k_{st}\, N_1 - \left[\sigma_a^{T1}(\lambda_L)\ \ I_L + \sigma_a^{T1}(\lambda_p)\ I_p \right] N_T - \qquad (3)$$

$$- N_T/\tau_T + N_{T*}/\tau_{T*}$$

for the (first) triplet state population, while

$$N_0 + N_1 + N_2 + N_T + N_{T*} = N = \text{constant} \qquad (4)$$

insures particle conservation.

The laser intensity, evolves according to the following photon gain equation in the active medium

$$\frac{dI_L}{dt} = I_L\ \{ N_1 \left[\sigma_e(\lambda_L) - \sigma_a^{S1}\ (\lambda_L) \right] - N_o\, \sigma_a^{So}(\lambda L) \ -$$

$$- N_o\, \sigma_a^{So}(\lambda_L) \ - N_T\, \sigma_a T_1(\lambda_L)\ \} \qquad (5)$$

or, $(dI_L/dz) = I_L\ g(\lambda)$, being $g(\lambda)$ the expression between curly brackets.

In equation 1, it is possible to set

$$(dI_L/dz) = (\eta/c)\,(dI_L/dt) \qquad (6)$$

The round trip gain, must account for losses in the particular cavity, other than the noted ones, If it is possible to dump all losses as reflective ones, the round trip gain $G(\lambda)$ will have the following expression,

$$G(\lambda) = 2\, l\ g(\lambda) - \ln[R(\lambda)] \qquad (7)$$

Being $R(\lambda)$ the effective reflectance of the cavity, and l the active length of the (laser) medium.

Working these equations out, it is possible to evaluate the minimum power needed for a Flash-Lamp- pumped system, to achieve oscillation.

From the population rate equation, the threshold power can be evaluated to be

$$P_{th} = A\ I_{th}\ (hc/\lambda_p) \qquad (8)$$

where

$$I_p^{th} = [\ \sigma_a^{So}(\lambda p)\ \tau_f\ ]^{-1}\ .\ (N_1/N)_{th}$$

Thus, for a 10 cm long Dye cell, with an area $A = 3\ cm^2$ and a concentration 2.0 10^{-4} M (corresponding to $N = 2\ 10^{17}\ cm^{-3}$ and with mirrors of $R_1 = 1$ and $R_2 = 0.1$ reflectivities, the population threshold is found to be $(N1/N)_{th} = 6\ 10^{-3}$, and the power to achieve it (considering $\sigma_a{}^{So}$ ($\lambda = 500$ nm) = $10^{-16}\ cm^2$) rounds the tens of kW.

In the case there is a possibility of matching the Lamp output to the narrow high absorption region of the Dye, this Figure is precisely 15 kW , and it is clearly a minimun.

Following the same procedure, but specializing the rate equations to short pulse pumping (Triplet losses negligible), the reader should easily convince himsef that for Nitrogen Laser pumping, such threshold power rounds the 6 kW range.

Variations of the above set of equations have been used to give the correct answer to several problems posed by Flash-Lamp-Pumped- Dye Lasers and even for CW laser pumped ones.[1]

Even in Flash-Lamp-Pumped - Dye Lasers, when one wishes to account for certain multimode phenomena (namely time evolution[1]), a variation of the above system can be used.[16,17].

In cases like short pump-pulse , or in short pulse generation, the evolution of the intensity as in (6) is incorrect, and a convective derivative should be used in (4), to account for propagarion effects.[9].

In short - pulse Dye Laser excitation, amplification of spontaneous emission (ASE) is a significative drawback in many designs, particularly in those which a very narrow emission line is required, or in amplifier- chain setups. Many ways to overcome ASE or to evaluate correctly its influence on the performance of a system, had been devised.[1,8-9].

In amplifiers chains it is seemingly possible to minimise its deletereous effect taking into account the delay between "laser oscillation" and ASE, by pumping the stages slightly delayed with respect to the geometrical one.

Tuning mechanisms for Laser pumped, Flash-Lamp pumped, and CW pumped Dye Lasers are reviewed in a number of publications.[1,12]. They include geometric, interferometric, or "slaved" selection of master wavelength . The last one include injection locked systems [2] (by other lasers or by locking to a particular well defined atomic transition) or effective double pumping of the Dye,[18].

Very recently, however, a new type of tuning mechanism has been discovered, based on degenerate four wave mixing in photorefractive crystals[19]. This system had prove to be a very efficient way of reducing the linewidth, and is self tuning- self repetitive under certain experimental conditions.

The lack of space precludes a detailed analysis of the last most significatives achievements in tuning mechanisms. The references will provede the reader with more precise information.

To coclude this section about Dye Lasers, a few words on wavelength stabilization in CW Dye Laser. In fact, for the many applications of this type of lasers, most requires an accurate control of the output wavelength. In this sense, one of the last original wxternal Dye- frequency stabilizer, combines an acousto-optic frequency shifter with a fast electro- optic phase modulator. The system also reduces the linewidth of a commercial system from 3GHz to 20 kHz [20].

V.- Other topics.

I think we have scanned a wide variety of topics concerning excitacion procedures, models, ASE reductions, tunability and frequency control of Dye Lasers, However, this is not a self- contained review, so the reader is encouraged to refer to the bibliography to complete the overlook, with some more details on each topic.

In addition to this, Laser Dyes has been presented and analized to understand the basic physics underlying the essential features of Dye Lasers.

Nevertheless - I insist - the file is still open.

Until very recently, dye Lasers had been a very useful tool (if not unique) in non - linear optics and in the study of many non- linear phenomena. Since very recently they became the subject of non - linear phenomena occurring at them.

I already mentioned the Optical Bistable behaviour, observed in Refs [10] , probably due to reabsorption in non-pumped regions or to triplet stated- energy storage. Other researchers had found a bistable loop in the wavelength output.

Chaotic behaviour in multimode Dye Lasers have been observed in a number of cases [11], while high dimensionality- chaos has been claimed [11] by Chyba et al. to be generated by flow turbulences in the Dye Jet.

VI.- Applications

Since the advent of Dye Lasers many applications became possible , due to the tuning capabilities they present in ranges of optical radiation useful for spectroscopy and, most important, the narrow bandwidths attained even by commercial systems. so just a list of them will be indefinitely long, out of the scope of these notes.

The reader should refer to the correspondig chapters in 1973 Edition of Schafer's book and its revised 1977 version, or the 1979 Wallenstein review, to get but a feeling of what is going on today in that field.

Certainly, spectroscopy (linear and non-linear) is one of the fields where Dye Lasers are used the most. Entirely new types of spectroscopic techniques were possible since the advevt of these lasers; and entirely new applications of spectroscopy were also possible.

To mention but a couple of the recently reported ones, I refer to one [21] in which the achievement of purally rotational CARS, simultaneously with Coherent Raman Spectroscopy, allows arbitrary separation between the pump and the broadband Stokes' lasers. A thermometric study through the rotational spectrum of O_2 obtained by a Dye Laser and its pump is there in presented. The reported experiment shculd be of interest in multiple wavelength- Dye Lasers [14]

The second one refers to an application of saturation spectroscopy to plasma diagnosis, [22] . The reported results are a direct measuremtnt of the cathode fall region of an electron- beam sustained- negative glow through the measurement of the quadratic Stark shift of the lines of the buffer gas.

A major channel of efforts in Dye laser applications is the ultra-short light pulse generation (USLPG). To understand that, one must bear in mind that only with dyes, were the shortest pulses ever obtained [6].

Other interest in are USLPG in the green region (towards shorter l's), as Passive Mode-Locking of Rhodamine 110 by HICI and DASBTI. [23]

Finally, it is worth noting the generation of subpicosecond light pulses in a colliding pulse mode-locking system, pumped by a visible, Xe ion laser. [24]

VII.- Outlook

According to the available bibliography, tendencies in the different subjects I mentioned in these notes, can be separated as follows:

In Laser Dyes, the efforts are in the direction of

1) obtain dyes with better photostability in the region beyond 1.5 μm and in the soft UV region [25];

2) Studies of molecular Dynamics that lead to photoisomerization understanding and its proper relation to USLPG;

3) extend the region of USPLG to the blue or even to the soft UV region of the spectrum.

The pumping arrangements follow the line of obtaining compactness, [26] which is rather difficult due to the need of recirculate the Dye solution. Tuning, Bandwidth reduction and Stability are major goals. [27]

Output stability, multistability and/or chaotic behavior are reviewed in recent articles by Fu and Haken [13]. It is still a field of research. Multimode operation and nonlinear interactions with the Dye host are taken into account; however, transverse mode effects are inevitable in Dye lasers.

Coupled cavities, and phase locking, after a period of rest, are becoming subjects of interest because of what had been said in the notes [28].

In the branch of applications, it is hard to see a simple tree. Nevertheless, in USLPG one can observe an increasing interest in dispersive effects of mirrors in the fs range [29].

The surface enhanced spectroscopy branch is another field in which Dye Lasers offer unique solution to experimental problems.

References

(The references are grouped when represents several aspects of the same problem. In Ref.1 I tried to group all reviews on Dye Lasers necessary to prepare these notes. All references prior to 1977 are likely to be cited in Ref.1, therefore they are not cited here.)

[1] "Dye Lasers", F. P. Schafer, Ed. (Springer-Verlag, Berlin, 1977,3rd. Edition; Liquid Lasers, F. P. Schafer, in "Laser Handbook" Vol I, F. T. Arecchi, Schultz Dubois, Eds. (North Holland Pub. 1972); "Laseres de Colorante y sus aplicaciones a la Espectroscopia Optoacústica", J. Tocho , unpublished (1988); "dye Lasers", enlarged version of these notes, H. F. Ranea Sandoval, unpublished (1988).

[2] "A tunable, single mode, Injection Locked Flash-lamp Pumped Dye Laser; Blitt, Ganiel, and Treves, Appl. Phys. **12**, 69 (1977); "Spectral narrowing of a Flash-lamp High Energy Dye Laser by two stage Injection Locking", Okada, Maeda, and Miyazoe, IEEE J. of Quantum Electron. **QE15**, 616 (1979).

[3] "A 1 MW p-Terphenyl Dye Laser", Bucher, and Chow, Appl. Phys. **13**, 267 (1977); "The XeCl* Excimer Laser: A powerful and Efficient UV Pumping source for Tunable Dye Lasers", Telle and Huffer, Opt. Commun. **38**, 402 (1981).

[4] "A Reliable High Average Power Dye Laser", Jethwa, Schafer, and Jasny, IEEE J. of Quantum Electron. **QE14**, 119 (1978)

[5] "A 200 W Average Power Narrow Bandwidth Tunable, Waveguide Dye Laser", Mazzinghi, Burlamacchi, Matera, Ranea-Sandoval, Salimbeni, and Vanni, IEEE J. of Quantum Electron. **QE17**, 2245 (1981).

[6] "A novel Pumping Arrangement for Tunable Single ps Pulse Generation with an N2 Laser-Pumped Distributed Feedback Dye Laser"; Bor, Opt. Commun. **29**, 103 (1979).

[7] "Tunable ps Pulse generation by an N2 Laser-pumped, self Q-switched, Distributed feedback Dye Laser, Bor, IEEE J. of Quantum Electron. **QE16**, 517 (1980).; Simultaneous Operation of Distributed Feedback Dye Laser and Grating Tuned Dye Lasers", Golub, Erez, and Shuker, Appl. Phys. **B31**, 75 (1983).

[8] "Stored Energy in Pulsed Dye Laser Amplifiers", A. Hnilo, O. Martinez, and E. J. Quel, IEEE J. of Quantum Electron. **QE22**, 20 (1986).

[9] "Amplified Spontaneous Emission in Narrow Band Pulsed Dye Laser Oscillators - Theory and Experiment", Nair, and Dasgupta, IEEE J. Of Quantum Electron. **QE21**, 1782 (1985).

[10] "Threshold Singularities of Optically Pumped Dye Laser" Heudorferand, and Marowsky, Appl. Phys. **17**, 171 (1978); "Bistability in the Operation of an Organic Dye Laser", Haag, Munz, and Marosky,IEEE J. Of Quantum Electron. **QE17**, 349 (1981); "Bistability and Self-sustained Intensity Oscillations in a Ring Laser with Optical Backscattering", Kühlke, Acta Physica Polonica **A61**, 547 (1982); "The influence of Triplet Losses on Optical Hysteresis in Pulsed Dye lasers", Koepke, Acta Physica Polonica **A66**, 741 (1984); "Observation of Higher Dynamical States of a Homogeneously Broadened Laser, Hillman, Krasinski, Boyd, and Stroud, Phys. Rev. Lett. **52**, 1605 (1984); "Bistability in Wavelength Using an Electro-optically Tuned Dye Laser"; Porte, and Goedburger Opt. Commun. **51**, 331 (1984).; "Optical Bistability and Hysteresis in a Colliding Pulse Mode-Locked Femtosecond Dye Laser", Jacobovitz, Brito Cruz, Mansur, and Scarparo, Opt. Commun. **59**, 233 (1986).

[11] "Deterministic Chaos and Dynamical Instabilities in a Multimode CW Dye Laser", Atmanspacher, and Scheingraber, Phys. Rev. **A34**, 253 (1986); "Instabilities and Chaos in a Multimode, Standing Wave, C.W. Dye Laser"; McMackin, C. Radzewicz, Beck, and Raymer, Phys. Rev. **A38**, 820 (1988) and references therein; "Chaos in a Good Cavity Single Mode Dye Laser due to Turbulent Dye Flow", Chyba, Gage, Ghosh, Lett, Mandel, and MacMackin, Opt. Lett. **12**, 422 (1987).

[12] "Pulsed Dye Lasers", Wallenstein, in Laser Handbook 3, M. L. Stitch, Ed. (North Holland Pub., 1979); "Laseres de Colorantes Excitados por un laser de N2", G. Bilmes, H. Ranea-Sandoval and J. Tocho, Monografias Cientificas del CIOp **#3**, (1985).

[13] "Semiclassical Dye Laser Equations and the Unidirectional Single-frequency Operation" , FU, and Haken, Phys. Rev. **A36**, 4802, (1987); "Multichromatic Operation in C. W. Dye Lasers", Fu, and Haken, Phys. Rev. Lett. **60**, 2614 (1988); "Semiclassical Theory of Dye Lasers: The Single frequency and Multifrequency Steady States of Operation", Fu, and Haken, JOSA **B5**, 899 (1988).

[14] "Characteristics of a MUlticolor Dye Laser", P. Burlamacchi, and H. Ranea Sandoval, Opt. Commun. **31**, 185 (1979); "Donor Concentration Dependence of the Gain of a Dye Mixture Laser", Govindanunny, Sivaram, and Srinivasan, Indian J. of Physics **54B** , 32 (1980).

[15] "Exciton Catalog of Laser Dyes", Exciton; "New Efficient Laser Dyes for Pulsed and CW Operation in the UV ", Hüffer, Schieder, Telle, Raue, and Brinkwerth, Opt. Commun. **33**, 85 (1980).

[16] "Laser de Colorante Bombeado por un laser de Xe Ionico" R. Duchowicz, Ph. D. Dissertation, Universidad nacional de La Plata, La Plata, B.A. , Argentina (1981), and references therein.

[17] "The Time resolved Spectrum of a Pulsed Dye Laser", Tocho, Ranea-Sandoval, Duchowicz, and Garavaglia, Optical & Quantum Electron. **16**, 393 (1984), and references therein.

[18] "Frequency Locking in Doubly Pumped Dye Laser", Tocho, Duchowicz, Gallardo, and Garavaglia, IEEE J. of Quantum Electron. **QE17**, 1591 (1981), and references therein.

[19] "Self Scanning of a Dye Laser due to Feedback from a BaTiO3 Phase Conjugate Reflector", Whitten, and Ramsey, Opt. Lett. **9**, 44 (1984); "Self Scanning of a CW Dye Laser Having a Phase Conjugate Resonator Cavity", Feinberg, and Bacher, Opt. Lett. **9**, 420 (1984); "Phase Conjugate Feedback into a CW Ring Dye Laser", Ramsey and Whitten, Opt. Lett. **10**, 362 (1985).

[20] "External dye Laser Frequency Stabilizer", Hall and Hansch, Opt. Lett. **9**, 502 (1984).

[21] "Simultaneous Rotational Coherent Anti-Stokes Raman Spectroscopy and Coherent Stokes Raman Spectroscopy with arbitrary Pump Stokes Separation"; Eckbreth and Anderson, Opt. Lett. **11**, 496 (1986).

[22] "Electric Field Distribution in the Cathode Sheath of an Electron Beam Glow Discharge", Lee, Andersen, Rocca, Marconi, and Reesor, Appl. Phys. Lett. **51**, 409 (1987).

[23] Passively Mode-Locked CW Rhodamine 110, Dye Laser", French and taylor, Opt. Lett. **11**, 297 (1986).

[24] "Sub-ps Pulse Development in a Colliding Pulse Mode-locking Dye Lasert Pumped by a Pulsed Xe Laser", Ansari, Dienes, and Whinnery, Opt. Lett. **10**, 19 (1985).

[25] "Tunable ps Pulses around 1.3 μm Generated by a Synchronously Pumped IR Dye Laser", Seilmeier, Kaiser, Sens, and Drexhage, Opt. Lett. **8**,205 (1983).

[26] "Dye Laser Pumped by a Nd:YAG Laser Pulses Frequency Doubled in a glass Optical Fiber", Osterberg, and Margulis, Opt. Lett. **11**, 516 (1986); "Tunable near IR ps pulses from a Short Cavity Dye Laser", Bohnert, Boggess,

Mansour, Maxson, and Smirl, IEEE J. of Quantum Electron. **QE22**, 2195 (1986).

[27] "Dye Laser Frequency Locking to the Hyperfine Structure ($3S_{1/2}$, F=2 - $3P_{1/2}$, F=2) of Sodium D_1 line by using Polarization Spectroscopy, Kim. Kong, and Lee, Appl. Phys. Lett. **52**,417 (1988).

[28] "Coupled Resonators Employing Phase Conjugate and Ordinary Mirrors", Anderson, Opt. Lett. **9**, 420 (1984). "Experimental Study of a Ring Dye Laser with an External Retroreflective Mirror", Lawandy and Afzal, IEEE J. of Quantum Electron. **QE22**, 2251 (1986); "Single mode CW Dye Laser by External Cavity Tuning", Schinca, Scaffardi, and Ranea-Sandoval, unpublished results.; "Effects of Backscattering on the Behavior of a Dye Ring Laser", R. Roy, and L. Mandel, Opt. Commun. **35**, 247 (1980); "Intensity Stabilization by Controlled Self-Injection in a Pulsed Ring Dye Laser", Tam, and Meyer, Opt. Commun. **49**, 349 (1984).

[29] "fs Pulse Laser Chirp Compensated by Cavity Mirror Dispersion", Yamashita, Ishikawa, Torizuka, and Sato, Opt. Lett. **11**, 516 (1986).

NITROGEN AND PULSED DYE LASERS

Francisco A. Manzano and Alejandro L. Peuriot
Ceilap (Citefa - Conicet)
Zufriategui 4380. (1603) Villa Martelli. Argentina

I PREFACE

These notes have been written with the scope of filling out the concepts exposed at the International School of Laser and Quantum Optics .

The aim of this work is to introduce and to help taking knowledge of two types of lasers which are relatively easy to construct and have several simple applications in chemistry and physics.

It was not our purpose to fill this subject , neverthless referances are included which contains the main information about this lasers.

At the same time some considerations on the design are included, which we wish to be useful for those the readers that are introducing themselves into this subjet.

II. THE NITROGEN LASER

II.1 INTRODUCTION

In 1963 Mathias and Parker [1] obtained for the first time laser emission in a molecular species in near infrared region of the spectrum from an electronic transition between excited electronic states of the Nitrogen molecule.

In this experiment they used a discharge tube axially excited by a short rise-time current pulse obtained from a high voltage pulsed discharge.

Later, Heard [2] using a similar excitation system with very high voltages (100 - 150 KV) obtained pulsed laser emission in the near ultraviolet region with peak powers of about 10 W in a low pressure discharge.

This emission was due to electronic transitions of the second positive (tripled) system of the molecular Nitrogen bands and its more intense line was λ= 337.1 nm .

Afterwards, in 1965 Leonard [3] produced ultraviolet laser pulses with peak powers of several tenths of KW from transversal discharges at lowers voltages (25 KV) making possible the use of Nitrogen lasers as an important and powerful luminous source in several fields of research and technology.

Since that year the Nitrogen laser has found increasing importance due to the emission in the ultraviolet, short pulse width, their high repetition rate and reliability. Many studies and designs were reported which allowed to understand the excitation mechanisms and the dynamics of generation of the ultraviolet laser emission.

Up to the moment three groups of identified lines and approximately a hundred of non identified ones were reported ranging the near infrared and the near ultraviolet regions of the spectrum [4,17].

The most important line is one at 337,1 nm due to the high peak power achieved and to the wide range of applications and these notes will be dedicated to discuss the N_2 UV laser.

The high gain of the transition (~50 dB /m)[3] allows this necessary to use mirrors if the length of the laser tubes exceeds a critical value [5]. However, in this case the beam quality is poor and highly divergent.

The importance of the work is justified by the peak power attained (higher than 10 MW), short pulse-width (100 ps -10 nsec) generated and the short wavelength obtained with relative simple devices [5, 6, 16] . Recentely new applications have appear like systems based on Nitrogen lasers for point surface measurements by laser induced fluorescence [18] demostration of multiple order Stokes frequency generation in optical fibers [19] and medical laser fluorimeter for tissue metabolism measurements [20].Additionally the Nitrogen laser had become the most popular pump source for dye lasers due your reliability and relative low cost.

II.2 THE NITROGEN LASER

The most important laser emissions bands and the most studied ones of the Nitrogen laser involve transitions between the electronic states $B^3\pi_g$ to $A^3\pi_u$ (first positive bands systems) from which an infrared emission is obtained and between the molecular levels $C^3\pi_u$ to $B^3\pi_g$ (second positive bands systems) giving rise to an ultraviolet emission (Fig.II 1) [4,8,9].

Both emissions consist of wide bands, roughly 0.2 nm, composed by a large number of very closely spaced rotational lines.

In the two systems the condition to establish a population inversion can only be fullfilled in a transient manner because the lifetime of both lower laser levels is substantially longer than those of the upper laser levels. Consequentely the populations tend to equalize in a time of the order of the upper laser level lifetime andthe laser emission stops. The laser emission is called self-terminating.

Therefore these lasers can only be operated by the use of a pulsed excitation so that the current pulse rise-time must be shorter than the effective lifetime of the upper laser level considered.

Several types of devices have been used. From fast pulsed axially or transversely electrical high voltage discharges to electron-beam excitation at reduced pressures (20 - 100 Torr) or at several atmospheres [4,10,15,21].

The accepted mechanism of excitation of the laser levels $C^3\pi_u$ and $B^3\pi_g$ is the direct electron impact from the ground state $X^1\Sigma_g$. This is strongly supported because it can be observed that the emission of spontaneous and laser radiation occurs on the onset of current pulse. This mechanism is also supported by the fact that the calculations for the gain distribution, based on this assumption, exhibited by the rotational structure of the 0-0 vibrational transition and the behavior of the power laser of different bands when the laser is cooled are in good agreement with the experimental data [4,8]. Therefore high energy electrons in sufficent numbers must be present to produce inelastic and selective

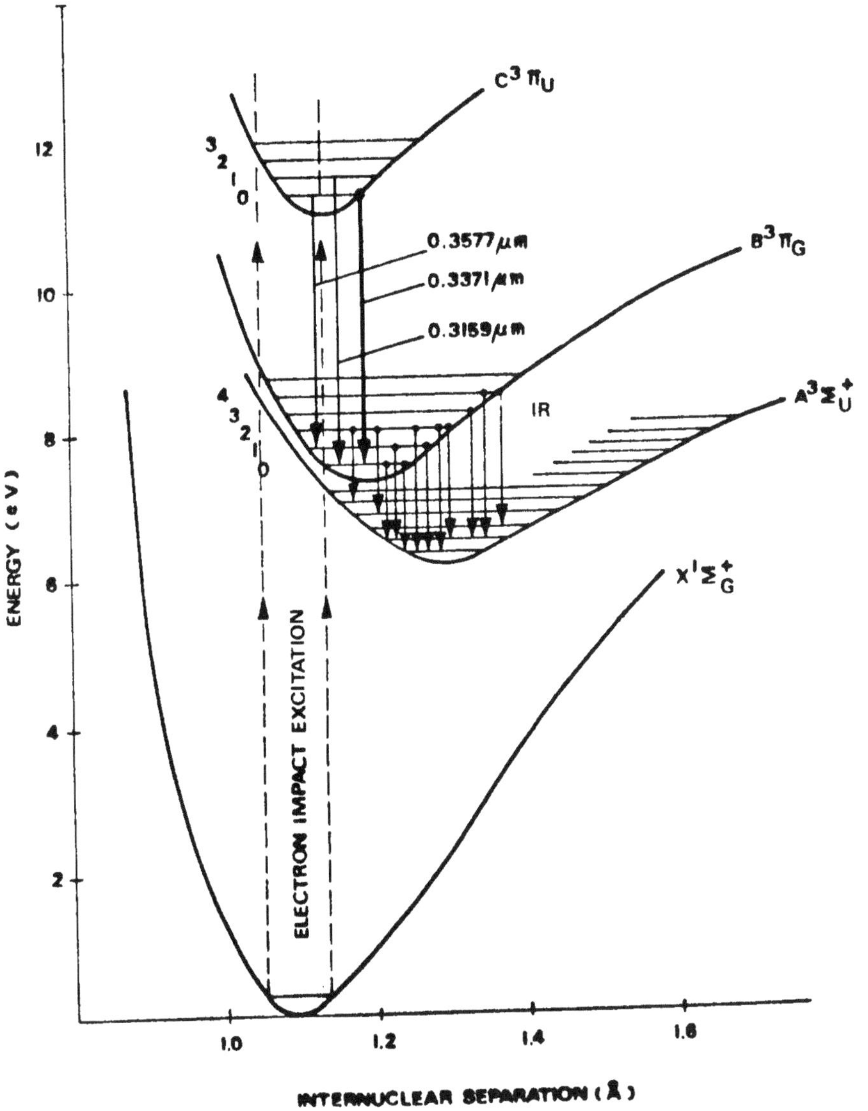

Fig.II.1 Energy levels of the Nitrogen molecule shows the ultraviolet laser transitions collisions with Nitrogen ground state molecules. It may be useful to discuss briefly this excitation mechanism.

The Fig.II 2 shows the total cross-sections of excitation for $B^3\pi_g$ and $C^3\pi_u$ states versus the electron energy [9].

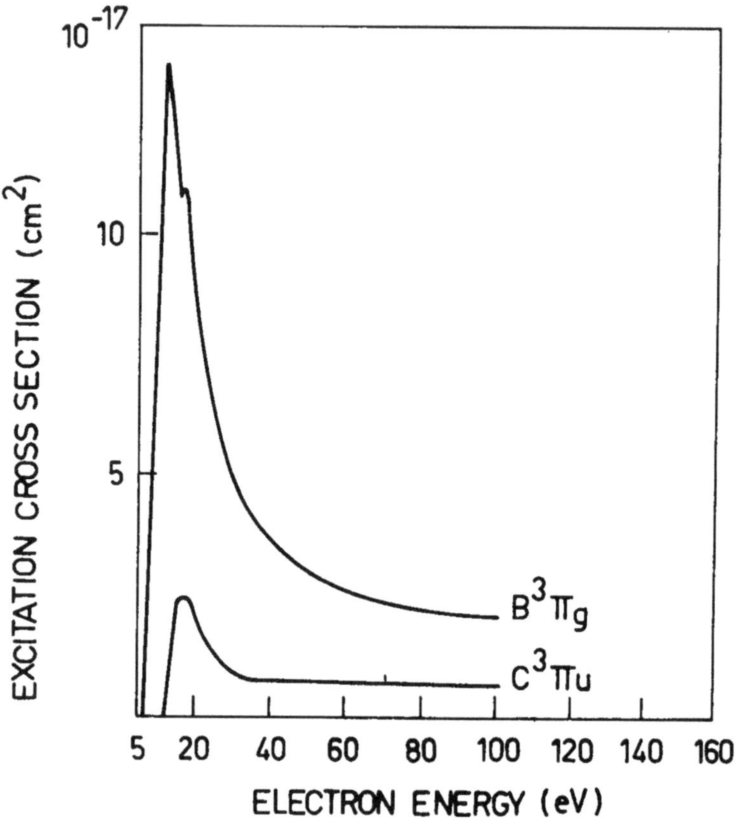

Fig.II.2 Excitation cross-sections by electron impact of the upper and lower laser levels of the Nitrogen molecule. (After C.S. Willet) [9]

It may seen that the maximum of the total cross-sections of $C^3\pi_u$ is several times smaller than that of the $B^3\pi_g$ level and both maxima appear aproximately at the same energy. Likewise the excitation function of $C^3\pi_u$ is always smaller than that of the $B^3\pi_g$.

It seems that the population inversion of levels cannot be attained by direct electron impact.

However, to decide if the inversion of population is posible cross- sections for each vibrational sublevel corresponding to an electronic term should be compared with those of the other electronicstate. Hence the relative positions of the potential curves and evaluation of the associated distribution of Franck-Condon factors play a decisive role.

On the other hand, from Fig. 1, if the excitation by direct electron impact with fast electrons is considered valid and applying the Franck-Condon principle it can be deduced that the lower vibrationals levels of $C^3\pi_u$ will be preferenteally populated with respect to the lower vibrationals levels of the $B^3\pi_g$.

This may be explained because the relative displacement to ground state of the internuclear separations is smaller for the $C^3\pi_u$ than the $B^3\pi_g$ state.

So, qualitatively speaking, it would appear possible to produce an inversion of population only for the lower vibrational levels of the electronic states.

If the excitation cross-sections of the vibrational levels are calculated from the Franck-Condon factors it may be deduced that it is only possible to produce inversion between the vibrational levels $v = 0, 1$ of $C^3\pi_u$ and the $v = 0, 1$ of $B^3\pi_g$ except the $v = 1 \dashrightarrow v = 1$ transition [9].

Three main laser bands have been reported in the ultraviolet region: $\lambda = 337{,}1$ nm ($v = 0$ to $v = 0$); $\lambda = 355{,}7$ nm ($v = 0$ to $v = 1$) and $\lambda = 315{,}9$ nm ($v = 1$ to $v = 0$) for the second positive bands system. This experimental data confirm the mechanism of excitation by direct electron impact.

Taking into account the above discussion the Nitrogen laser at the 337,1 nm may be considered in the frame of a three level laser theory with the upper laser level ($C^3\pi_u$) and the lower one ($B^3\pi_g$) excited from the ground state level ($X^1\Sigma_u$) by direct electronimpact [11, 12].

The laser rate equations set governing the populations densities N_0, N_c, N_b of the fundamental, upper and lower levels can be written taking into account that for the Nitrogen molecule the following relations between the lifetimes are valid:

$$t_{cb} \sim 40 \text{ nsec} >> t_{bo} \sim 10\ \mu\text{sec}; \quad t_{co} > t_{cb}$$

and utilizing the Franck -Condon principle:

$$dN_c/dt = \langle r_{oc} \rangle N_e N_0 + \langle r_{bc} \rangle N_e N_b - (L_{cb} + t_{cb}^{-1}) N_c - \Phi(\nu_{cb}) B^i{}_{cb} (N_c - N_b)$$

$$dN_b/dt = \langle r_{ob} \rangle N_b N_0 + (L_{cb} + t_{cb}^{-1}) N_c + \Phi(\nu_{cb}) B^i{}_{cb} (N_c - N_b)$$

The photon number evolves as:

$$(1/h\nu_{cb})d\Phi(\nu_{cb})/dt = N_c t_{cb}^{-1} + \Phi(\nu_{cb})B^i_{cb}(N_c - N_b)$$

and the ions are created according to:

$$dN_e/dt = N_e N_0 <r_i>$$

where $<r_{ij}>$ denotes excitation cross-sections from level i to j averaged over the electrons velocities distribution, N_e is the electron density, B^i_{cb} is the rate of induced emission, L_{cb} is the rate of collisional deexcitation, $\Phi(\nu_{cb})$ is the energy density at the ν_{cb}frequency and $< r_i>$ is the ionization cross-section.

The Nitrogen molecules are also associated to an excitation circuit as a variable resistivity R(t) thatis mainly originated by the electron-molecules collisions. Thus thegain of energy of the electrons as function of their electron temperature T_e is given by [11]:

$$d(kN_eT_e)/dt = I^2R(t)-N_e[N_0(<r_{oc}>E_{oc}+<r_{ob}>E_{ob})-N_o<r_i>E_i-$$
$$-N_o<r_vE_v>-E_{bc}(N_b<r_{bc}>-N_cL_{cb})]$$

where $<r_vE_v>$ is an averaged excitation cross-section between vibrational levels of ground state and E_i is the ionization energy of the Nitrogen molecule.

The equation for the electrical circuit, which consists of a capacitor C charged at a voltage V_0 connected suddenly to a laser tube through an stray inductance L_s and resistance R, is:

$$Ls(dI/dt) + [R(t) + R]\, I = V0 - \int_0^t (I/C)\, dt'$$

The equations system takes the form of rate equations coupled to the excitation circuit equations. They have been established and numerically solved by several authors [11,12,14] obtaining a good agreement with the experimental data.

The conditions for the population inversion, the laser power density and its temporal evolution are obtained from the solution of the equations system.

In first approximation the condition for inversion is the same as for a pulsed laser with a bottleneck lower laser level so say the inversion exists for times shorter than the

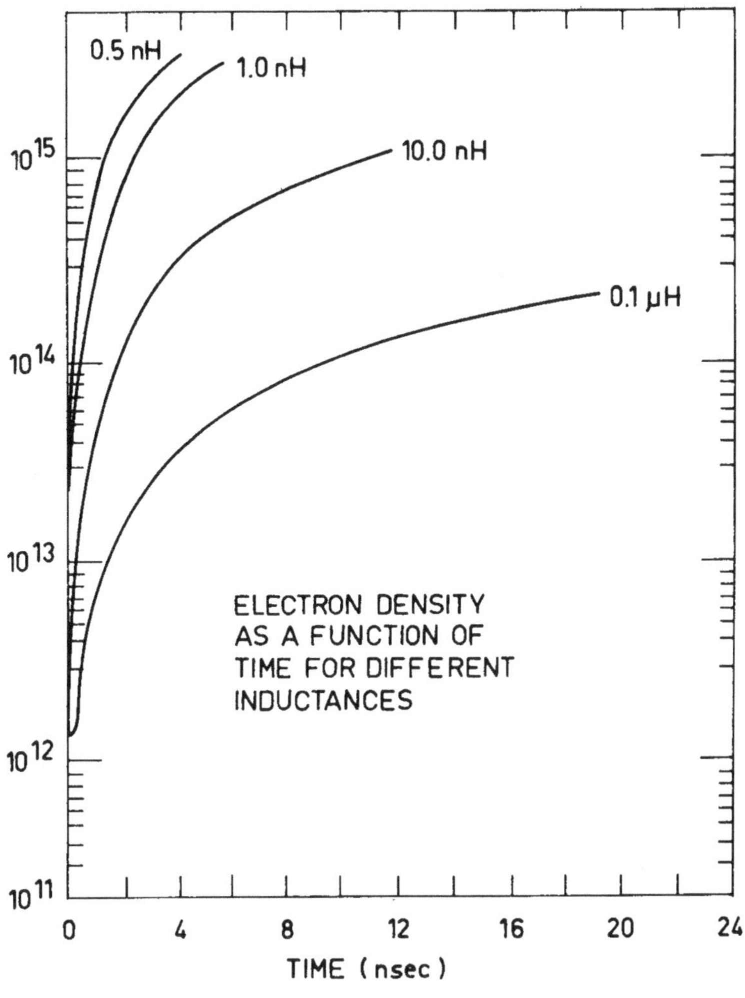

Fig.II.3 Time history of the electrondensity for different inductances the excitation circuit (After A.W. Ali) [11]

upper lifetime. However, for high electron densities ($Ne \geq 10^{14}cm^{-3}$) the rate of collisional deexcitation must be taken into account.

Therefore, the inversion time is reduced and the laser emission is reduced.

The laser power density W is essentially proportional to an excitation rate of the upper laser level [13],

$W \propto Ne \langle r_{oc} \rangle$

The mean value must be calculated from the velocity distribution of the electrons which is a function of the electron temperature Te .

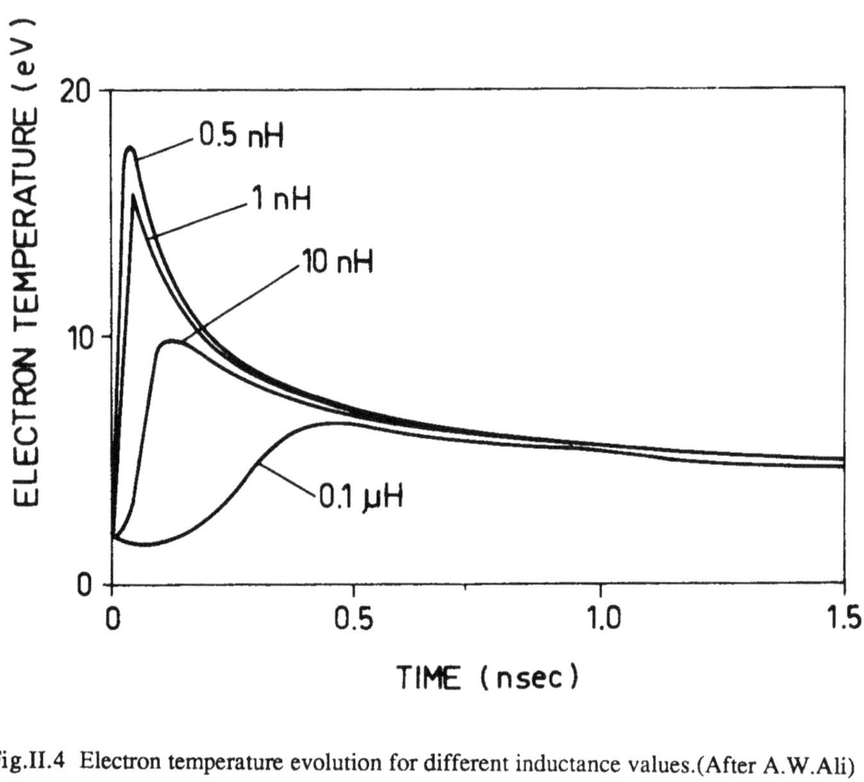

Fig.II.4 Electron temperature evolution for different inductance values.(After A.W.Ali) [11]

The electronic density is dependent on the capacitor voltage and mainly on the inductance circuit that limits the rise rate of the electron number. Fig.II 3 and Fig.II 4 show that at lower inductances faster risetimes of the current pulses and higher electrons densities are obtained. Simultaneously the electron temperature is also increased. Taking into account all these effects higher power densities are expected from lower inductance circuits.

Fig.II 5 shows the effect on the power density when the inductance is reduced. The voltage dependence can be understood because increasing the high voltages leads to more intense electrical fields and hence to higher electron velocities.

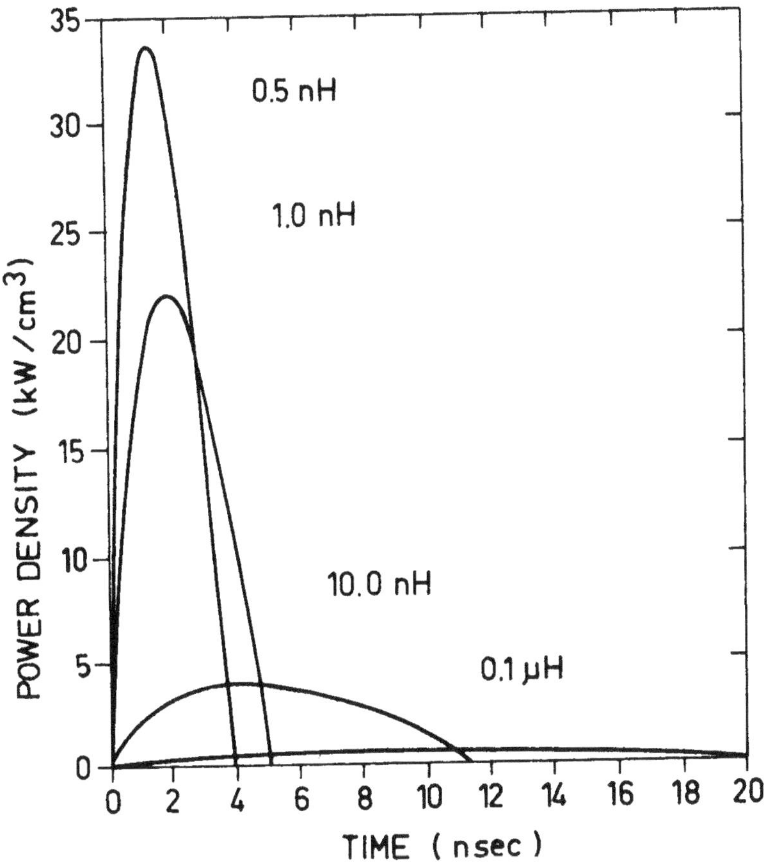

Fig.II.5 Time history of power density for different inductances. (After A.W. Ali)[11]

In addition when the electron velocities are increased the ionization rate is increased too. Therefore, high voltage and low inductance circuits are primordial for the Nitrogen lasers.

The Fig.II 6 shows the electron energy loss as a function of electron temperature for the cited levels [11].

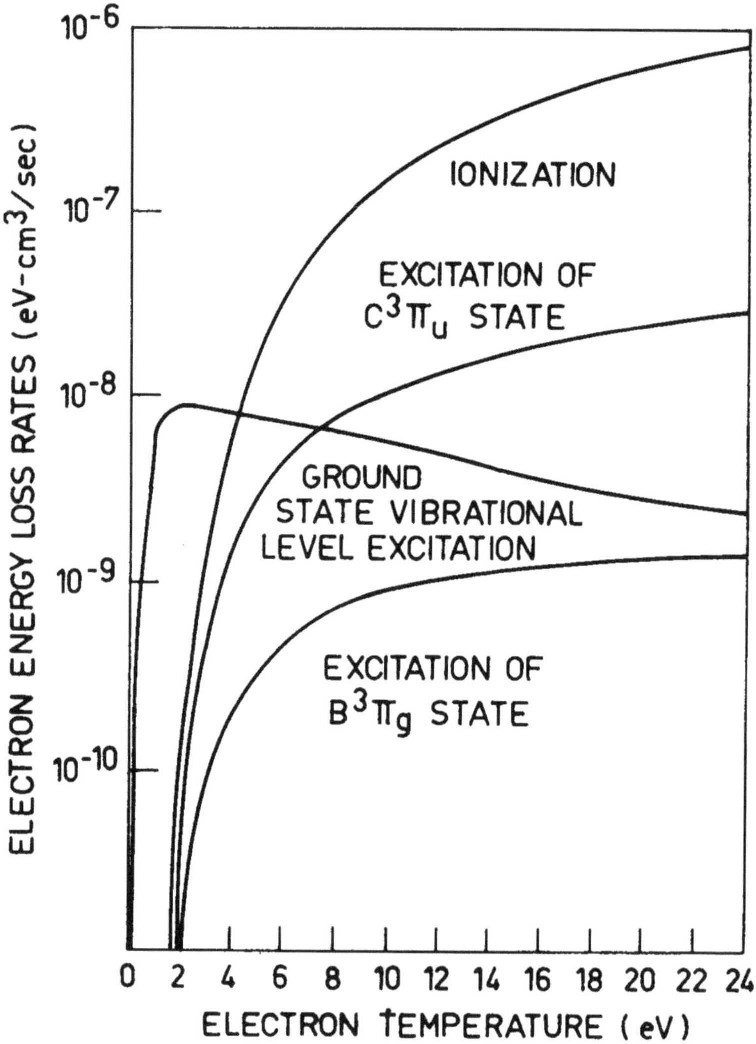

Fig.II.6 Energy deposition at the different levels of the Nitrogen molecule as function the electron temperature. (After A.W. Ali) [11]

It is evident that the rate of energy transfer from the electrons to the upper level and of the electron production by ionization depends strongly on the electron temperature. It should be noted that for electron temperatures below 4 eV the main effect is exciting vibrationally the ground state.

Consequently, in order to obtain high efficiency, electron temperatures higher than 4 eV must be attained.

In the time scale in which population inversion takes place, the electrons can be considered in steady state with the instanta-neous electric field [14] and it has been shown that the distribution function is closely a Maxwellian function when $E/P \geq 40$ V/cm.Torr where E is the applied field and P is the pressure.

Therefore, it is possible to calculate the electron temperature as a function of the E/P value, equating the ionization velocities from the Townsend ionization theory to those calculated from the kinetic model for the electron movement. The expression obtained is [14]:

$$k\,Te = 0.11\,(E/P)^{0.80}\ \mathrm{eV}$$

Consequently, the electron temperature increases with E/P; so to higher capacitor voltages correspond higher laser power density. Therefore it is desirable to reach E/P values as high as possible under the condition of obtaining arc free discharges.

Resuming, high laser power densities should be associated to very low inductance electrical circuits.

II.3 PRACTICAL CIRCUITS

Several types of excitation methods have been used to obtain laser emission in molecular Nitrogen.

These methods include pulsed high-voltage electrical discharges at reduced pressure (2 - 100 Torr) or at high pressures (5000 Torr) in pure Nitrogen or with additives and electron beam excitation.

The discharge circuits of Nitrogen lasers are characterized by a very low inductance, a very fast rise time of voltage and current pulses, and a very high peak current.

Two different high voltage pulse generators are effective, efficient and commonly used for Nitrogen lasers : the Blumlein transmission lines and capacitive transfer circuits using trans-mission lines or discrete high voltage ceramic capacitors.

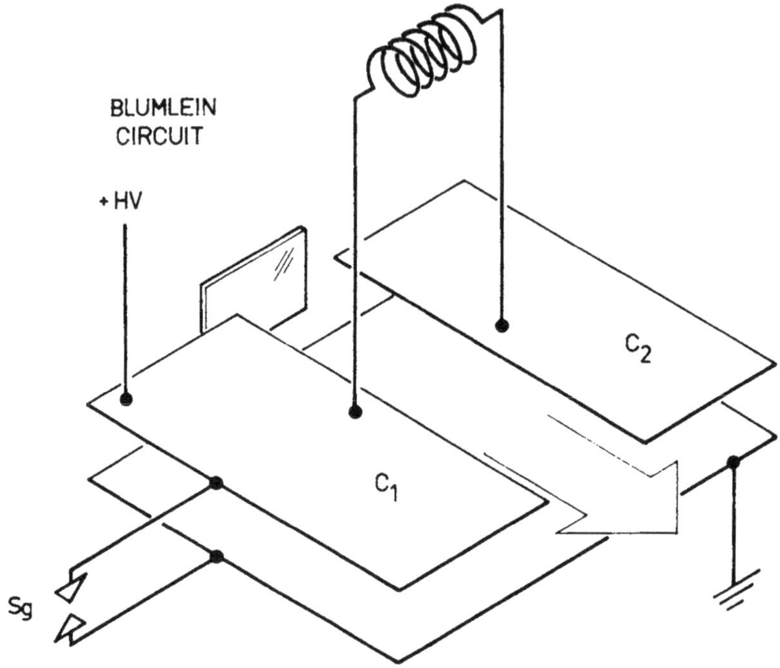

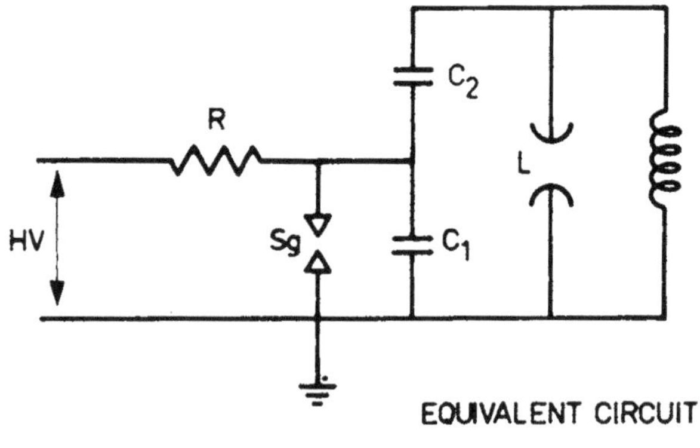

Fig.II.7 a) Schema of the Blumlein circuit.
b) Equivalent circuit.

In both circuits the output voltage is provided by the RCL oscillations of the network that start when a gaseous switch is closed.

The Blumlein circuit is composed by three flat metallic plates (Cu or Al is commonly used) insulated by dielectric sheets arranged to form two transmission lines and two capacitors (see Fig.II 7) with a common side and a gap where the discharge will be occur.

The energy storage capacitors are connected by a coil for charging them from a high voltage supply.

When the gaseosous switch (spark-gap or thyratron) is fired one capacitor is short-circuited and the voltage across the electrodes begins to oscillate rising theoretically a voltage twice respect to the power supply.

This value is not generally reached because the break-down in the gas occurs at lower voltages for the typical Nitrogen pressure range normally used for optimal emission.These circuits are more efficient for the excitation of the Nitrogen molecules because the damping effect in the resistance of the switch is less important and have a lower inductance value than those using discrete capacitors but they are voluminous and difficult to build [14].

In the CEILAP we have studied and developed Nitrogen lasers based on capacitive transfer circuits using discret ceramic high voltage capacitors whose diagram is shown in the Fig.II 8.

The operation principle is the following: the capacitor C_A is charged to the voltage V_0 through the limiting resistance R_L and the inductance L_c, which keeps null the voltage between the electrodes but does not take part in the dynamics of the circuit when the thyratron is closed ($L_c >> L_s$, see below).

When a trigger pulse closes the thyratron (Th) the stored energy in C_A is transferred to a group of N capacitors C_d connected in parallel to the electrodes, and as long as the break-down in the gas is not produced the equivalent circuit (Fig.II 8) begins to oscillate with a pulsation:

$$\omega^{-1} = (L_s C_A C_D/(C_A + C_D))^{1/2}$$

where $C_D = N\ C_d$.

At the same time the voltage V_e between the electrodes oscillates at equal frecuency until the breakdown occurs.

The voltage between the electrodes is given by :

a)

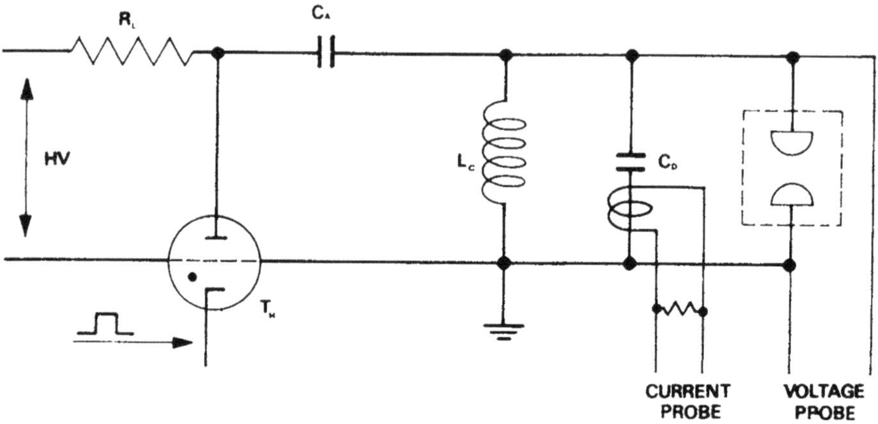

b)

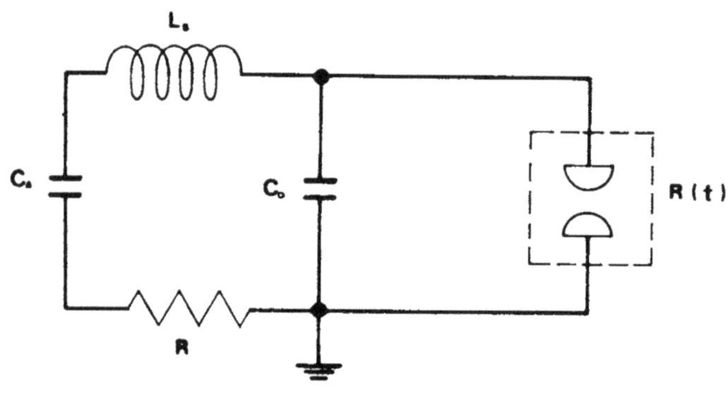

Fig.II.8 a) Diagram of the tranfer capacitive circuit build in our laboratory
b) Equivalent circuit after the thyratron is switched.

$$V_e(t) = - V_0 (C_A / (C_A + C_D)) e^{-(R/2L_sC)t} (1 - \cos \omega t)$$

where

L_s is the stray inductance of the circuit.

R is the stray and switch resistance
C is $(1/C_A + 1/C_D)^{-1}$

According the above expression the voltage V_e between the electrodes can rise 2 V_0 (C_A / C_A+C_D) but the breakdown in the gas generally, is produced at voltages slightly lowers than this value.

In order to assure good efficiency in the energy transfer and compensation of loss of energy in the stray resistance in our design we choose $C_A = 2C_D = 50\ \eta F$.

The voltage and current waveforms as measured by us are shown superposed to a laser pulse in Fig.II 9. The timeorigin is the switching of the thyratron, the current probe is

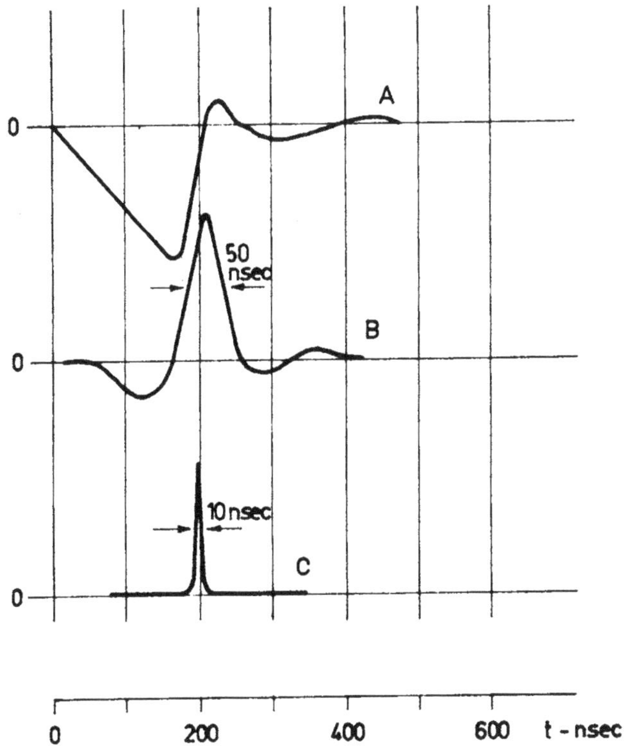

Fig.II.9 a) Voltage waveform between the electrodes.
b) Current waveform.
c) Laser pulse.
Vertical scale arbitrary. Horizontal scale 100 nsec/cm

placed on the ground side of one of the distribution capacitors and the voltage probe is on the electrodes. The time scale is 100 nsec . It can be seen that the voltage between the electrodes rises up to the voltage supply in 200 nsec and then the breakdown is produced.At this moment the potential difference falls down suddenly in 50 nsec originating the current pulse as shown.As is predicted by the theory the laser pulse appears in the rise of current pulse.

The laser tube is made from an acrylic block, Fig.II 10, 660 mm long and a cross-section of (60 x 40) mm^2 laterally emptied to form a rectangular cavity of 50 mm x 550 mm . Two centered holes are drilled in the front and the rear of the block to mount the mirrors. The electrodes were carved with a Chang type profile (33 mm width, 500 mm long) in the middle of an aluminium plate with the same lateral dimensions of the acrylic block. These plates close laterally the laser tube with an O'ring seal (see Fig.II 10). The discharge gap is 25 mm and his width is roughly 10 mm . A dielectric coated mirror and a quartz substrate forming a resonant cavity close the ends allowing the evacuation of the tube and assuring minimum vacuum losses.

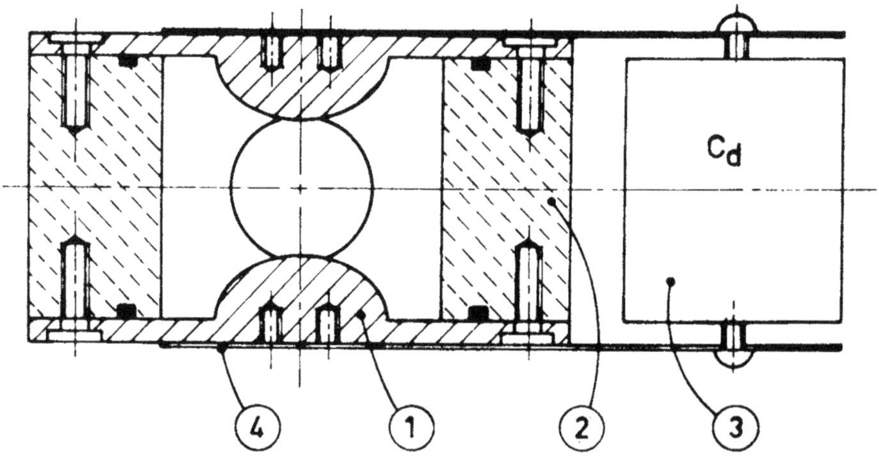

Fig.II 10 Transversal view of the laser head (scale 1:1)
1.- Al electrodes (gap 25 mm) Chang type
2.- Lucite box
3.- Distribution capacitors C_d
4.- Copper sheets

With this device pulses of 6 - 8 nsec (FWHM) length are obtained with 6 mJ of energy per pulse at a repetition of 4 pps. The laser can be operated with gas flowing at pressures of 50-120 Torr voltages up to 20 Kv and at repetition rates up to 15 pps.

The beam is rectangular (25 x 10) mm very homogeneous and with a very low divergence.

III. PULSED LASER PUMPED DYE LASERS

A general overview of the excitation and deexcitation mechanisms in a dye medium and the equations describing these processes are found in a separated chapter of this book(Dye lasers. H Ranea-Sandoval), and we will not deal with them in this section.

III.1 CAVITY CONFIGURATION

In laser-pumped dye lasers, the laser medium can be excited in either a longitudinal or transverse pumping arrangement. Which of these pumping schemes is more appropriate depends mainly on the beam profile of the pump laser. For pump sources with an extended beam profile of nearly rectangular cross section such as N_2 and excimer lasers, the transverse pumping scheme is more adequate.In this scheme the laser light is focused by a cylindrical lens (or cylindrical and spherical combination) through a side window into the cell which contains the dye solution. The focusing and dye concentration are adjusted so that the excited region of the dye solution forms a pencil-shaped volume whith nearly circular cross section on the inner cell wall of about 0.15 mm width for a typically 10 mm cell length. This scheme also is used for pumping with copper vapour and frequency doubled Neodymium lasers.

In laser-pumped dye lasers the number of round-trips of the light across the cavity is very limited during the short excitation time of 5-20 ns; therefore the linewidth of the laser radiation will not be much smaller than the bandwidth of any spectral filter inserted into the laser cavity. Since an unsaturated single-pass gain of 30 dB over a pathlength of a few millimeters is easily achieved with dye media pumped by powerful pulsed lasers, thus very narrow band wavelength selectors can be tolerated in the laser cavity despite their losses, as high-order diffraction gratings and high finesse Fabry-Perot etalons.

The use of a grating in autocollimation (Littrow mount) as cavity end reflector, provides both a mean of tuning the wavelength and narrowing the linewidth; but a linewidth bellow a few Angstroms is difficult to obtain due to the small diameter of the laser beam covering only a small area of the grating. This difficulty is overcome introducing some means of beam expansion inside the cavity. Telescopes and near grazing incidence prisms and gratings are used with this objective.

According to these ideas, different approaches have been reported to produce tunable, narrow linewidth emission from laser-pumped pulsed dye lasers. The more important ones are displayed in Fig.III 1- III 4.

A dye laser design which has been successfully used for many years was reported by T.W. Hansch [22], Fig.III 1. A high optical quality inverted telescope is placed inside the cavity to ensure collimation of the light striking the grating and producing a two-dimension beam expansion. The cavity is closed by a high order echelle grating (316 lines/mm, 7th order) blazed at 63° as selective reflector and for high gain dyes, such as

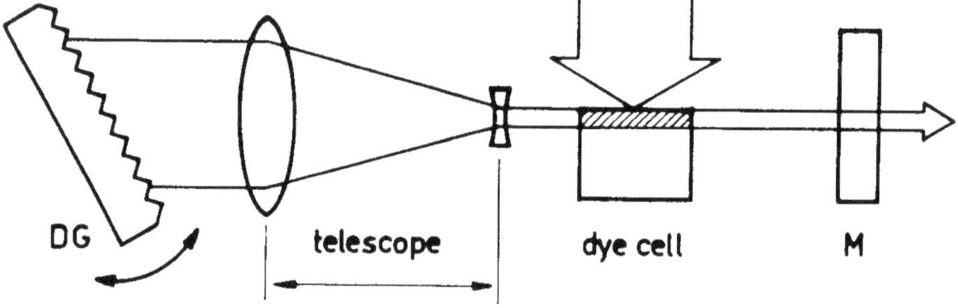

Fig.III.1: Hansch type dye laser cavity. DG: echelle diffraction grating, M: outcoupling mirror (optical flat wedge).

some rodamines and coumarines, an uncoated quartz surface with 4% reflection is sufficient as outcoupling mirror. Linewidths of 0.08-0.1 cm^{-1} with energy efficiency up to 10% in rodamine 6G can be obtained by a careful alignment of the telescope. The necessary cavity length is about 40 cm for a 20x magnification telescope. This length approaches the coherence length of the emitted laser light, then not discrete axial modes structure is expected. This allows smooth continuous tuning of the laser wavelength without tracking of the laser cavity length.

In some cases the use of shorter cavities is important, by example when very short pump pulses are employed or in applications where a single axial cavity mode is required. Folded mirror [23] and Cassegrain telescope [24] beam expanders have been reported to reduce cavity length (20-47 cm). Shorter cavities and higher magnifications are possible by near grazing incidence on prisms and gratings. Additionally only one-dimension beam expansion is produced.

The use of near grazing incidence prisms have the advantages of simplicity and compactness (10-15 cm cavity). Prisms-expander cavities generally take one of two forms, Fig.III 2: in one configuration [25,26] a 100% reflecting mirror forms one end of

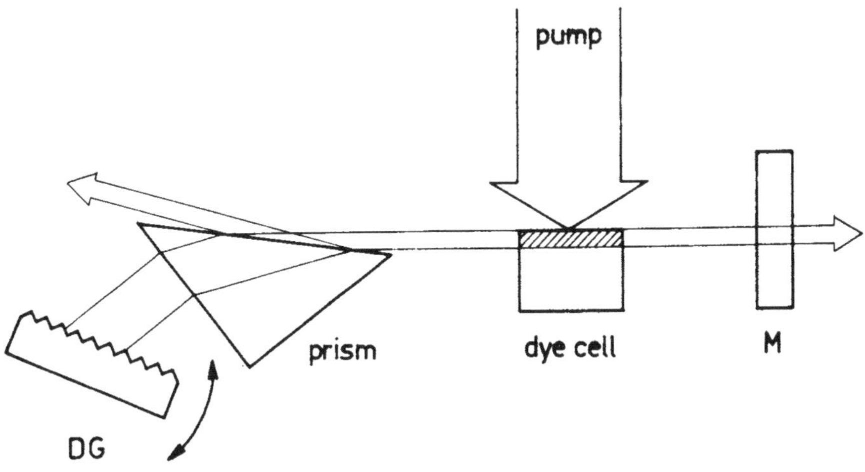

Fig.III.2: Prism type beam-expander dye laser cavity. DG: echelle diffraction grating, M: partially or full reflecting mirror.

the cavity and the laser output is obtained from the reflective losses at the incidence face of the prims. For this arrangement spectral linewidths of 0.1-0.2 cm^{-1} have been achieved at energy conversion efficiencies up to 8% in rodamine 6G pumped with a nitrogen laser [26]. It is found, however, that the output of such cavities includes a large component of non-tunable super-radiant background (one passage amplification of spontaneous emission: ASE). The effect is greatly diminished when the high reflecting mirror is replaced by a coupling mirror from which the output beam is taken [27], but in this case laser power is substantially reduced. This configuration was improved using a double-prism-expander as shown schematically in Fig.III 3. The double prism approach enables to obtain the same degree of beam expansion as in single-prism cavity, while minimising the reflection losses and overcoming the dispersive effect of the prism with an adecuated achromatic design. Linewidths of 0.03 cm^{-1} with efficiencies ≥ 10% are possible [28]. Multi-prism achromatic-expanders have been studied exhaustively [29,30]; a design of four prisms at low angle of incidence approaching Brewster angle, configured in achromatic pairs has been used in commercial dye lasers. Their performance exceeds that of conventional Hansch design [31].

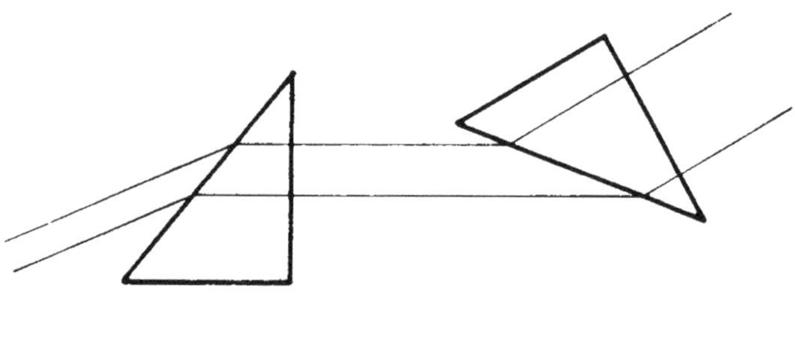

Fig.III.3: Achromatic pair prism-expander.

Another simple and popular now-a-day used solution, was introduced independently by Shoshan et al. [32] and by Littman and Metcalf [33]. In this new design a diffraction grating is illuminated across its length-perpendicular to the grating grooves with the incident unexpanded dye laser beam by selection of an incidence angle near 90 (grazing angle) as shown in Fig.III 4. The cavity resonator is completed by a tuning mirror that retroreflects the diffracted beam back towards the grating where it is diffracted again and returned to the dye cell. As in the prism-expansor two output coupling configurations are possible: if the mirror M_1 is a full reflector the laser output is taken from the zero order reflection in the grating [32,39]; and if M_1 is partially reflecting, often a quartz window as in the Hansch design, then this is used for output coupling [33]. Typically 1800 or 2400 grooves/mm blazed holographic gratings are used with a cavity length of 10-15 cm. Both configurations achieve narrow band operation, with linewidth ≤ 0.08 cm^{-1} and energy efficiency of 3% in rodamine 6G dye at grazing angle of 89° -89.5°. This scheme is also very suitable for longitudinal pumping with near Gaussian beam pump lasers. Pumping with the second harmonic of a TEMoo Nd:YAG and using a short cavity (5 cm), a stable single longitudinal mode dye laser was constructed [34]. An improved design with scanning over a range larger than 30 cm^{-1} without mode hopping is available in a commercial model [35].

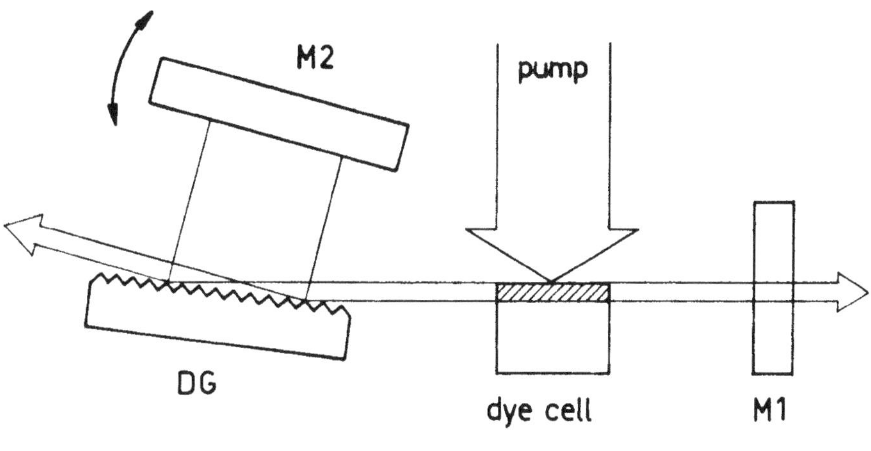

Fig.III.4: Grazing incidence type dye laser cavity. DG: first order blazed diffraction grating, M_1: full reflector tuning mirror, M_2: partially or totally reflecting mirror.

Generally further linewidth reduction and additional fine tuning in the above (transverse pumped) described designs can be obtained introducing an intracavity tilted Fabry-Perot etalon in the expanded collimated light field [22,36,40], but laser power and energy conversion efficiency are reduced too. In grazing incidence grating dye lasers, with transverse pumping, a profitable solution is to replace the tuning mirror M_2 of Fig. 4 by an echelle grating in Littrow mount [37-39]. Two posible orientations of the grating satisfy Littrow condition but for improved operation the gratings dispersion must be added. Single axial mode emission with time-averaged linewidths of 0.02 cm^{-1} in a nitrogen laser pumped 15 cm cavity, was reported by [39]. Additionally, single mode dye laser was obtained in a rather more complicated grazing incidence ring geometry [41] and with a grazing incidence folded cavity [55].Table I resumes comparative data for various dye laser cavities.

TABLE III.I Comparative data for various dye laser cavities. Mx: beam-expander magnification, LG: Littrow-grating and M: mirror

Beam Expander Type	Mx	Tuning Element	Etalon Finesse	Linewidth (cm^{-1})	Reference
------	---	LG	---	1~10	----
telescope	20	LG	---	0.08-0.1	[22]
telescope	20	LG	20	0.01	[22]
prism	40	LG	---	0.1-0.2	[26]
prism	40	LG	10	0.01	[26]
two-prisms	60	LG	---	0.03	[28]
four-prisms	40	LG	---	0.1	[31]
grating	50-60	M	---	0.1-0.05	[32,33]
grating	50-60	LG	---	0.02-0.025	[38,39]
grating	20	M	---	0.023	[55]
grating	28	M	7	0.012	[40]

III.2 WAVELENGTH SCAN

Linear wavelength scan in pulsed dye lasers, over the full range covered by a laser dye (~50 nm), is carried out with a mechanism similar to the one used in optical grating spectrometers. This mechanism known as "sine-bar" drive is shown schematically in Fig.III 5a. Briefly one end of the sine-bar can slide over a flat surface FS which is fixed on a drive nut DN. A stepper motor rotates a precision screw PS moving the drive nut. The other extreme of the sine-bar is fixed to a pivot point P (precision rotation stage)

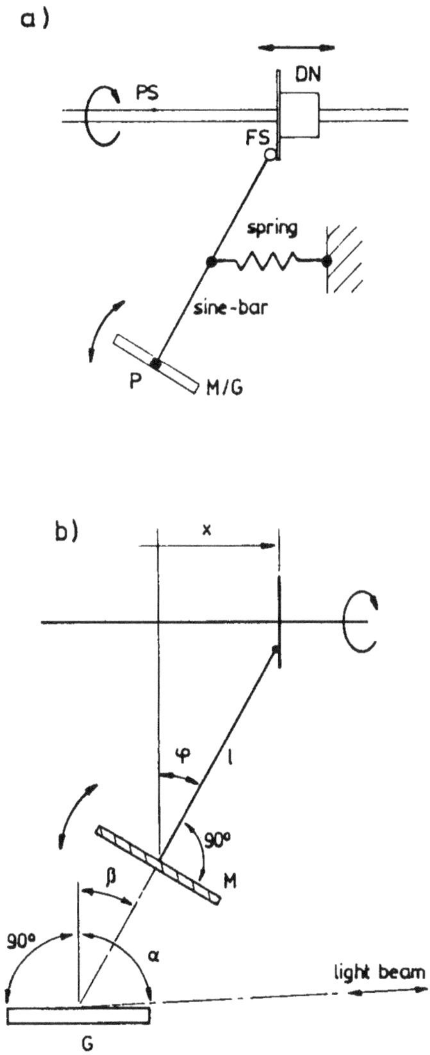

Fig.III 5: Sine-bar drive mechanism for linear wavelength tuning.
a) Scheme; PS: precision screw, DN: drive nut, FS: flat surface, P: pivot point and M/G: mirror or grating.
b) Sine-bar drive geometry in a grazing incidence cavity. M: tuning mirror, G: grating.

where the laser tuning mirror or grating is mounted. The Fig.III 5b shows an idealized geometric diagram for a grazing incidence grating configuration. The grating G is supposed parallel to the precision screw and the sine-bar is perpendicular to the mirror M. If the grazing angle α = constant, then from the grating equation,

$$d\lambda = [\, d/m \,] \cos \beta \, d\beta. \qquad (1)$$

where d is the grating groove spacing and m the diffracted order. From the geometric diagram,

$$dx = l.\cos \varphi \, d\varphi \qquad (2)$$

been l the length of the sine-bar and $\varphi = \beta$; from (1)and (2) results:

$$d\lambda/dx = [\, d/l.m \,] = \text{const.} \qquad (3)$$

Expression (3) shows that linear scanning of laser wavelength is obtained by rotation of the precision screw at constant velocity.

In practice, errors in construction and alignment are unavoidable and must be considered in a precise linear scanning mechanism. Actually $\beta = \varphi + \Delta$, where Δ is composed of two terms: first some small angle ε, which exists between the grazing incidence grating G and the precision screw; and second some angle δ between the sine-bar and the light beam striking the mirror M. Nevertheless Δ can be suppressed approaching $\delta = -\varepsilon$. The lack of perpendicularity between the precision screw and the flat surface mounted on the drive nut is another source of nonlinearity. If γ is the deviation of perpendicularity, it can be demonstrated that,

$$d\varphi /dx = 1/\, l(\cos\varphi + \tan\gamma.\sin\varphi) \qquad (4)$$

and let be $\Delta = \varepsilon + \delta$ any residual error, then:

$$d\lambda /dx = [d/l.m] \cos(\varphi + \Delta)/(\cos\varphi + \tan\gamma .\sin\varphi)$$

$$\approx [d/l.m] \, [1 - (\Delta + \gamma)\tan\varphi] \qquad (5)$$

Equation (5) gives the first order corrections to equation (3), and shows that sine-bar drives are actually tangent drives. In the Hansch type cavity, equation (5) also holds with $\varphi = 2\beta$. In this case nonlinear terms becomes important owing to high φ values, since $\alpha = \beta \approx 63$ in a typical echelle grating. A different sine-bar design, for this particular case, was used in [42].

Finer and more precise wavelength tuning is achieved by pressure tuning [43]. This is accomplished enclosing the laser grating and air spaced etalons in a pressure chamber where gas and pressure can be changed. The grating (and etalons) selects a fixed wavelength inside the gas medium, dependent only on its geometry. If the light wavelegth λ remains constant in a medium of variable index of refraction n, the corresponding vacuum wavelength changes linearly with n,

$$\lambda_{vac} = \lambda \, . \, n \qquad (6)$$

The refractive index of a gas depends on its density o according to dispersion relation (Gladstone and Dale law)[44]:

$$(n^2 - 1)/(n^2 + 2) = \rho . \text{const.} \qquad (7)$$

Since gas density is proportinal to gas pressure and n is usually very close to 1, eq.(7) becomes, to a good approximation,

$$n - 1 = p \, . \, \text{const.} \qquad (8)$$

and a pressure change Δp will result in a proportional change of vacuum wavelength,

$$\Delta\lambda_{vac} = \kappa \, \Delta n = \lambda \, \Delta p \, . \, \text{const.} \qquad (9)$$

Also the light frequency $\nu = c/\lambda_{vac}$changesapproximately linearly with pressure,

$$\Delta\nu \approx -(c/\lambda) \, \Delta p \, . \, \text{const.} \qquad (10)$$

Scanning ranges of about 0.15 nm/atm in nitrogen and 0.25 nm/atm in CO_2 can be obtained at the yellow wavelengths.

III.3 DYE AMPLIFIERS

In a high gain medium as dyes, with strong pumping conditions (N_2, excimer and Nd:YAG), a single passage Amplification of Spontaneous Emision (ASE) can produce an excessive non-tuned background noise on the dye laser output. Increasing pumping rate broadens the spectrum of emitted fluorescence and increases the ASE noise. Operation of dye solutions in this ASE mode has been reported by [45,46]. The effect of ASE must also be consideredin dye oscillator-amplifier systems, where ASE can interfere with the amplification of the real laser radiation.

A good laser system with high spectral purity is constructed distributing pumping power among a laser oscillator and one or two amplifier stages to increase substatially the laser output and reducing ASE background [47-49]. This considerations show the importance of studying ASE and signal amplification in a dye medium, as much for constructing dye systems in the ns range as for generation and amplification of ultra-short (ps or less) pulses.

Laser pumped dye amplifiers have been extensively studied in the past years [50-52]. Analytical solutions of a dye amplifier equations have been obtained under restrictive conditions in some special cases. General quantitative results (ASE and pulse evolution) are possible only by numerical resolution of the integro-differential equations set [53,56].

In a recent paper Hnilo et al. [54] verify a formula which allows the calculation of the small signal gain of a dye laser amplifier from single measurements of the ASE. In the case of transverse pumping, this formula indicates that (besides the dye parameters) the gain is determined by only two variables: the ratio between length L and radius r of the active dye volume (assumed cylindrical) and the pump rate. This makes it possible to draw a set of curves giving the unsaturated gain as a function of pumping rate for any possible combination of the geometrical parameters. On the other hand, when the amplifier is saturated the value of interest is the stored energy in the medium. A second set of curves of the available energy E_{av} vs pumping energy E_p can be derived from the former, where the available energy is understood as the difference between the stored energy and the energy which rests in equilibrium in the medium after the passage of the amplified pulse.

The Fig.III 6 shows a typical plot for a transverse second harmonic Nd:YAG pumped dye amplifier. The straight (dotted) line gives the energy efficiency asymptote which depends on the ratio between amplified to pump frequencies, the slope of pumping pulse and the decay time of the dye. Fixing E_p and keeping constant one of the geometrical parameters (L or r) an optimum working point can be searched. It is important to take into account that the ASE increases quick above the elbow of the curves, then the working point of the amplifier must be placed as close as possible to the maximun efficiency asymptote.

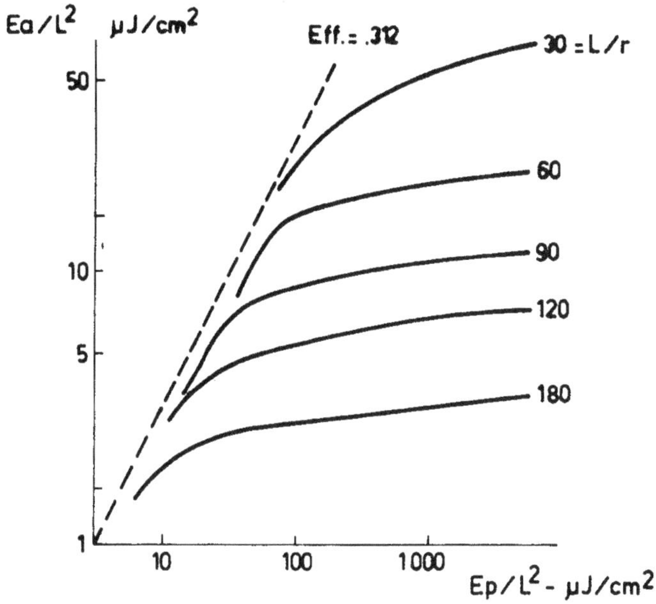

Fig.III.6: Available energy (E_a) in a dye laser amplifier as a function of pumping energy (E_p) for different values of the parameter L/r. From reference [54].

When the output pulse energy of the dye oscillator, as generally happens, is not sufficient to saturate the amplifier a chain of two or more amplifiers may be designed. Some examples applied to the design of chains of utra-short pulse dye laser amplifiers were given in [54].

IV REFERENCES

[1] L.E.S. Mathias and J.T. Parker, Appl. Phys. Lett. 3 (1963) 16.

[2] H.G. Heard, Nature 200 (1963) 667.

[3] D.A. Leonard, Appl. Phys. Lett. 3 (1965) 16.

[4] V.M. Kashlin and G.G. Petrash, Pulsed Gas-Discharge Atomic and Molecular Lasers.Proc.P. N. Lebedev Physics Institute Vol 81 , N.G.Basov Ed. Moscow (1976).

[5] P.Ritchter,J.D. Kimel and G.C. Moulton, Appl.Optics 15 (1976) 756. [6] J.D. Shipman and A.C. Kolb, IEEE J. Quantum Elect. QE-2 (1967) 298.

[7] H.M.von Bergman, V. Hasson and D. Preussler,J. Phys. E Scientific Instruments 9 (1976) 73.

[8] C.A. Massone, M. Garavaglia, M. Gallardo, J.A.E. Calatroni and A.A. Tagliaferri, Appl. Optics 11 (1972) 1317.

[9] Colin S. Willet, Introduction to Gas Lasers:Population Inversion Mechanisms Pergamon Press (1972).

[10] B. Godart, IEEE J. Quantum Elect. QE-10 (1974) 147.

[11] A.W. Ali A.C. Kolb and A.D. Anderson, Appl.Optics6(1967) 2115.

[12] E.T. Gerry, Appl. Phys. Lett. 7 (1965) 6.

[13] A.W.Ali, Appl. Optics 6 (1969) 5.

[14] W.A. Fitzsimmons, L.W. Anderson, C.E. Riedhauser and J.M. Vrtilek, IEEE J. Quamtun Elect. QE-12 (1976) 624.

[15] A.Dias Tavares, C.E. Fellows and C.A. Massone, Appl.Phys.B 38 (1985) 259.

[16] A.J. Schwab and F.W. Hollinger, IEEE J. Quantum Elect. QE-12 (1976) 183.

[17] N.Subhash, Sudha C. Kartha, and K. Sathianandan, Appl. Optics 22 (1983) 3612.

[18] S. Mont_n and S. Svanberg, Appl.Phys B 38 (1985) 241.

[19] Toru Mizunami and Keiji Takagi, IEEE J.Quamtum Elect. QE-22 (1986) 227.

[20] Roland Roux, Laser Focus pag.51-52 January 1986.

[21] H.Strohwaldand H.Salzmann, Appl. Phys. Lett. 28 (1976) 272.

[22] T.W. Hansch, Appl. Optics 11(1972)895.

[23] R. Koning, G. Minkwitz and B. Christov, Ootics Comm.32(1980)301.

[24] E.J. Beiting, Optics Comm. 29(1979)209.

[25] S.A. Myers, Optics Comm. 4(1971)187.

[26] D.C. Hanna, P.A. Karkkainen and R. Wyatt, Optical & Quantum Electronics 7(1975)115.

[27] E.D. Stokes et al., Optics Comm. 5(1972)267.
[28] F.J. Duarte and J.A. Piper, Optics Comm. 35(1980)100.
[29] J.R.M. Barr, Optics Comm. 51(1984)41.
[30] F.J. Duarte, Optics Comm. 53(1985)259.
[31] G.K. Klauminzer, IEEE J. Quan. Elec. QE-13(1977)920.
[32] I. Shoshan, N.N. Danon and U.P. Oppenheim, J. Appl. Physics48(1977)4495.
[33] M.G. Littman and H.J. Metcalf, Appl. Optics 17(1978)2224.
[34] M.G. Littman, Appl. Optics 23(1984)4465.
[35] M.G. Littman and J. Montgoomery, Laser Focus: February 1988, pag.70.
[36] T. Chang and F.I. Li, Appl. Optics 19(1980)3651.
[37] I. Shoshan and U.P. Oppenheim, Optics Comm. 25(1978)375.
[38] M.G. Littman, Optics Lett. 3(1978)138.
[39] M.K. Iles, A.P. D'Silva and V.A. Fassel, Optics Comm. 35(1980)133.
[40] S. Mory, A. Rosenfeld, S. Polze and G. Korn, Optics Comm. 36(1981)342.
[41] S.G. Dinev, I.G. Koprinkov, K.V. Stamenov and K.A. Stankov, Optics Comm. 35(1980)403.
[42] T.D. Rymond, S.T. Walsh and J.W. Keto, Appl. Optics 23(1984)2062.
[43] R. Wallenstein and T.W. Hansch, Appl. Optics 13(1974)1625.
[44] R.S.Longhurst: Geometrical and Physical Optics,Longmans Group(1973) U.K.
[45] M.E. Mack, Appl. Phys. Lett. 15(1969)166.
[46] H.P. Grieneisen, R.E. Francke and A. Lago, Appl. Phys.15(1978)281.
[47] I. Itzcan and F.W. Cunningham, IEEE J. Quan. Elec. QE-8 (1972)101.
[48] R. Wallenstein and T.W. Hansch, Optics Comm. 14(1975)353.
[49] J.E. Lawler, W.A. Fitzsimmons and L.W. Anderson, Appl. Optics 15(1976)1083.
[50] U. Ganiel, A. Hardy, G. Neumann and D. Treves, IEEE J. Quan. Elec. QE-11(1975)881.
[51] G. Dujardin and P. Elamant, Optics Comm. 24(1978)243.
[52] A.A. Hnilo, O.E. Martinez and E.J. Quel, IEEE J. Quan. Elec.QE-22(1986)20.
[53] A.A. Hnilo, F.A. Manzano and O.E. Martinez, J. Opt. Soc. Am. B4(1987)629.
[54] A.A. Hnilo and O.E. Martinez, IEEE J. Quan. Elec. QE-23 (1987)593.
[55] A. Lago, G. Wochl Jr. and R. Riva, Appl. Phys. B49(1989)73.
[56] L.G. Nair and K. Dasgupta, IEEE J. Quan. Elec. QE-21 (1985)1782.

CO2 LASERS : PHYSICS, TECHNOLOGY AND APPLICATIONS

E. J. Quel, C. A. Rosito, and V. B. Slezak
CEILAP (CITEFA-CONICET)
Zufriategui y Varela, Villa Martelli
Buenos Aires, Argentina

Outline

1. **Introduction**

2. **Historical review**

3. **The CO_2 molecule: its vibration-rotational spectrum**

4. **Different types of CO_2 lasers**
4.1. Electrical excited lasers
4.2. Gas dynamic lasers
4.3. Transfer chemical lasers.
4.4. Optically pumped CO_2 lasers

5. **Laser gain and linewidth.**

6. **Mechanisms of excitation of CO_2 lasers**

7. **Optical generation and extraction in homogeneously broadened lasers**

8. **Electrically excited diffusion cooled CO_2 C.W. lasers.**
8.1. General description
8.2. Cooling mechanism
8.3. Electrical excitation
8.4. Power extraction

9. **Description of diffusion cooled devices**
9.1. Conventional slow flow DC excited lasers
9.2. Sealed-off lasers
9.3. Waveguide lasers
9.4. Second generation diffusion-cooled lasers

10. **Electrically excited convection cooled CO_2 lasers**

1. Introduction

The main scope of this work is to provide a general description of the CO_2 laser and its applications. Special emphasis is on the the electrically excited types which are the most widely used.

The development of this laser has grown enormously from the early beginnings in 1964. Starting from longitudinal DC excited low pressure devices, many variations using pulsed or RF excitation, sealed-off or fast flow systems have been developed and have found wide diffusion in industrial, medical, military and research environments.

After a short historical review of the most important results on CO_2 lasers from 1964, this work goes on with a description of the CO_2 molecule and the active medium. Different types of CO_2 lasers are presented and classified by the excitation methods. Power generation and extraction are analyzed for homogeneous broadening. Electrically excited CO_2 lasers are distinguished by the cooling mechanisms, namely diffusion or convection.

Next, high pressure pulsed lasers are described with their different characteristics. The last section contains an overview of the main applications of CO_2 lasers.

2. Historical review

The first laser emission from CO_2 at 10.6 μm was obtained by C.K.N. Patel et al. [1] in 1964, and later [2], [3], the continuous and pulsed emission was carefully studied. Legay and Legay-Sommaire [4] made an important contribution about the addition of N_2

to CO_2. Patel published also similar works [5], [6],[7] where he informs about the increase of the output power upon adding N_2, which is excited in the CO_2 discharge. Moeller and Ridgen [8] showed that adding He to the CO_2 the laser power was increased noticeably, and Patel et al. [9] obtained more than 100 Watts c.w. by means of the mixture of the 3 gases: CO_2 -N_2-He.

Witteman [10] showed that it is possible to construct reliable sealed-off CO_2 lasers. With this type of lasers, Freed [11] obtained a frequency stabilization around 9 KHz and 20 KHz for observation periods of 3 x 10^{-2} and 1 sec.

Q-switched techniques were applied [12] and due to the long lifetimes of the transitions, repetitive short pulses of high intensity could be obtained (10 KW/m and pulse width around tenths of μs).

Hill [13] obtained a good performance of a CO_2 laser pulsed by a fast electrical discharge. (5 μs, 100 pps, pressure 50 Torr, and peak power of 200 KW).

Another important progress was the use of a fast gas flow (300 m/s) in c.w. that allowed passing from a value of 50 W/m for low flow to several KW/m. In [14] and [15] 2300 W and 27,200 W were obtained.

Dumanchin and Rocca-Serra [16] obtained interesting results related with the increase of energy and power per unit volume in a pulsed CO_2 laser: 3 mJ/cm^3 and 240 mJ/pulse.

Atmospheric pressure was used by Beaulieu [17] obtaining 2J/pulse and 17% efficiency from 1 m active length and La Flamme [18] used later the preionization which favours the uniformity and stability of the main discharge.

Basov and Oraevskii [19] suggested the possibility of using the dependence of relaxation times on temperature in order to achieve population inversion by means of a brusque change of temperature. Gerry [20] obtained a power of 60 KW multimode and 30 KW monomode in a gas dynamic laser. In this type of laser instable cavities are used [21] due to the possibility of a big active volume for the fundamental mode.

Continuous laser emission at 10.6 um from a CO_2 molecule was obtained by means of chemical pumping initiated in HCl-CO_2 and HBr-CO_2 and HF-CO_2 [22].

Moreover, by means of emission from a HBr TEA laser at 4.23 μm that is strongly absorbed by the CO_2 R(20) line, Chang and Wood [23] obtained optical pumping of this molecule at high pressure, where the overlapping of levels allows for almost continuous tunability. Bridges et al. [24] showed the possibility of using dielectric waveguides in CO_2 lasers.

Some good reviews on the fundamental processes in CO_2 lasers are those by Tychinskii [25], Sovolev and Sokovikov [26], Cheo [27], DeMaria [28] and Willett [29]. High power c.w. lasers have been reviewed by Abilsiitov et al [30] and pulsed lasers by Wood [31]. The books by Duley [32] and Witteman [33] cover the subject of CO_2 lasers and applications.

3.The CO_2 molecule: its vibration-rotational spectrum

The CO_2 molecule is linear symmetric with an axis of symmetry along the nuclei and a plane of symmetry perpendicular to this axis. It has thus only two degrees of rotational freedom, and four vibrational degrees of freedom as shown in fig. 1.
Vibration ν_1 , is the symmetric stretch mode; ν_2 .is a bending two-fold degenerate vibration since oscillations can take place in two different planes containing the molecular axis; ν_3 is the asymmetric stretch mode. The wave numbers associated with the four vibrations are [27]:

$$\nu_1 = 1388.3 \text{ cm}^{-1};\ \nu_2 = 667.3 \text{ cm}^{-1};\ \nu_3 = 2349.3 \text{ cm}^{-1}.$$

If each vibration is treated separately as a harmonic oscillator, the total vibrational energy will be given by:

$$E_v = h\,c\,[\,\nu_1\,(v_1+1/2) + \nu_2\,(v_2+1) + \nu_3\,(v_3+1/2)\,] \tag{1}$$

where v_i is the occupation number of mode ν_i .

The vibrational state of the molecule can be described with three numbers representing the number of vibrational quanta in each mode, i.e.:

$$(\,v_1\,,\,v_2^{\,l}\,,\,v_3\,)$$

The superscript l associated with the bending mode is a quantum number and determines the angular momentum (in units of h) associated with the vibration and directed along the molecular axis C_∞ . It takes the values v_2, v_2-2 , ...,0 for even v_2 and v_2, v_2-2 , ...,1 for odd v_2 . Levels with $l = 0$ are not degenerate and levels with $l > 0$ are double degenerate [34].

If the anharmonicity of vibrations is taken into account the vibrational energy is given by [34]:

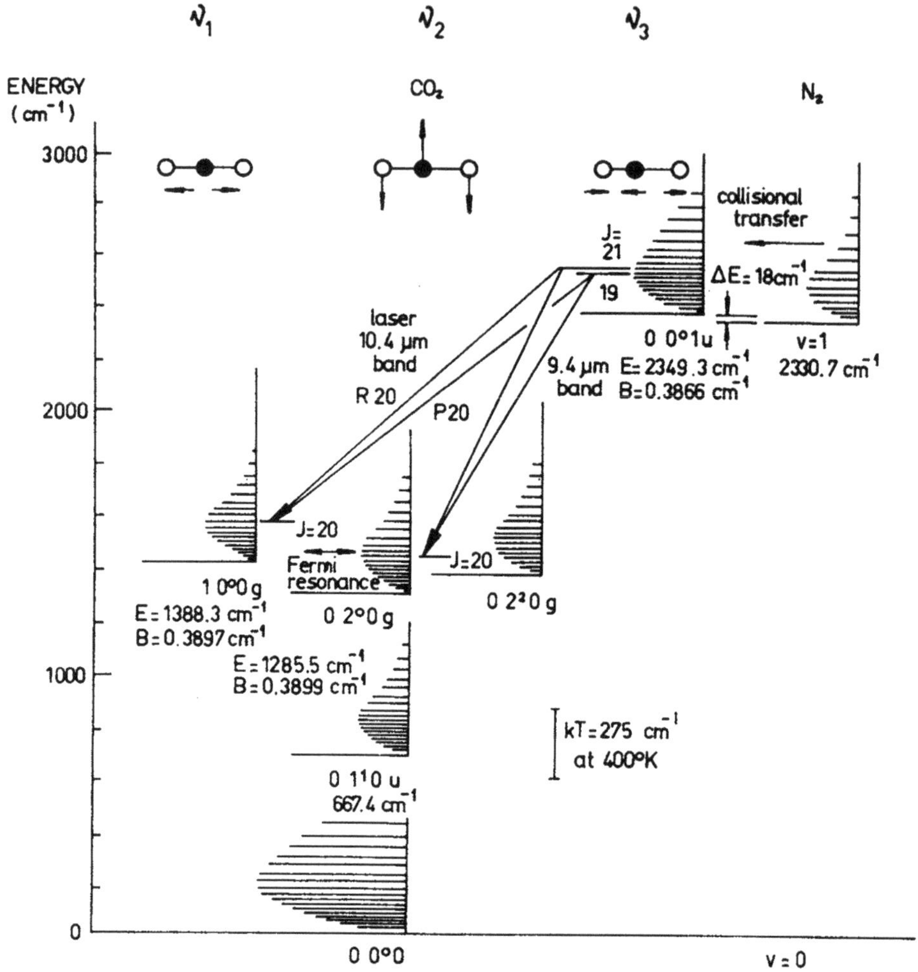

Fig. 1. Lower energy levels of CO_2 and N_2 molecules (electronic ground state). The horizontal lines are proportional to the population of each rotational level.

$$\frac{E_v}{hc} = \sum_{i=1}^{3} \omega_i \,(v_i + d_i/2) + \sum_i \sum_k x_{ik} (v_i + d_i/2) (v_k + d_k/2) + \sum_i g_{ii} \, l_i^2 \tag{2}$$

where $d_i = 1$ for nondegenerate and 2 for doubly degenerate vibrations. The double sum arises from cross coupling between different modes of vibration through the anharmonic force constants x_{ik} ; the last summation is called the l-type doubling.

The rotational energy E_r of the CO_2 molecule is given by the formula:

$$\frac{Er}{hc} = F(J) = B\,J\,(J+1) - D\,J^2\,(J+1)^2 + \ldots \tag{3}$$

where J is the rotational quantum number and B, related to the inertia moment of the molecule I through $B = h / (8 \pi^2 I c)$, is called the rotational constant and depends slightly on the vibrational state. B values wich lie in the range of 0.38-0.39 cm^{-1} are given in fig 1. The term $D\,J^2\,(J+1)$ accounts for the non rigidity of the molecule and will be neglected in what follows.

The population n_{Jv} of the rotational level J associated with the vibrational level v (v represents the three numbers v_1, v_2, v_3) is given by:

$$n_{Jv} = \frac{n_v}{Z_r(T)} (2J+1)\, e^{-Er/kT} \tag{4}$$

where n_v is the total population of the vibrational level v and $Z_r(T)$ is the rotational partition function given by:

$$Z_r(T) = \sum_J (2J+1)\, e^{-\frac{BhcJ(J+1)}{kT}} \tag{5}$$

When the rotational energy spacing is much less than kT, as in this case, the sum (5) can be approximated by an integration with the result:

$$Z_r(T) = \frac{1}{2} \frac{kT}{Bhc} \tag{6}$$

The value of J corresponding to the maximum of the distribution (4) is obtained by equating (dn_{Jv} / dJ) to zero. The result is:

$$J\,max \cong \sqrt{\frac{kT}{2hcB}} - \frac{1}{2} \cong \sqrt{T} \qquad (7)$$

which means that for laser temperatures of 300 - 400 K Jmax is in the range 17-20.

When the two 0 atoms have the same atomic weight, due to symmetry considerations only even (odd) "J" numbers appear in the rotational spectrum if the vibrational wave funtion is symmetric (asymmetric).

The infrared spectrum of CO_2 consists of transitions between vibrational-rotational levels, subject to the selection rules: $\Delta v = \pm 1$ and $\Delta J = \pm 1,0$. The transitions with $\Delta J = -1, 0, +1$ are called the P,Q and R branches respectively.

In the dipole approximation, transitions between rotational levels belonging to vibrational states of the same parity are forbidden.

The 10^00 and 02^00 levels lie relatively close (1338.3 and 1285.5 cm^{-1} respectively) and are in Fermi resonance. These two states are therefore mixed and have been more accurately labeled $(10^00, 02^00)_I$ and $(10^00, 02^00)_{II}$ respectively. They are the terminal levels of the 10.4 μm and 9.6 μm laser bands, their common upper level being the (00^01) at 2349.3 cm^{-1}. For brevity we shall keep labelling them (10^00) and (02^00).

The CO_2 laser lines are identified by the vibrational band edge and the type of rotational transition indicating the J value of the lower level. By example 10 P (20) is the transition $00^01 - 10^00$ ($J=19 \rightarrow J=20$).

Patel [1] using a laser with only CO_2 as the active medium observed 14 simultaneously oscillating lines in the P-branch of the $00^01 - 10^00$ band. When the mixture CO_2-N_2-He is used only a few laser lines oscillate at one time (P(18), P(20) and P(22)).

Other laser transitions in the range 11 to 18 μm have been reported [35], [36] but are much weaker than those oscillating in the 00^01-10^00 and 00^01-02^00 bands.

4. Different types of CO_2 lasers

In this section a brief description of different types of CO_2 lasers will be given. The devices differ widely in excitation methods, operating pressure, optical cavities and in general output characteristics.

4.1. Electrical excited lasers

These include conventional c.w. lasers using longitudinal electrical excitation and longitudinal slow gas flow, c.w. lasers using fast gas flows with electrical excitation either longitudinal or transversal to the gas flow, sealed-off lasers and pulsed transversal discharge high pressure lasers. Energy can be extracted by means of open or waveguide resonators. In some cases RF excitation is used.

Fig. 2a and 2b show elementary longitudinally DC excited devices. Other types will be shown in the corresponding sections (Sec. 8 to 11).

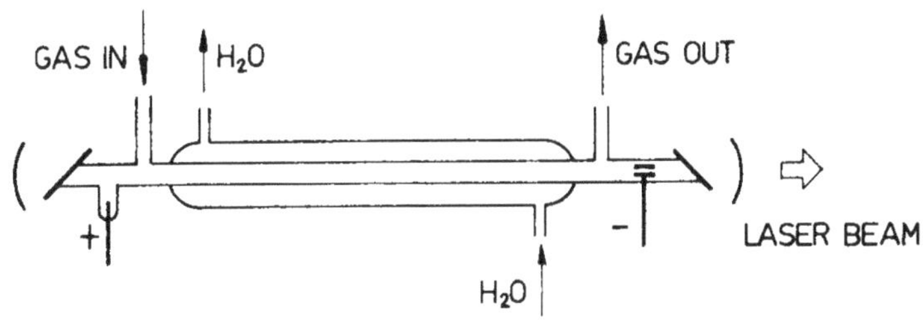

Fig. 2a. CO_2 laser with slow longitudinal flow of gases and longitudinal discharge.

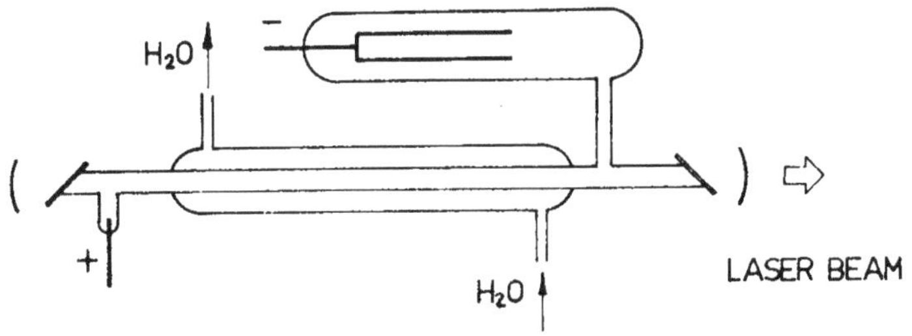

Fig. 2b. Scheme of a sealed-off laser.

4.2. Gas dynamic lasers

A preheated mixture of the working gases (CO_2, N_2, He or H_20) is delivered to a nozzle where it is accelerated and translationally cooled. The different V-T relaxation rates of molecules with different energies lead to a population inversion; then the gas enters into a cavity where energy is transformed in laser radiation. A schematic diagram is shown in fig.3.

The development of GDL is described in several reviews [37], [38]. The advantage of these lasers consists in the possibility of creating continuously operating high-power devices, without electrical excitation. Furthermore these systems offer the possibility of obtaining laser radiation using combustion products of common fuels.

The output power of GDL depends on certain parameters like mixture composition, initial conditions of pressure and temperature (P_0, T_0) and nozzle geometry.

As an example a power up to 60 KW was reported [39] on a CO_2-N_2-H_20 mixture for $T_0 = 1200$ K.

4.3. Transfer chemical lasers

In this type of lasers a high population inversion is produced by a chemical reaction which propagates after being initiated by a "trigger" [40], [41]. This process is very efficient.

The antisymmetrical stretching mode of the CO_2 molecule is excited by fast quasi-resonant v-v energy transfer from chemically pumped HCl(v) or DF(v). High rates of v-v transfer minimize the collisional deactivation of HX molecules.

The vibrational energy gap between HX and CO_2 is around 500 cm^{-1}, but this difference is compensated by the rotational excitation of HX molecules.

The first model of a chemically pumped laser without electrical excitation is the continuous subsonic DF-CO_2 laser. High power is obtained from transversal gas flow devices, which can be used also with HF and DF exclusively. Fig. 4 shows a scheme of a typical laser system.

Combining the GDL system with the TCL one, very high power is obtained in continuous emission.

It is also possible to obtain pulsed emission from a TCL, producing the chemical reaction either with flash lamps placed along a longitudinal tube when a mixture of gases enters into the system through one end, or with pulsed electrical longitudinal discharges. The latter method is the most efficient one.

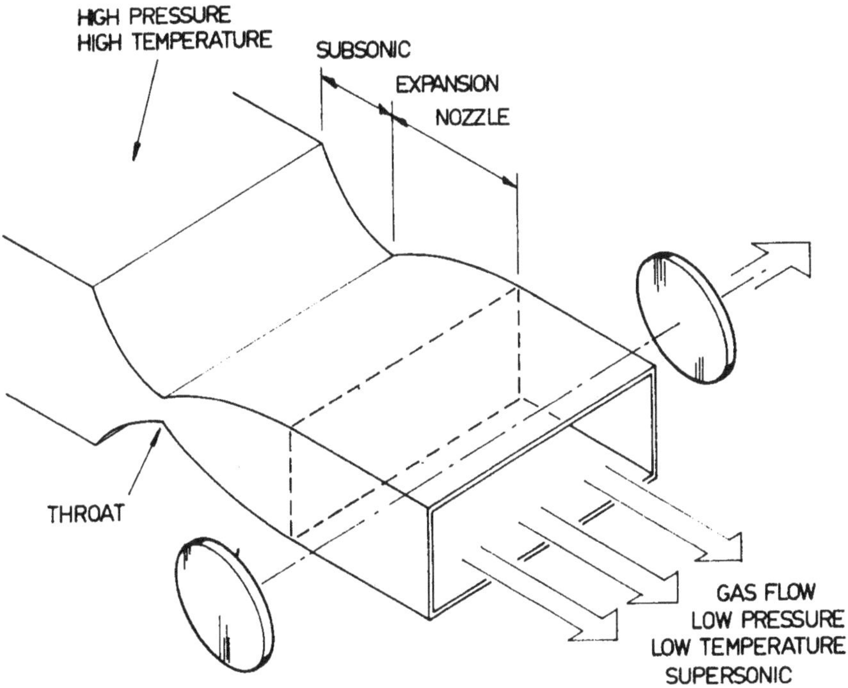

Fig. 3. Scheme of a gas dynamic laser.

4.4. Optically pumped CO_2 lasers

The CO_2 laser can be also optically pumped. This has been achieved exciting by blackbody radiation [42], [43] and by the 4.23 μm line of a HBr laser [29].

In [43] for the first time a c.w. CO_2 laser pumped by a blackbody cavity was reported. Fig. 5a shows schematically a blackbody cavity surrounding the laser tube. The configuration shown permits the thermalization of the radiation transmitted by the laser

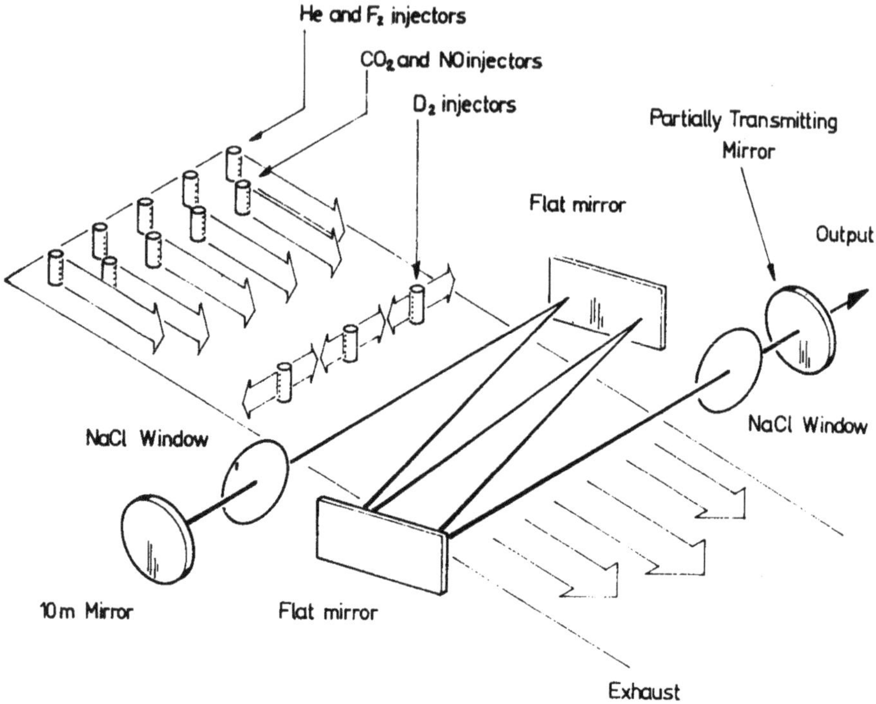

Fig. 4. Scheme of a transverse flow chemically excited laser.

with a hole in the spectral energy distribution, when reemission by the walls occurs. In this way all of the energy source is utilized.

Fig. 5b shows a simplified scheme of the blackbody pumped CO_2 laser.

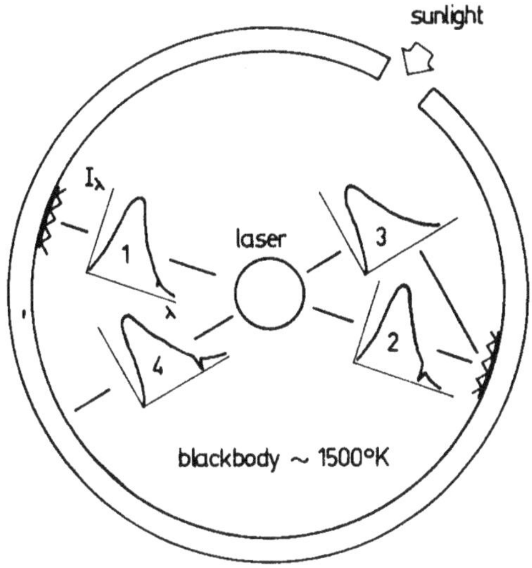

Fig. 5a. Thermalization of a nonequilibrium radiation field by a black body cavity.

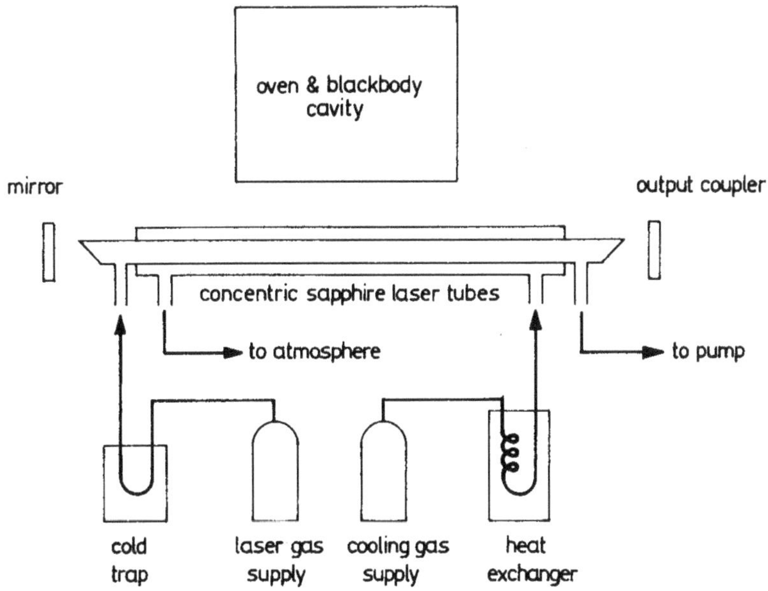

Fig. 5b. Set-up of a black-body-pumped CO_2 laser.

5. Laser gain and linewidth

The gain coefficient of the CO_2 laser is given by:

$$\alpha(\nu) = \frac{A_{21} c^2}{8 \pi \nu^2} \left(n_{v''J''} - n_{v'J'} \, g_{J''} / g_{J'} \right) g(\nu) \tag{8}$$

where A_{21} is the spontaneous emission Einstein coefficient, $g(\nu)$ is the normalized lineshape function

$$\int_0^\infty g(\nu)\, d\nu = 1$$

and $n_{v''J''}$, $n_{v'J'}$, $g_{J''}$ and $g_{J'}$ are the upper and lower level populations and degeneracies respectively. Since $g_{J''} = 2J'' + 1$, $g_{J'} = 2J' + 1$ and $J' = J'' \pm 1$ and typical values of J' and J" are around 20 (see eq (7)), $g_{J''}/g_{J'} \cong 1$ and this factor is frequently ommitted. In thermodynamic equilibrium the vibro rotational populations $n_{v'J'}$ and $n_{v''J''}$ are related to the total vibrational populations $n_{v'}$ and $n_{v''}$ by equation 4. However under lasing conditions, if the probability of stimulated transitions becomes comparable to the rotational relaxation rate a hole will be created in the upper rotational population distributions and a peak in the lower one (see fig.6).

It should be noted that the rotational relaxation rate is of the order of the collision rate which can be calculated by kinetic theory.

$g(\nu)$ describes the spectral distribution of the radiative spontaneous decay, which is broadened due to lifetime effects, Doppler shifts or collisions.

At low gas pressures,Doppler shifts of the transition frequencies determine the line shape which is Gaussian. Doppler width is given by:

$$\Delta\nu_D = 2 \nu_0 \sqrt{\frac{2kT \ln 2}{Mc^2}} \tag{9}$$

where M is the molecular mass; T: temperature and ν_0 is the central frequency of the line. At higher gas pressures the line shape is collision broadened (Lorentzian) and the linewidth is given by [44]:

$$\Delta\nu_P = 7.58 \left(\Psi_{CO2} + 0.73\, \Psi_{N2} + 0.64\, \Psi_{He} \right) P \left(300/T \right)^{1/2} \text{MHz} \tag{10}$$

where P: total gas pressure (in torr) of the laser .mixture; Ψ_x is the proportion of the gas x in the laser mixture where x is CO_2,N_2 or He. Fig. 7 shows the linewidth vs. pressure of typical gas mixtures. In the region around 10 torr the transitions are partially homogeneously and partially inhomogeneously broadened and the gain is expressed in terms of the Voigt function and related functions [45].

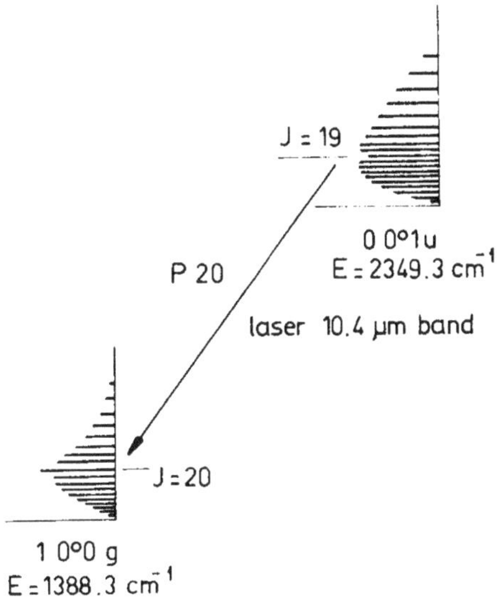

Fig. 6. Hole and peak created in the upper and lower laser level when the probability of stimulated transitions is comparable to the rotational relaxation rate.

Now we can calculate the gain coefficient for a normalized Lorentzian lineshape, which is usual in power lasers, from the expression (8) taking (for the $(00^01)\rightarrow(10^00)$ transition) [46], $g(\nu_0) = 2/\pi\,\Delta\nu_p$:

$$\alpha(\nu) = \frac{A_{21}c^2 2}{8\pi\,\nu^2\Delta\nu_p}\,(n_{v''J''} - n_{v'J'})$$

$$\alpha(\nu) = 6.88\ 10^{-15}\ (n_{v''J''} - n_{v'J'})\ [1/\Delta\nu_p] \tag{11}$$

where n is in cm^{-3} and $\Delta\nu_p$ in MHz.

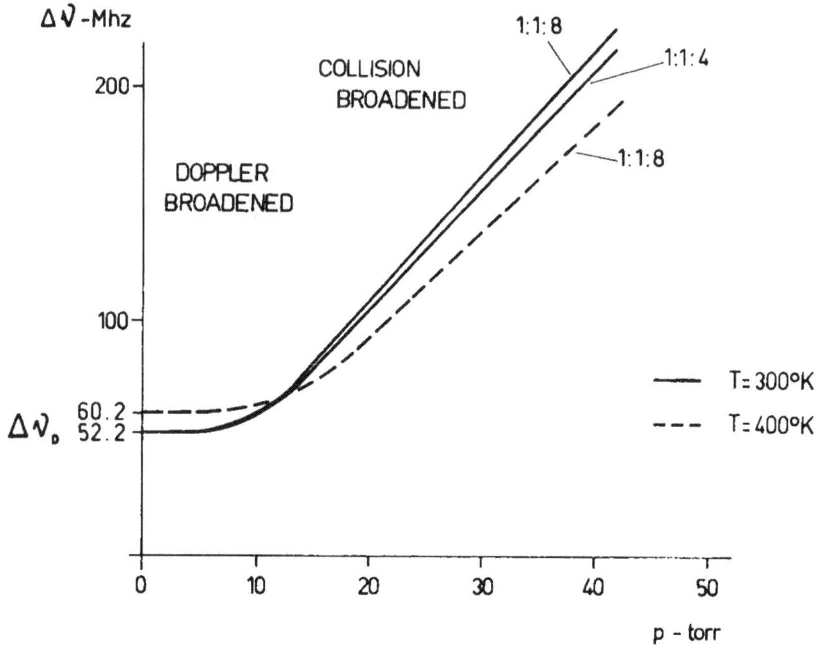

Fig. 7. Doppler and collisional line broadening for a typical CO_2-N_2-He mixture as a function of total pressure.

For a 1:1:8 CO_2-N_2-He mixture at 25 torr total pressure and 400^0K working temperature we obtain from fig. 7 $\Delta v = 115$ MHz and assuming population inversion of the order of 0.1% of the total CO_2 molecules (7.92×10^{16} cm^{-3}) we obtain from (11) $\alpha = 0.47 \times 10^{-2}$ cm^{-1} (i.e. 0.47%/cm or 2 db/m), typical value for a conventional low pressure CO_2 laser. It should be pointed out that in the pressure broadened regime the gain is basically independent of pressure because both the linewidth and the population inversion increase linearly with pressure (assuming the fraction of excited molecules remains constant).

In a Doppler broadened CO_2 laser below 10 torr, supposing Boltzmann distribution of population over rotational states, we obtain the gain coefficients for P and R branches as a function of the ratio a, where $a = N00^01 / N10^00$; $N00^01$ and $N10^00$ population of v_3

and ν_1 respectively as shown in fig.8 [47]. It should be noted that the gain in the P-branch can be positive even if there is not population inversion. Furthermore the amplification in the P-branch is always higher than in the R-branch.

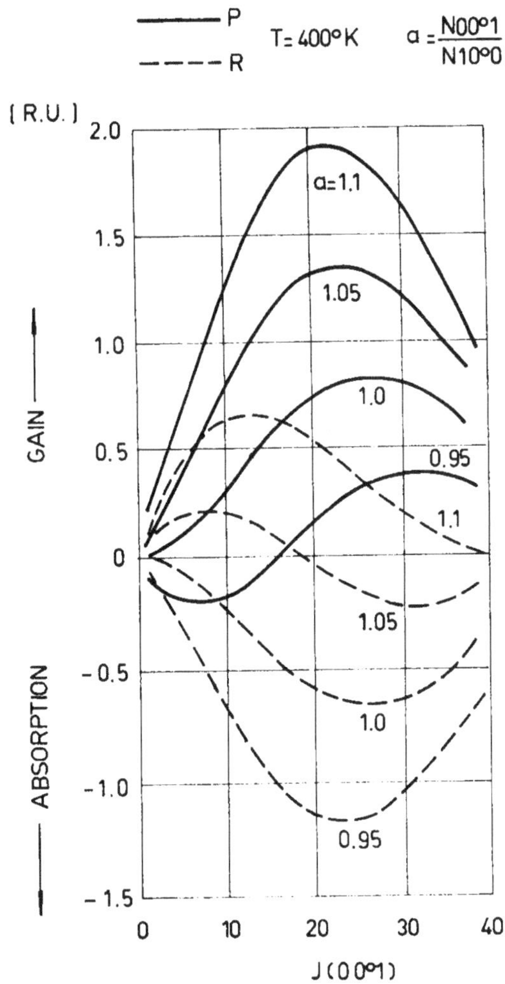

Fig. 8. Theoretical gain curves for the transition of the P and R branch for different values of $N_{00°1}/N_{10°0}$.

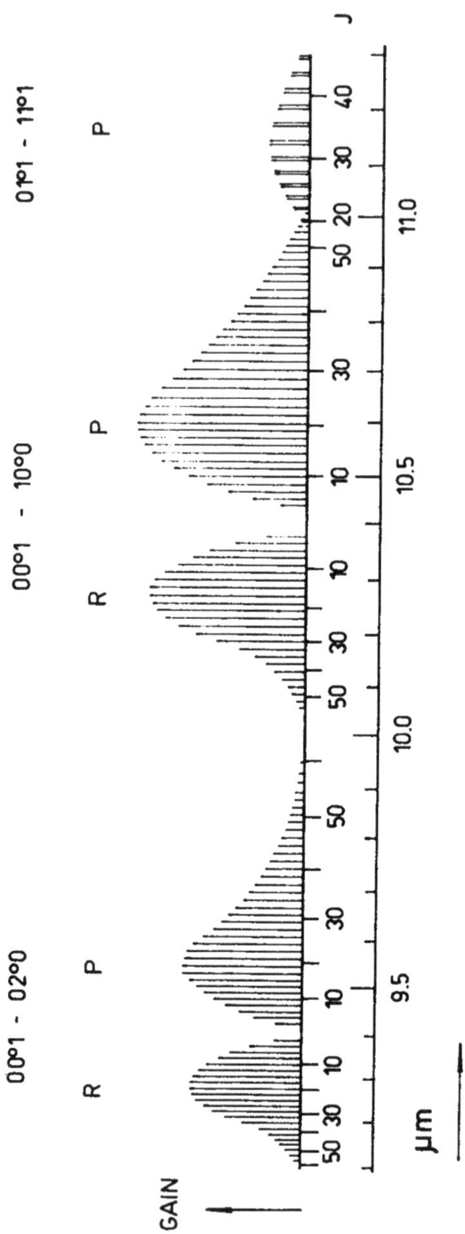

Fig. 9. Vibro-rotational CO_2 laser lines.

The CO_2 laser can be tuned to oscillate on over 100 different lines in the 9 to 11 μm range. Fig. 9 shows the different transitions and the relative intensities.

6. Mechanisms of excitation of CO_2 lasers

Population inversion in CO_2 can be achieved by electric discharge, chemical reactions, optical pumping, gas dynamic processes, electron-beam and energy transfer from preheated N_2. In this section we shall concentrate in electrically pumped lasers which are the most widely used.

The excitation of the upper laser level in pure CO_2 discharges is produced by direct electronic impact. The addition of N_2 substantially favours the laser action. Since N_2 is a homonuclear diatomic molecule, its dipole radiation is forbidden and, once excited to the first vibrational state, it can only decay by collisions with the wall or with other molecules. The (00^01) level of CO_2 is only $\Delta E = 18\ cm^{-1}$ higher than the vibrational level v=1 of N_2. Therefore N_2 de-excitation via collisions with unexcited CO_2 molecules is a high probability process [48] resulting in very efficient laser pumping. The necessary degree of ionization for a self-sustained discharge calls for electron energies above 1 eV.

Nighan [49], [50] has calculated the fractional power transferred to the vibrational and electronic level of CO_2 and N_2 as a function of the average electron energy; his results are shown in fig 10a and 10b. It can be deduced that the 1-3 eV range is generally necessary in actual lasers as a compromise between efficient CO_2 pumping and an adequate rate of charge generation for the self-sustained discharge.

Studies on the paths of electronic excitation of N_2 in the discharge have been performed [51]: observing the sidelight emission from a CO_2-N_2-He laser it was concluded that some N_2 molecules are excited to the $C^3\Pi_u$ state by electronic impact from the ground state and by collisional energy transfer from excited He. Lowering of the vibrational temperature when passing from non-lasing to lasing conditions indicated an effective transfer of energy to CO_2 in the latter case.

Since the CO_2 (00^0v) and N_2 (v) "ladders" are almost coincident, an appreciable fraction of the energy transferred to the high order modes of N_2 ends up in the upper laser level. Adding up all these contributions it is observed that between 60% and 70 % of the input energy can be transferred to the upper laser level. Therefore, as the quantum efficiency of the 10.4 μm band is about 40%, laser efficiencies of 24-28% should be attained. These figures are in fair agreement with the experimental value for a well optimised c.w. CO_2 low pressure laser.

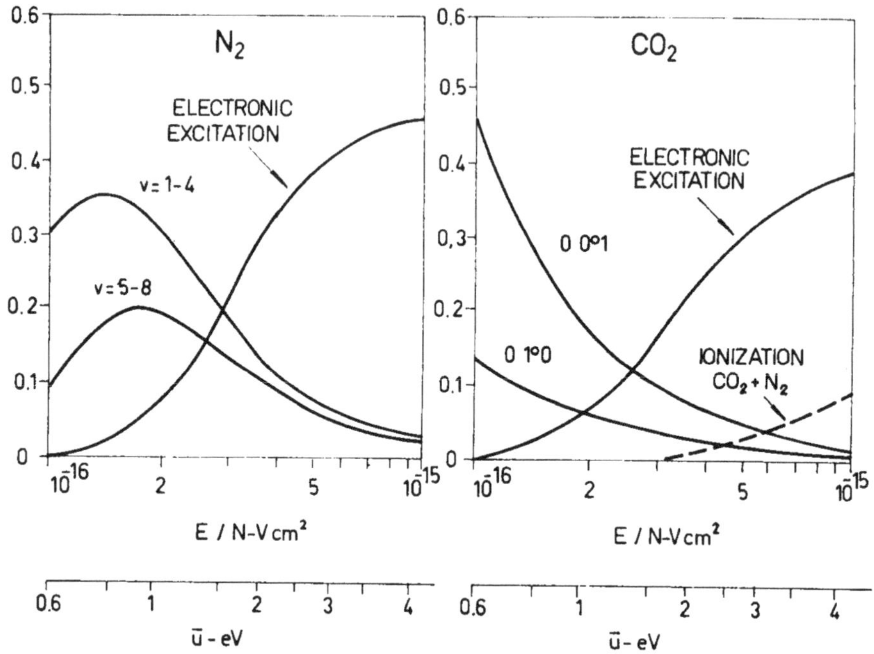

Fig. 10a. Fractional electron power transfer to N_2 in 1:1:8 CO_2-N_2-He mixtures.

Fig. 10b. Fractional electron power transfer to CO_2 in 1:1:8 CO_2-N_2-He mixtures.

The temperature of the discharge must be kept low to reduce the thermal population of the lower laser level (which is only 5 kT above the ground state at room temperature). Significant pumping of the 10^00 and 02^00 levels (which are strongly coupled by a Fermi resonance) may result in the range $u = 3$ - 5 eV where u is the electron average energy, due to resonance effects in the cross-sections for electronic impact excitation [31]. However, this detrimental effect is more important in high pressure lasers.

The addition of helium cools the gas owing to its high thermal conductivity, thus reducing the thermal population of the lower laser level. Furthermore, as most electron-helium collisions are elastic due to the high ionization energy of helium, the average electronic energy tends to be reduced for a given value of E/N. A third benefical effect resulting from the addition of helium is an increase in the relaxation rate of the lower laser level.

Since the dissociation energy of CO_2 is 2.8 eV, the dissociation of CO_2 can be as high as 50%. Consequently during the discharge an appreciable density of CO molecules can be produced. The total cross section for vibrational excitation of CO by electronic impact is similar to N_2 [26], [52], [53] and the difference between the energies of the vibrational level of CO and the 00^01 level is 170 cm^{-1}, which is smaller than the average kinetic energy kT. So CO can help the upper laser level excitation, but it is not as effective as N_2 in exciting the vibration of CO_2. Furthermore the CO_2 concentration necessary for laser action must be maintained; this can be achieved by the addition of H_2O which helps to reduce the degree of dissociation of CO_2 into CO and O. This is mainly helpful in sealed off systems (see page 104 of ref [33]).

7. Optical generation and extraction in homogeneously broadened lasers

In the following section we will deal with collision broadened CO_2 lasers since these are the most widely used.
In this case the gain coefficient follows the known relation [54]:

$$\alpha = \frac{\alpha_0}{1 + I / I_s} \tag{12}$$

where α_0 is the unsaturated or small signal gain, and I_s is the saturation intensity, defined as the power density at which the gain falls to $\alpha_0 / 2$.
Combining (12) with the definition of α and integrating we obtain:

$$\alpha_0 l = \ln \frac{I_{out}}{I_{in}} + \frac{1}{I_s} (I_{out} - I_{in}) \tag{13}$$

where I_{in} and I_{out} are the input and output intensities respectively and l the active medium length. When the medium is well saturated ($I \gg I_s$) eq. (13) shows that the product α_0 l is the available power per unit volume and should be maximized in a good design. However, not all this power is available as laser output because of cavity losses and because the condition $I \gg I_s$ is in some cases difficult to fulfil. If we define the optical extraction efficiency η_{opt} as the ratio of the laser output power (W_{out}) to the power available in the laser medium, the former quantity will be given by:

$$W_{out} = \eta_{opt}\, \alpha_o\, I_s\, l\, A_s \qquad (14)$$

where A_s is the cross section of the active volume and we have assumed all magnitudes constant over this volume.

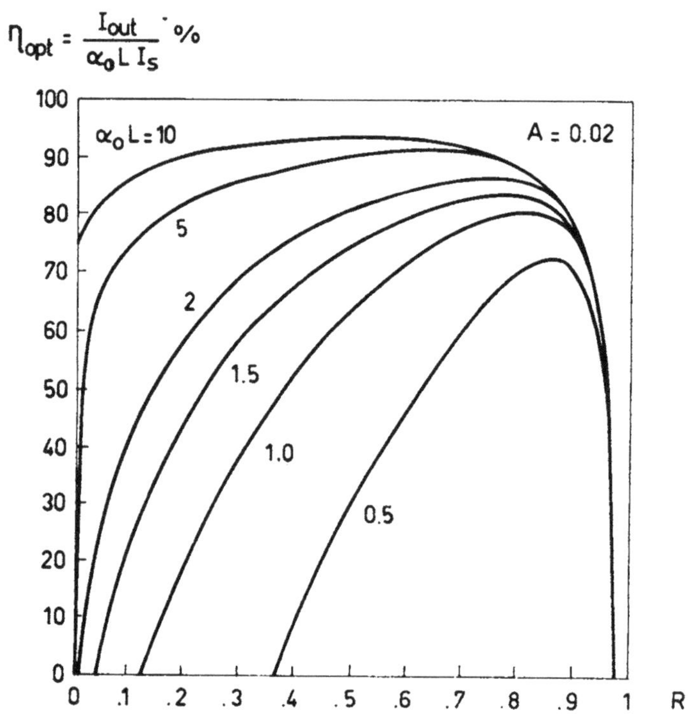

Fig. 11. Optical efficiency for a conventional longitudinal cavity, vs. reflectivity R, for losses of 2%.

η_{opt} as a function of the output mirror reflectivity, gain α_0 l and relative cavity losses A was calculated in ref [55] under the assumption of saturation due to the sum of the two opposite intracavity waves. It is shown in Fig. 11 for the particular value A = 0.02.

The values of α_0 and I_s are dependent on excitation mechanisms, total pressure and translational temperature of the gas. The influence of temperature is of paramount importance. The lower laser level 02^00 lies around 1400 cm^{-1} (depending on the final rotational number J), wich is only about 5 kT above the ground state at 400 K. (See fig. 1). Therefore, the increase of the translational temperature of the gas will produce significant population of the lower laser level and this may even decrease to negative values, i.e. turn the medium into an absorber instead of an amplifier. This has been corroborated by measurements of cells containing hot CO_2 [56], [57] in wich absorption coefficients of the order of 0,02 cm^{-1} are obtained at 600 K. Even in the smaller electrically excited devices of the types we will discuss in the following sections kilowatts of waste power must be exctrated from the discharge. This can be done in two different ways, leading to two well differentiated CO_2 laser categories:

1) Diffusion-cooled or "slow flow" CO_2 lasers.
2) Convection-cooled or "fast flow" CO_2 lasers.

These will be studied in the following sections.

8. Electrically excited diffusion-cooled CO_2 C.W. lasers

8.1. General description

The active medium of c.w. diffusion-cooled lasers is the positive column of a CO_2-N_2-He discharge. The mixture either flows slowly along the tube or is confined in sealed off devices. The cooling is provided by a water or oil jacket. The heat generated in the electric discharge diffuses radially (hence the name) to the tube wall and is carried away by the cooling fluid. Fig. (2a and 2b) shows typical flowing and sealed-off elementary low power devices (<100 W).

8.2. Cooling mechanism

This has been discussed by Smith and McCoy [58] and DeMaria [28]. Let's estimate the characteristic time in which the waste energy is extracted from the discharge.

The mean time between collisions of a CO_2 molecule is $\tau_c = \lambda / v_t$ where λ is the mean free path and v_t the thermal velocity of the molecules. Being a random walk process, the distance between the initial and final positions of the molecule after n collisions will be $(n)^{1/2}\,\lambda$ (and not $n\,\lambda$). As $n = \tau_d / \tau_c$ where τ_d is the travelling time, we obtain

$$\tau_d = \frac{R^2}{\lambda^2}\,\tau_c = \frac{R^2}{v_t\,\lambda} \qquad (15)$$

Using typical values for λ and v_t from kinetic theory and for R = 0.5 - 1 cm and T = 350 K we obtain $\tau_d \sim 0.1$ sec. or longer for slow flow lasers. We thus conclude that most of the heat is extracted via diffusion to the cooled wall.

8.3. Electrical excitation

The electric discharge is confined in a glass or quartz tube, giving origin to a well developed and stable glow. In the positive column the longitudinal field is constant and therefore there is no net volumetric charge [59], [60]. The densities of electrons and positive ions are equal: $n_e = n_i = n$. Ionizing collisions are produced in the volume, whereas volume recombination is negligible. The dominant loss mechanism is the diffusion of the charges to the wall where they recombine. Charge particle density follows the law:

$$n = n_o\, J_o\,(2.4\, r / R) \qquad (16)$$

where n_o is the electron density at the tube axis; r is the radial distance from the tube axis; R is the tube inner radius, and J_o is the zero order Bessel function.

Current density **J** roughly follows the same radial Bessel distribution. At high current levels the distributions tend to be flattened departing from the Bessel curve. Since E is constant, the power dissipated per unit volume **J.E** also follows the same law, which is not favourable to heat extraction. In fact, at high current levels the gain can be reduced or even become negative at the tube center, favouring ring type combination of modes instead of TEMoo.

The kinetics of excitation of laser levels in the discharge has already been described in sec. 6. Fig. 12 shows a typical dependence of gain on current [61], [27]. The decrease at high currents is due to heating of the gas and consequent thermal population of the lower laser level 02^00.

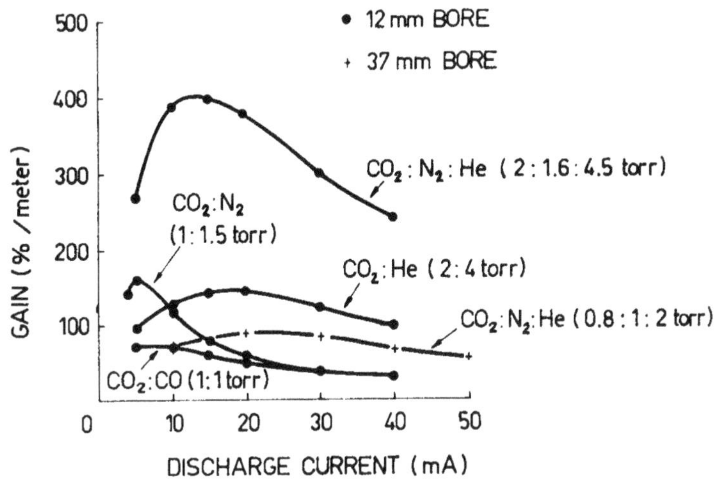

Fig. 12. Gain as a function of current. From (27).

Slow flow, longitudinal discharges of this type can be obtained with no stability problems up to about 100 torr of total pressure. However, as we shall see in the following section, there is no advantage in increasing the working pressure beyond 20 - 30 torr since this does not lead to higher power outputs in this type of lasers.

In DC excited lasers, particularly at high operating pressures, the cathode current density increases causing depletion of oxygen from the CO_2 gas mixture through oxydation of the cathode. This affects the operational lifetime of the laser. During mid 1970's research was brought about which resulted in the development of RF excited CO_2 lasers [62]. The advantages of RF relative to DC excitation are:

. possibility of producing electrodeless discharges thus avoiding the above effect and problems of sputtering and contamination;

. possibility of using transverse excitation, which permits reduced operating voltages;

. simpler and faster modulation and regulation, and

. better volume-filling capabilities.

However, the complexity of the power supply increases and the impedance matching of the excitation source with the discharge must be carefully considered. Some details of this are given in sec. 9.3 and 9.4.

8.4. Power extraction

The output power is governed by eq. (14), in which η_{opt} and α_0 have already been discussed and I_s can be deduced from a simple two level model.

Saturation intensity of diffusion-cooled lasers has a somewhat intriguing behaviour. First measurements showed values ranging from about 20 to 100 W/cm^2 for similar discharge tube bores, discharge currents, flow rates and gas mixtures [58], [63].

A simple model prepared by Smith and McCoy [58] explains the dispersion in the results: saturation intensity depends on the radius of the probe beam due to diffusion effects. A simplified set of rate equations can be written for a two level model:

$$\frac{\partial n_2}{\partial t} = R_2 - \frac{n_2}{\tau_2} - (n_2 - n_1)\frac{\sigma I}{h\nu} - \frac{n_2 - n_{20}}{\tau_{d2}} \qquad (17)$$

$$\frac{\partial n_1}{\partial t} = R_1 - \frac{n_1}{\tau_1} + (n_2 - n_1)\frac{\sigma I}{h\nu} - \frac{n_1 - n_{10}}{\tau_{d2}} \qquad (18)$$

where n_2 is the upper level population, n_{20} is the upper level population in absence of stimulated emmision, R_2 is the upper level volume pumping ratio, n_1, n_{10} and R_1 are the corresponding values for the lower level, and σ is the cross section for stimulated emission. The time τ_{d2} is the characteristic time for diffusion of excited CO_2 molecules into the beam and τ_{d1} is the characteristic time for diffusion out of the beam of molecules in the lower laser level.

Solving for $\alpha = \sigma(n_2 - n_1)$ in steady state:

$$\alpha = \frac{\sigma(R_2\tau_2 - R_1\tau_1)}{1 + \frac{\sigma I}{h\nu}\left(\frac{\tau_2\tau_{d2}}{\tau_2 + \tau_{d2}} + \frac{\tau_1\tau_{d1}}{\tau_1 + \tau_{d1}}\right)} \qquad (19)$$

($n_2 >> n_{20}$ and $n_1 >> n_{10}$ also assumed) which has the form of eq. (12) with

$$\alpha_0 = \sigma(R_2\tau_2 - R_1\tau_1) \qquad (20)$$

being the small signal gain and

$$I_s = \frac{h\nu}{\sigma}\left(\frac{\tau_2\tau_{d2}}{\tau_2 + \tau_{d2}} + \frac{\tau_1\tau_{d1}}{\tau_1 + \tau_{d1}}\right)^{-1} \qquad (20)$$

the saturation intensity.

In a CO_2 laser the collisional relaxation time for the upper laser level is an order of magnitude greater than the lower level, while τ_{d1} and τ_{d2} should be of comparable magnitude. Thus the second term in the denominator can be neglected and we obtain:

$$I_s = \frac{h\nu}{\sigma}\left(\frac{\tau_2 + \tau_{d2}}{\tau_2\ \tau_{d2}}\right) \tag{22}$$

τ_{d2}can be evaluated using the same arguments of section 8.2. using eq. (15), but replacing the tube radius by the beam radius. For 2.5 mm radius, v_t = 400 m/sec. and $\lambda = 3 \times 10^{-3}$ cm, typical values for a diffusion-cooled laser, we obtain $\tau_{d2} \sim 0.5$ m sec. which is somewhat shorter than the upper relaxation time τ_2 . Clearly for small radii τ_{d2} will dominate and I_s will be

$$I_s = h\nu / \sigma\tau_{d2} = h\nu v_t \lambda / \sigma R^2 \tag{23}$$

whereas for large radii:

$$I_s = h\ \nu / \sigma\tau_2 \tag{24}$$

By substituting the values $\tau_2 = 10^{-3}$ sec and σ for collisional broadening with $\Delta\nu$ = 115 MHz (see section 5) we obtain I_s = 0.3 w/cm^2, which is much smaller than the experimental values. Christiansen [63] points out that the discrepancy lies on the fact that the CO_2 laser is a multilevel system, therefore τ_2 must be replaced by a higher relaxation rate due to strongly coupled levels within the 00^01 band, each of which can relax at the rate τ_2. Fig.13 shows the experimental values of the saturation parameters as a function of the average beam radius, which behave as stated by eq.23. Besides Ersoy [64] evaluated the effect of diffusion on gain saturation for homogeneous and mixed broadening.

From the above we conclude that the power per unit length given by $\pi R^2 \alpha_0 I_s$ remains roughly constant because the increase in mode volume is compensated by the decrease of saturation intensity. The fact that the maximum power output per unit length does not depend on tube diameter can be established using general thermal arguments. The available output power of the laser will be $W_{out} = \eta_D W_D$ where W_D is the discharge input power and η_D is the conversion efficiency from electrically to optically stored energy. By energy balance the waste power $W_D (1 - \eta_D)$ must be equal to the power flowing out of the gas, radially to the wall

$$W_D\,(1-\eta_D) = K\,(\partial T/\partial r)|_R\; 2\pi R l \qquad (25)$$

where K is the thermal conductivity of the gas evaluated near the tube wall (K depends on temperature). For a given temperature profile we can always write:

$$T - T_W = (T_o - T_W)\, f(\xi) \qquad (26)$$

were T_o is the maximum temperature at tube axis, T_W is the wall temperature, $\xi = r/R$ and $f(\xi)$ is the adimensional temperature profile. (All convective effects have been neglected).

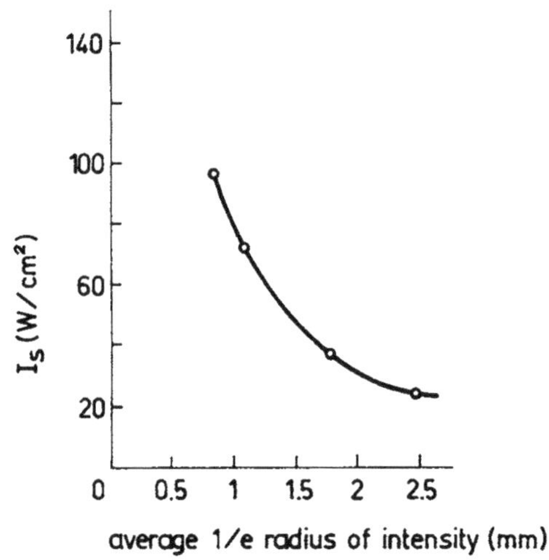

Fig. 13. **Experimentally measured saturation parameter as a function of the average beam radius. CO_2: 1.4 torr, N_2: 1.7 torr, He: 7 torr, H_2: 0.2 torr, I: 26 mA.**(63)

Replacing (26) in (25), using $W_{out} = \eta_D W_D$, we obtain for the available power per unit lenght:

$$\frac{W_{out}}{l} = \frac{2\pi\,\eta_D}{1-\eta_D}\, K\,(T_o - T_W) \left| \frac{d f(\xi)}{d\xi} \right|_{\xi=1} \qquad (27)$$

This is independent of tube radius and operating pressure (K is independent of pressure) and only depends on discharge current through the excitation efficiency η_D , which, when optimized, is again essentially independent of tube parameters.
For a Bessel temperature profile $f(\xi) = J_o\,(2.4\,\xi\,)$ the derivative in eq. (27) is 1.25 and for a parabolic profile $f(\xi) = 1 - \xi^2$ is 2. Actual profiles are difficult to obtain, but fall around this range.
Using the Bessel value and assuming $T_o - T_w = 200$ K , K = 1.2 10^{-3} W/cm °K (for a typical gas mixture), $\eta_D = 0.24$ as explained in section 6, we obtain $W_{out} / l = 59$ W/m , which, assuming an optical extraction efficiency of 0.9 (see fig. 11) gives 53 W/m. Experimental values are between 40 and 70 W/m for a wide variation of experimental situations.

The conservative value of 50 W/m is generally accepted although somewhat higher values can be obtained by flowing the gas at linear velocities of a few meters per second to provide some extra convective cooling.

9. Description of diffusion cooled devices

9.1. Conventional slow flow D.C. excited lasers

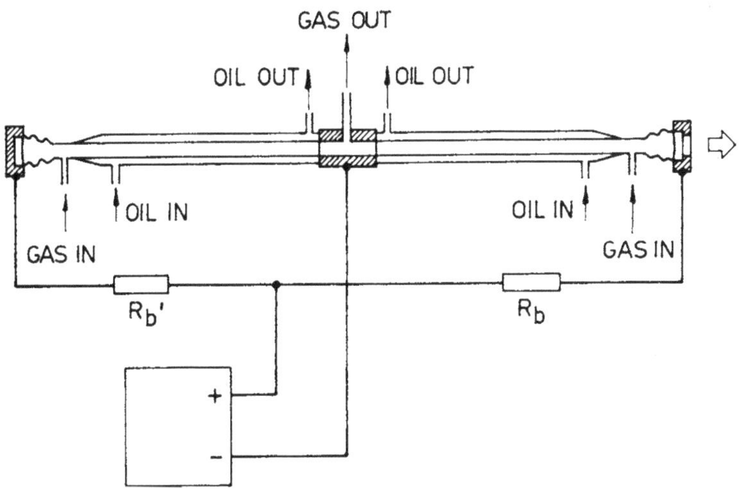

Fig. 14. Typical CO_2 axial slow flow laser configuration.

A typical device of 210 W output power (design and construction at CITEFA-CEILAP) is schematically shown in fig. 14 and its main characteristics are resumed in table I.

TABLE I

Technical data of a flowing c.w. diffusion cooled CO_2 laser

Tube diameter (int.)	16	mm
Tube length (each section)	1.6	m
Discharge current (each section)	44	mA
Discharge voltage	12	kV
Total pressure	10	torr
CO_2:N_2:He fractions	9:18:73	
Linear velocity of gas	4	m/sec.
Power input (two sections)	1.056	kW
Power output (two sections)	210	W
Efficiency (electrical to optical)	19.8	%
Power per unit length	65.6	W/m
Cooling fluid	YPF-64 oil	

As has been described in sec. 8, the only way of increasing the power output of a diffusion cooled laser is to increase the length of the active volume. Since about 8 - 10 meters length is necessary for 500 W output for practical reasons different folded configurations,as shown in fig. 15, are used. Fractioning the total length also helps avoiding the use of very high voltages. The sections are conected optically in series and are electrically fed in parallel via suitable ballast resistors. Above 1 kW output power these lasers become impractical because of the increasing complexity associated with the many foldings and the corresponding increase of the intracavity losses.

In summary, these are well understood, reliable devices, useful in the subkilowatt range.

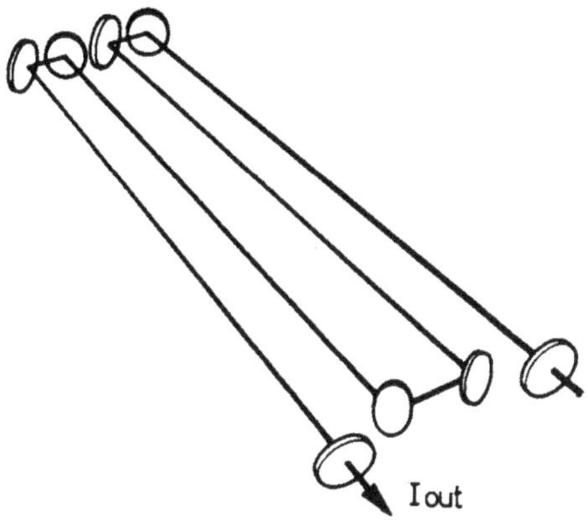

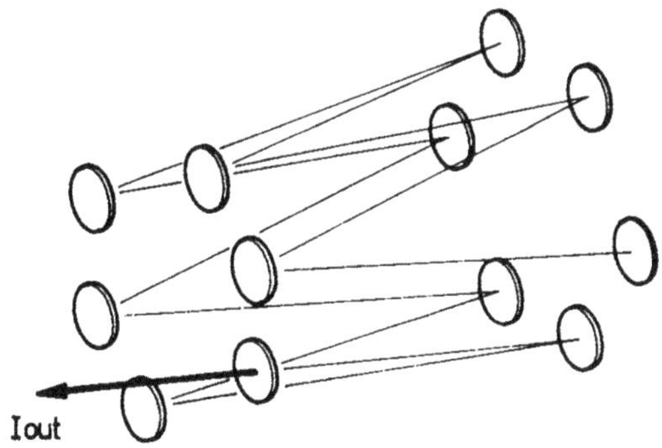

Fig. 15. Different folded configurations for diffusion-cooled lasers.

9.2. Sealed-off Lasers

The sealed off CO_2 lasers have a better stability and output power than gas flowing lasers due to lack of turbulence. Also they are more economic and easy to operate since there is no gas consumption. The maximum power output of these lasers is around 100 Watts.

The main problem is the chemical decomposition of CO_2 in the electrical discharge. It was found that small quantities of H_2O vapor (or H_2), added to the gas mixture of CO_2-N_2-He, avoid the consumption of CO_2. By means of mass spectrometry and lateral radiation analysis it was demonstrated that the CO concentration diminished very clearly in presence of H_2O. This is due to the CO recombination with HO.

The fig. 16 shows the significant increase of lifetime of this laser with the use of H_2O vapor.

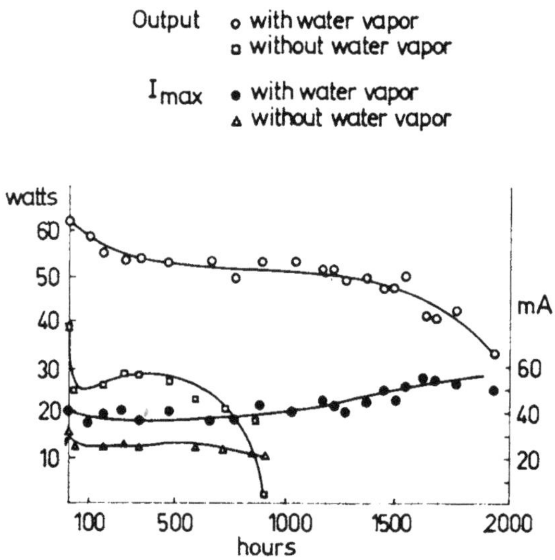

Fig. 16. The maximum radiation output, and the corresponding current, I_{max}, vs. operation time with a gas mixture of 1 torr CO_2, 2.5 torr N_2 and 11 He, and with a gas mixture that contains also 0.2 torr H_2 and 0.1 torr O_2.

It is more convenient to use fused silica tubes instead of pyrex glass, because this material generates problems of water absorption in the first hour of lasing [65]. The electrodes should be inert metals, like gold or platinum. Hollow electrodes of nickel are also good because they avoid sputtering. The outer surface of the cathode must be close to the glass or quartz tube to avoid the formation of carbonyls in the discharge.

One possible scheme of a sealed off CO_2 laser is shown in fig. 2b. The tube is in pyrex (30 cm length and 4 cm internal diameter) and has a nickel hollow cathode and Brewster ZnSe windows. With 6 Kv, 6 mA a power of 8 Watts was obtained from a 1:1:7 mixture with .a total pressure of 10 Torr.

9.3. Waveguide Lasers

The losses of a dielectric waveguide are very small if the inner diameter d is larger than $10^2\ \lambda$. For $\lambda = 10.6\ \mu m$ this implies d > 1 mm. The waveguide resonator can be terminated with flat mirrors; the coupling losses will be a function of the distance of the mirror to the waveguide and the confocal parameter of the gaussian beam.

The materials commonly used for the hollow waveguide are Pyrex glass, fused SiO_2, and ceramics like Al_2O_3, BeO, BN, which have higher heat conductivity.

Waveguide CO_2 lasers have the following advantages:

The emission is on the TEMoo mode, since the waveguide attenuates higher order modes.

The small signal gain is 2-3 times larger than in conventional lasers, allowing for a short cavity length.

The pressure of the active medium can be larger, increasing the homogeneous linewidth. The combination of a short cavity and relatively high pressure allows an increase of the frequency tunability around each CO_2 laser line.

The small signal gain α_0 increases with gas pressure for a Doppler broadened line and becomes independent of P when the broadening is homogeneous (Sec. 5). By increasing further the pressure at constant excitation, diffusion towards the wall decreases and also population inversion. It may be concluded that α_0 has a maximum which occurs for a pressure between 15 or 75 torr depending on the WG diameter and gas mixture. It was found that the optimum electron temperature is obtained for $E/N \sim 2 - 3 \times 10^{-16}\ Vcm^2$.

A typical c.w. CO_2 laser [66] using 3 to 5 mm inner diameter Pyrex or fused SiO tubes is shown in fig. 17. An output power of 40-45 W per meter of discharge length was obtained with a grating tuned cavity and longitudinal DC excitation. Cooling is obtained by circulating a water solution of ethylene-glycol; the most convenient temperature was found between -10 an 0°C.

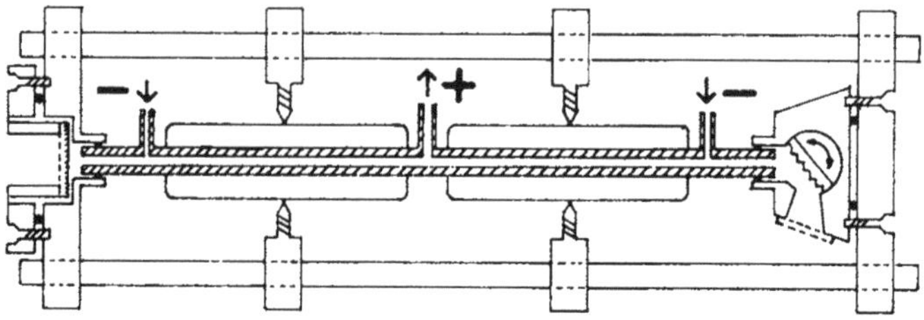

Fig. 17. A waveguide laser [66].

More advanced waveguide lasers are transversally RF excited. Two approaches are shown in fig. 18a and 18b [67], [68], [69]. The hollow-bore waveguide in fig. 18a is formed by a pair of metal electrodes, sandwiched between two ceramic pieces. Several shunt inductors placed along the laser length help to uniformize the discharge. In fig. 18b appears a technique of parallel-resonant distributed-inductance: here ceramic pieces are sandwiched between the electrodes to form a high-capacitance structure. The parallel inductance resonates with the capacitance of the waveguide structure at the desired frequency. This technique allows efficient coupling of RF power into the active medium. Using this RF excitation technique high power sealed-off waveguide lasers were conceived [70]. To achieve high output power from a compact laser, also a multitube system in the waveguide regime can be used.

An approach, proposed as early as 1970 [71] has been more recently used in the IGLAN 3, a compact diffusion-cooled device generating more than 4 kW c.w. output power [72]. As shown in the simplified scheme of fig. 19, it consists of a large number of small diameter (0.5 cm) discharge tubes inmersed in a common coolant and sharing the same optical cavity.

Heat removal conditions are very favourable because of the smaller tube radius. However, the oscillations of the different tubes are non-coherent, and the divergence of the entire beam is determined by the diameter of each tube and not the diameter of the entire bundle. Correspondingly, the focusing capabilities of this type of laser will be poor.

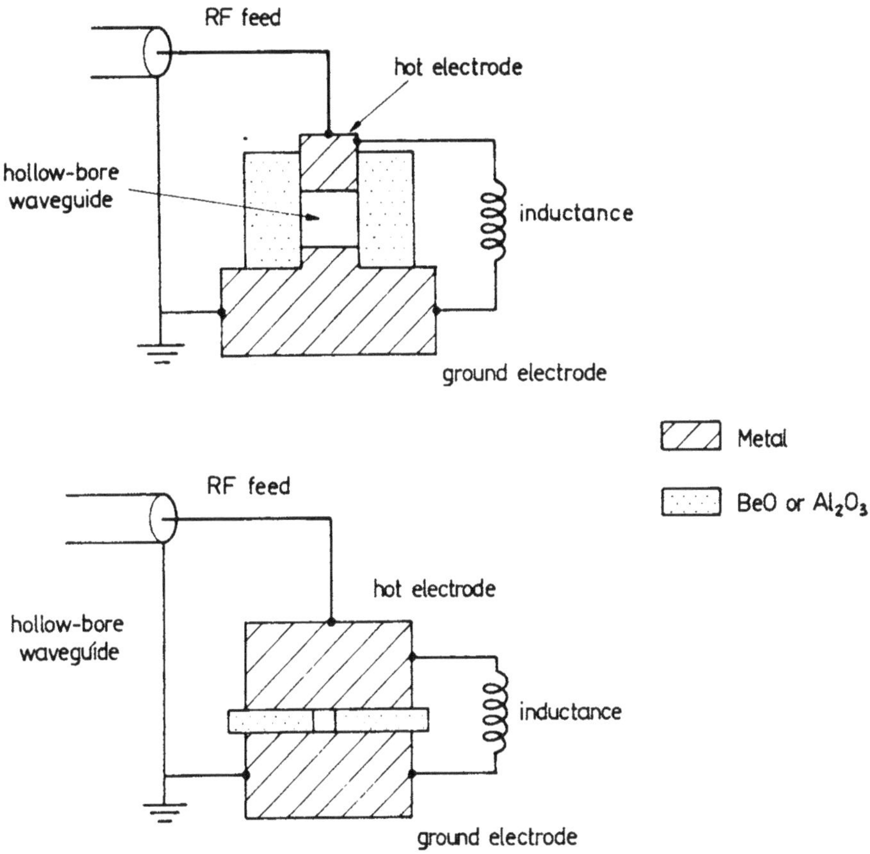

Fig. 18a. Radio-frequency excitation circuits for CO_2 waveguides: a pair of metal electrodes sandwiched between ceramic pieces.

Fig. 18b. Radio-frequency excitation circuits for CO_2 waveguides: ceramic pieces sandwiched between the metal electrodes.

In order to obtain single frequency operation the different waveguide elements must be at the same frequency and phase locked: different methods have been studied with this scope. [73]-[75].

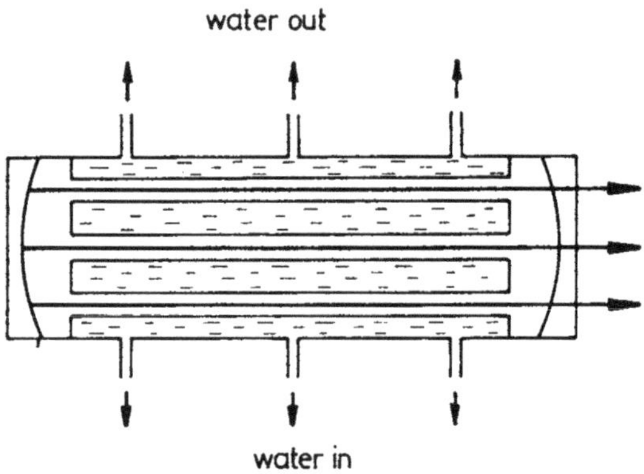

Fig. 19. Multitube diffusion-cooled laser.

9.4. Second generation diffusion-cooled lasers

The rigid power output limits of conventional cylindric discharge diffusion-cooled lasers impulsed research on new configurations with better waste power extraction capabilities.

Thin planar or anular active regions as shown in fig. 20, in which the active gas is in close contact with cooled walls are used. Reference [76] describes an experimental laser in which the active region is confined between two water cooled concentric electrodes and excitation is provided by a 42 Mhz RF discharge. Up to hundreds of Watts can be obtained this way in compact dimensions.
However, optical power extraction becomes complicated and mode quality is degraded.

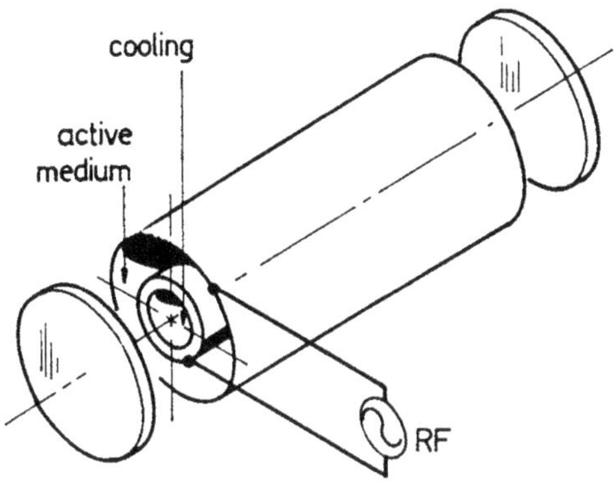

Fig. 20. Thin active region diffusion-cooled lasers.

10. Electrically excited convection-cooled CO_2 lasers

10.1. General principles

It is seen from the above that power output of diffusion-cooled CO_2 is limited by waste heat dissipation problems, and scaling up to multikilowatt regime is very difficult. The obvious answer to this problem is to increase heat dissipation by convective cooling, i.e. flowing the gas. If the flowing gas transit time is much faster than the characteristic time (eq. (15)) for diffusion to the walls, the heat will be extracted by convection and not diffusion. There is no upper limit, in principle, for the power levels obtained by means of this technique.

There are basically two geometries to obtain convection-cooling, flowing the gas along the optical axis (axial flow) (fig. 21a) and transverse to the optical axis (transverse flow) (fig. 21b and 21c). In the latter case the electric field of the discharge may either be or perpendicular (fig. 21c) along the flow (fig. 21b).

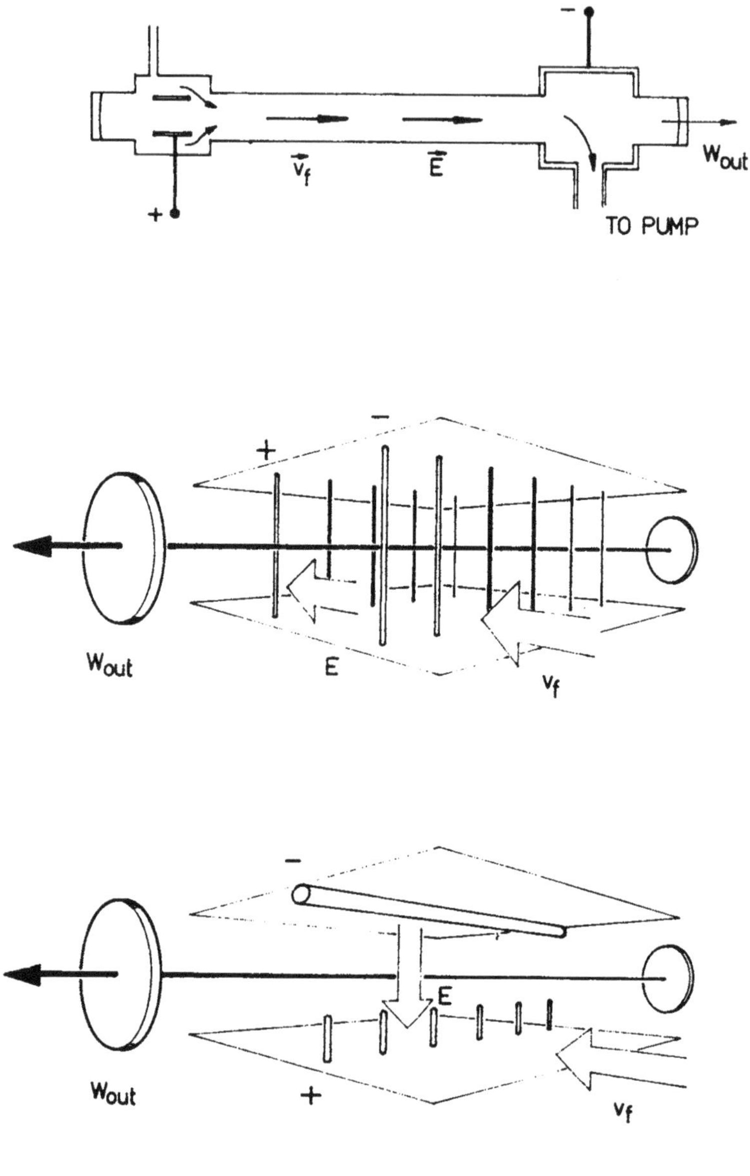

Fig. 21. Convection-cooled CO_2 lasers. A: axial flow, B: and C: transverse flow.

10.2. Cooling mechanisms

A general energy balance equation may be written for all types. Using the symbols defined in section 8, the waste power in terms of laser power output will be:
$W_{out} = (1 - \eta_D) / \eta_D$ and this should equal the heat absorbed by the gas and carried away : $m\ c_p\, \Delta T$

Hence:

$$W_{out} = \frac{\eta_D}{1 - \eta_D}\, m\ cp\, \Delta T \qquad (28)$$

where m is the mass flow, cp is the specific heat and ΔT the temperature rise.

Thus we can calculate the output power per unit mass flow W_{out}/m . For maximum $\Delta T = 250$ K, $\eta_D = 0.15$ (in fast flow discharges it is typically lower than in slow flow ones due to turbulence and other secondary effects) and cp = 2.3 Joule/ ^{o}K g for a typical mixture we obtain specific powers of about 100 W per g/sec of mass flow.

Although open cycle lasers are constructed for experimental studies, closed cycle devices with circulating gas and heat exchangers are mandatory for any industrial application. Detailed descriptions of these systems are given in sections 10.5 and 10.6.

10.3. Electric excitation

As relatively high pressures (typically 30 - 100 torr, up to one atmosphere in some cases) and high linear flow velocities (30 m/sec up to near sonic) are used, there are severe problems of discharge stability. If no special care is taken in design and operation, glow discharges tend to degenerate into arcs that fail to provide appropriate volumetric excitation of the active medium and may even cause permanent damage to the electrodes.
Optimum E/N corresponds, according to fig. 10, to average electron energy u of between 0.6 and 0.8 eV. On the other hand, the necessary degree of ionization (dotted line in fig. 10b) for a self-sustained high pressure discharge calls for higher values of E/N and u, thus lowering the excitation efficiency. To overcome this difficulty, some degree of preionization by UV photons, RF or DC auxiliary discharges are advantageous [28], [30], allowing the use of lower E/N values and consequently higher excitation efficiency. Apart from that, either the anode or the cathode is generally sectionalized in order to provide a stable discharge. A more detailed description of the devices are given in sections 10.5 and

10.6. It should be pointed out, however, that in many cases the limiting factor in the output power is the discharge stability and not the heat dissipation as implied in eq. (28).

10.4. Gain and saturation intensity

Let us write the rate equations for a small volume with a flowing medium in the z direction [28]. These are similar to eqs. (17) and (18), but the last terms, representing diffusion, are now replaced by terms accounting for the local change due to flow.

$$\frac{\partial n_2}{\partial t} = R_2 - \frac{n_2}{\tau_2} - (n_2 - n_1)\frac{\sigma I}{h\nu} - \upsilon\frac{\partial n_2}{\partial z} \tag{29}$$

$$\frac{\partial n_1}{\partial t} = R_1 - \frac{n_1}{\tau_1} + (n_2 - n_1)\frac{\sigma I}{h\nu} - \upsilon\frac{\partial n_1}{\partial z} \tag{30}$$

as an approximation we assume:

$$\upsilon\frac{\partial n_2}{\partial z} = \upsilon\,\frac{n_2 - n_{20}}{L} = \frac{n_2 - n_{20}}{\tau_F} \tag{31}$$

$$\upsilon\frac{\partial n_1}{\partial z} = \upsilon\,\frac{n_1 - n_{10}}{L} = \frac{n_1 - n_{10}}{\tau_F} \tag{32}$$

where L is the length of the laser volume in the direction of the flow and $\tau_F = L/\upsilon$ is the transit time. Replacing (31) and (32) in (29) and (30), averaging over the distance L, neglecting n_{10} and n_{20} because of the low temperature of the input gas, and assuming steady state, after solving for $\alpha = \tau\,(n_2 - n_1)$, we obtain a homogeneous saturation law as given by eq. (12), with small signal gain:

$$\alpha_o = \sigma\left(R_2\,\frac{\tau_2\,\tau_F}{\tau_2 + \tau_F} - R_1\,\frac{\tau_1\,\tau_F}{\tau_1 + \tau_F}\right) \tag{33}$$

and saturation intensity:

$$I_s = \frac{h\nu}{\sigma}\left(\frac{\tau_2\,\tau_F}{\tau_2 + \tau_F} + \frac{\tau_1\,\tau_F}{\tau_1 + \tau_F}\right)^{-1} \tag{34}$$

For small flow velocities $\tau_F \gg \tau_1, \tau_2$ and we obtain:

$$\alpha_o = \sigma\,(R_2\,\tau_2 - R_1\,\tau_1) \equiv \sigma\,R_2\,\tau_2 \tag{35}$$

and

$$I_s = h\nu\,/\,\sigma\tau_2 \tag{36}$$

for large flow velocities $\tau_F << \tau_1, \tau_2$

$$\alpha o = \sigma \tau_F (R_2 - R_1) \qquad (37)$$

and

$$I_s = h \nu / 2 \sigma \tau_F \qquad (38)$$

It should be taken on account that here also I_s takes smaller values than the measured ones, as stated in section 8.4.

At the low flow range α_0 increases with flow velocity because of the increase of R_1 and R_2, but only if pumping is correspondingly increased (eq. (35)). For fast flows, even for optimized pumping, the increase in R_1 and R_2 is compensated by the decrease in τ_F and the gain remains constant (eq. (37)). All this ideal behaviour is somewhat obscured in actual devices by discharge instability, losses associated with turbulence, insufficient pumping and/or heat extraction.

Fig. 22 taken from ref. [77], shows the increase of power with flow velocities and total pressure. We see a qualitative agreement with the elementary discussions given in this section.

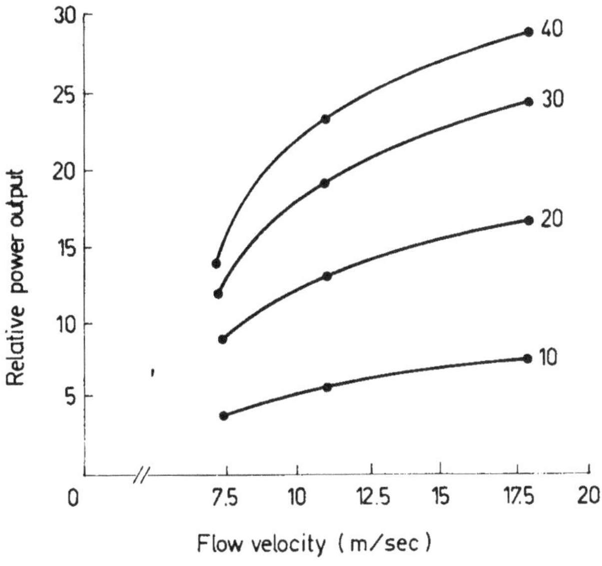

Fig. 22. Relative power output as a function of flow velocity and total pressure.

10.5 Axial flow devices

Fig. 23 shows a fast axial flow (FAF) laser. As all the mass flow must be forced through a narrow bord tube, relatively large pressure drops of the order of 50 torr are produced and large compressors of the Roots type are necessary. The pumping power is large and may become comparable to the electric input to the discharge, thus reducing the global efficiency appreciably. The extra energy given by the compressor to the gas must be extracted by means of a second heat exchanger.

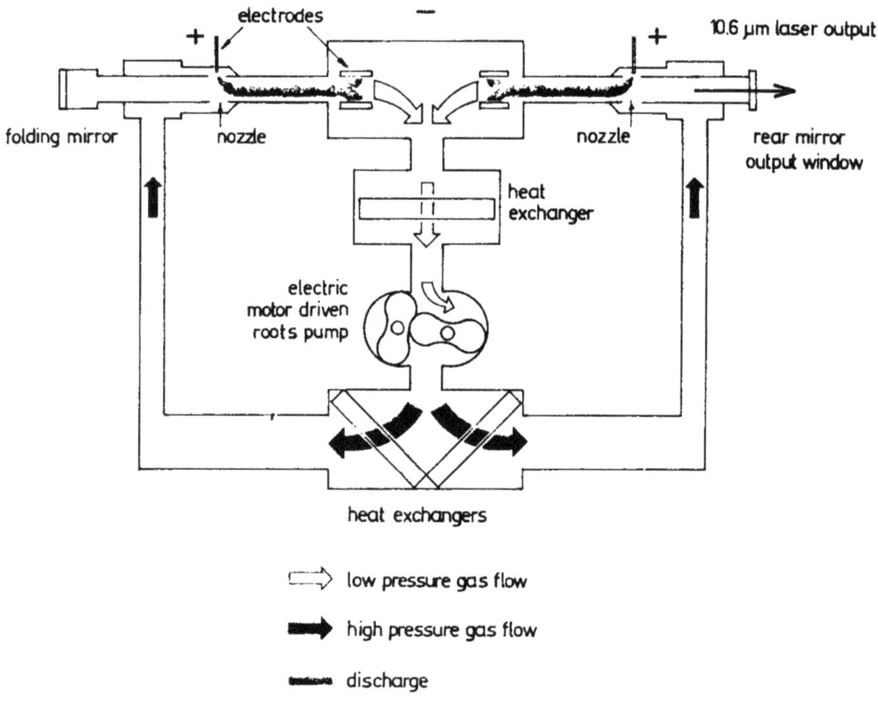

Fig. 23. Fast Axial Flow laser.

In most devices supersonic velocities are not allowed. Therefore, care should be taken in the design because the expansion of the gas due to temperature increase along the flow direction produces a corresponding velocity increase and the gas may reach the speed of sound. For the same reasons and in order to avoid very high temperatures near the gas outlet, discharge lengths must be kept low, of the order of 30-50 cm. Typical powers are of the order of 1 kW per meter of discharge at nominal mass flow. The main advantage of this configuration lies in the good matching between the cylindrically symmetric gain profile and the low order cavity modes, giving a good optical quality output beam.

A different approach to fast axial flow is shown in fig. 24 [78]. It employs a coaxial design which integrates discharge tube, blower and gas cooling in a single cylindrical housing, resulting in a very compact unit. The discharge is RF excited at 13.56 Mhz with a specific input power of 21 W/cm^3, and an output of 800 W is generated using four sections with an efficiency of 16% . The impeller is fitted with gas bearings, fed with the gas mixture and operated at 20.000 r.p.m. The main advantage of this design is its compactness.

10.6. Cross flow devices

The main advantage of cross flow devices is the relatively low speeds necessary to obtain the required mass flows as given by equation (28). For the typical values of length along the flow direction L = 5 cm and υ= 50 m/sec we obtain τ_F = 1 msec which is of the order of the upper level relaxation time at the typical operation pressure of 40 torr. These modest velocities also require modest circulating power. Pressure drops in the laser are small, and may be made less than 1 torr for operating pressures of the order of 50 torr. Axial or cross flow can be used provided the aerodynamic design is optimized with respect to the losses. Fig. 25 shows the gas flow strcture of a CITEFA-FIBA design [79]. In this 4 kW CW device care has been taken to minimize losses and velocities of up to 100 m/sec across a 1x0.1 m laser channel have been obtained using two contrarotatory axial impellers consuming less than 2 HP pumping power.

Fig. 26 shows a commercial transverse flow design [80] with a 1.5 kW power output. Segmented anodes and a common cooled cathode are used to refrigerate the discharge. In this type of devices α_0 is maximum at the initiation of the discharge and decreases downstream [81]. As the active region is typically 5 - 20 cm long in this direction, multipass optical cavities are required for efficient energy extraction. The measured saturation intensity in ref. [81] is 246 W/cm^2, much higher than diffusion-cooled typical values of 20 - 80 W/cm^2. This agress with theory of section 10.4 using the τ_F value of 1 msec and $\tau_2 \cong 2$ msec at the operating pressure of this laser.

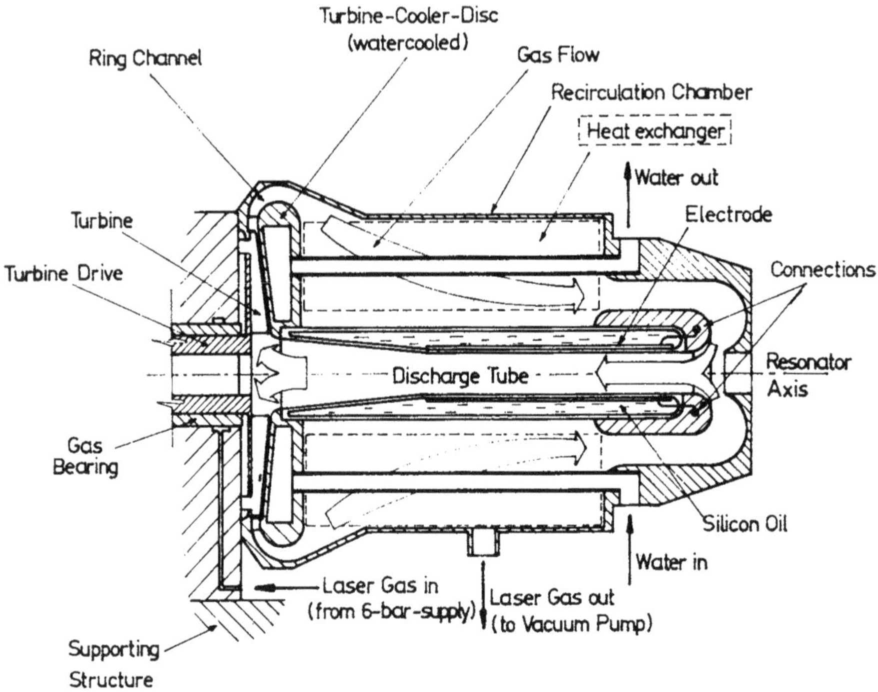

Fig. 24. Principle of laser with integrated turbo blower.

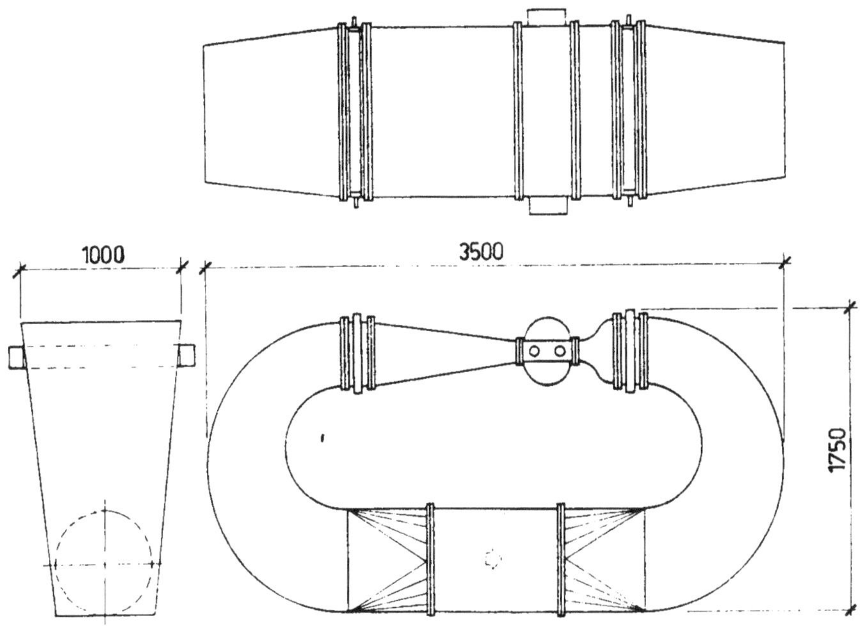

Fig. 25. Gas-flow structure of a CEILAP-Universidad de Buenos Aires design.

Considerable effort has been spent devising electrode configurations favouring stable glow discharges. Segmented anodes, segmented cathodes, plane, bar, cylindric with rounded or conic points in a variety of configurations impossible to describe here have been used [82]-[88]. Some configurations are shown in fig. 27.

DC and RF [82] auxiliary discharges and magnetic fields [89] have also been used for stabilization of the discharge.

Finally, two more sophisticated methods have been used, not only to stabilize the discharge, but also to obtain a more efficient electric pumping by independent control of E/N and the necessary degree of initial ionization (sections 6, 10.3). One is the electron beam ionization technique, in which electrons are injected into the discharge through a thin metallic foil and a more recent method is known as PIE laser [90]-[93] (Photoinitiated, Impulse enhanced, Electrically excited). In this method the active volume is initially photopreionized by an extended array of UV spark sources. A pulsed avalanching field of short duration is then applied to provide enhanced ionization through Townsend multiplication and then vibrational excitation is achieved by a continous non-avalanching DC discharge. Pulse repetition frequency is not critical and is around 10 khz.

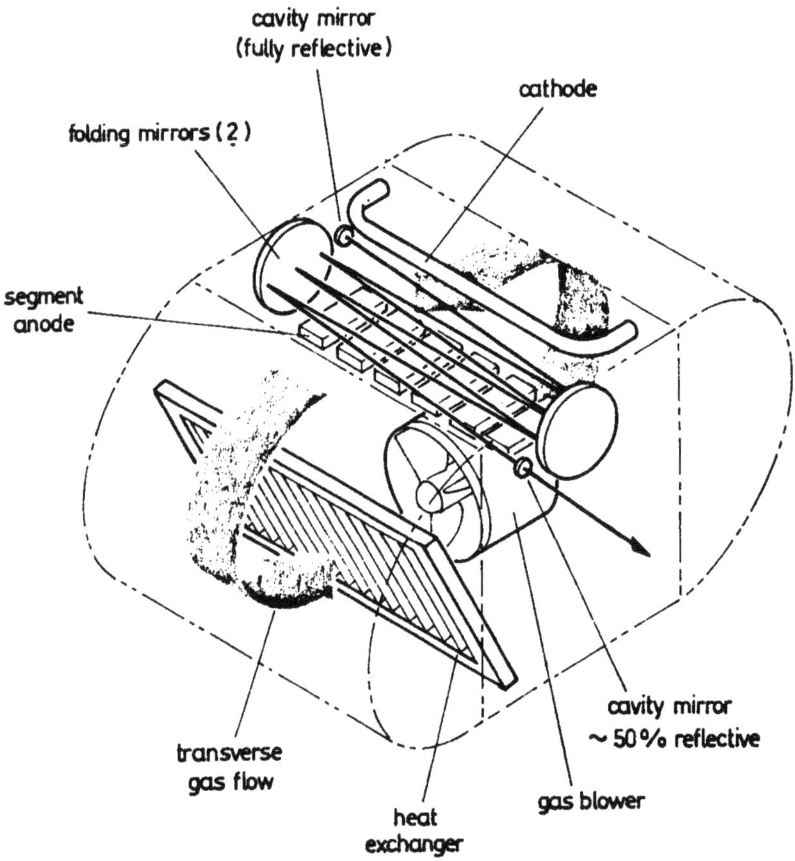

Fig. 26. Commercial transverse flow design.

Power outputs of up to 25 kW CW [93] with 19% efficiency have been obtained with this family of devices.

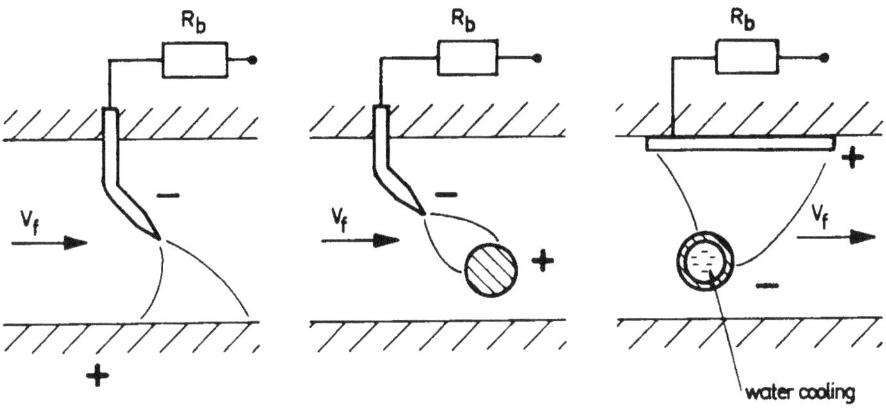

Fig. 27. Some electrode configurations for transverse flow.

11. Pulsed high pressure lasers

11.1. Introduction

A new family of lasers resulted from the work of Dumanchin and Rocca-Serra (16) and Beaulieu (17) in 1969 and 1970, who reported pulsed high pressure operation using transverse rather than longitudinal excitation. In the first case, uniform excitation up to 0.7 atm was obtained using two parallel rounded edges electrodes a few centimeters apart and four external electrodes to provide some initial ionisation near the main ones. In the second case, operation at atmospheric pressure was obtained using a large number of transverse discharges in parallel between resistor loaded pins acting as cathodes and a common anode. In both cases submicrosecond rise time pulses were used. These lasers are called TEA (Transversely Excited Atmospheric pressure). Subsequently, many other electrode configurations have been proposed and demostrated in order to increase the cross-section and uniformity of the discharge.

Dumanchin et al. [94], Laflamme [18] and Lamberton and Pearson [95] devised systems in which auxiliary discharges of limited energy provided preionisation of large cross-sectional areas. Richardson et al. [96] used volumetric preionisation by UV radiation from an auxiliary multiple arc discharge and obtained 300 Joules, 3 GW pulses at 8% efficiency.

TEA lasers are relatively simple and compact devices and represent a considerable advance over the earlier gain or Q-switched low pressure laser. However, the trend continued and operation of similar devices at pressures as high as 14 atm have been reported [97]. A distinct category of pulsed CO_2 lasers started with the work of Basov et al. [98] and Fenstermacher et al. [99], who used electron beams injected into the laser volume in order to produce substantial preionisation. In this mode of operation the electron beam controls the charge production while the transverse electric field can be adjusted to provide the optimum E/N for efficient pumping of the upper laser level.

11.2. Pulsed high pressure discharge

Other factors being equal, the energy output per unit volume should increase with the pressure due to the increase in the number of active molecules. Two types of problems are encountered, however, in trying to increase the working pressure.

Firstly, all the de-excitation rates which are collision-dominated are faster. The upper laser level of each CO_2 molecule must be pumped more times per second to maintain the population inversion. As the pumping process is not 100% efficient, a given amount of energy goes to other degrees of freedom and is finally thermalised. The resulting increase of temperature produces the detrimental effects already described.

Secondly, glow discharges tend to constrict into arcs above pressures of the order of 100 torr, making uniform excitation of the gas very difficult.

The first problem can be circumvented using either fast flow to provide convective cooling of the gas or using pulsed excitation, or both. The second approach is used in TEA lasers, in which the current pulse is generally shorter than 1 μsec.

The problem of uniformity of the discharge is a complex one and has received considerable attention. The mechanism of streamer formation in high voltage atmospheric, pulsed discharges is different from that of Townsend discharges [31]. Streamers begin with single electrons and grow as avalanches propagating towards the anode. In the initial stages, free electrons are added via collisions, but after about 15 nsec. light is emitted from the head and the avalanche progresses at much higher speed (about 10^9 cm/sec) due to photoionisation. Within a few nanoseconds these avalanches form a uniformly ionised region extending from very near the cathode to the anode. If the external supply of energy is maintained over a period of the order of several hundred nanoseconds, the discharge will become an arc. Therefore, uniform excitation can be obtained either using current pulses shorter than the arc formation time or providing some external way of limiting and spreading the current, or both.

Some important conclusions concerning high pressure pulsed CO_2 lasers can be obtained from Nighan's curves (fig. 10a and 10 b) which, although calculated for steady state, are also applicable to each successive value of E/N (or u) taken on by the pulsed discharge (this is because the discharge parameters change in a time scale which is slow compared with the electron collision rate). In the initial part of the pulse only fig. 10 b needs to be considered because the collisional transfer from N_2 to CO_2 (00^01) is comparatively slow. We observe that the compromise between good CO_2 (00^01) level pumping and an adequate rate of ionisation is more stringent than in low pressure discharges. It is clear from the curves that TEA lasers without preionisation tend to work with higher E/N and are therefore less efficient than CO_2 CW or slow pulsed low pressure lasers. It is also apparent that external preionisation will favour working at lower E/N and therefore more efficiently.

Having dealt with these excitation and efficiency problems, we shall now describe the ideal parametric behaviour which is to be expected from a pulsed CO_2 laser as a function of pressure; E/N and gas composition will be assumed constant [31] [100]:

- The fraction of energy delivered to the various levels is constant and therefore the output energy per unit volume increases linearly with pressure (provided, obviously, that the pumping, which is proportional to the current at constant E/N, is also increased accordingly).
- All the relaxation times, which are collision-dominated are inversely proportional to the pressure. This causes the pulsewidth to follow the same law.
- As a result of the above two conclusions the peak power increases quadratically with pressure.
- The maximum permissible repetition ratio is limited by the de-excitation of the lower laser level and therefore increases linearly with pressure.
- The maximum attainable average power per unit volume is proportional to the energy per pulse and the repetition ratio and as a consequence increases quadratically with pressure.

- The linewidth increases linearly with pressure. This is valid from pressures at which Doppler broadening is negligible (about 20 - 50 torr) to the point at which overlapping of adjacent rotational lines become appreciable (8 - 10 atm). (See sec. 5). The small signal gain is independent of pressure. (See sec. 5).

This ideal behaviour is roughly in agreement with experimental results regardless of the specific excitation configuration being used (provided E/N and gas composition are held constant and the discharges are arcfree). In a given laser, however, power and energy finally drop with increasing pressure due to discharge deterioration. Discharge quality of UV preionised TEA lasers is improved by the aggregate of small quantities (<< 1%) low ionisation (6 eV) organic substances as Triethylamine, Trimethylamine or Tri-n-propylamine [101]. This is called "seeding" the gas mixture and permits better energy deposition and consequently higher output pulse energies.

11.3. Output characteristics

Output pulses of TEA lasers have an initial high intensity spike and a secondary long-tailed one (fig. 28). The initial spike is delayed from the initiation of the current pulse. This delay is of the order of 1 μsec and decreases with increasing power output. The initial spike is due to the directly pumped CO_2 whereas the secondary one is due to the relatively slow collisional transfer rate from N_2 (v = 1) to CO_2 (00^01) which is about 2 x 10^4/sec torr [102] at room temperature. This is not noticeable in CW lasers. The delay is due to a relatively slow gain build-up which can be explained as follows [31]: as mentioned in connection with low pressure lasers, significant population of the 10^00 and 02^00 levels can occur due to electronic impact, particularly if u is in the range of 3 - 5 eV.

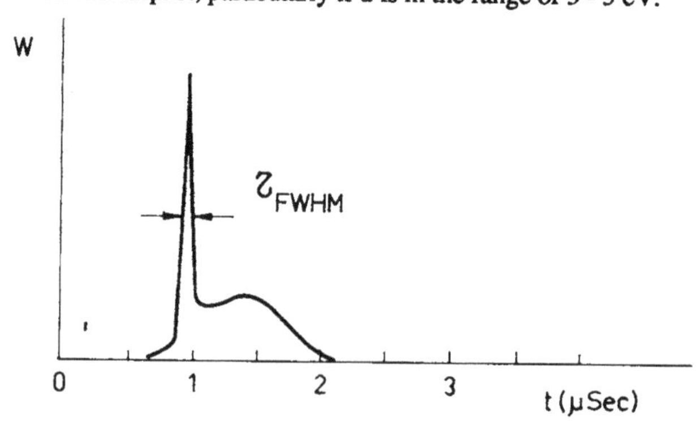

Fig.28 : Typical TEA CO2 - N2-He laser pulse

Therefore, the gain in the early stages is low and only increases after relaxation of the (coupled) 10^00 and 02^00 levels.

Manes and Seguin [103], based on previous work by Witteman [104], developed a so called five temperature model of the TEA laser which is in fair agreement with the experimental results. The different terms of a given mode of vibration were supposed to be exactly equidistant and tightly coupled (and in Boltzmann equilibrium) due to fast v - v interactions. Consequently, an effective temperature was associated with each of the four CO_2"ladders" (00v), (0v0), (v00) and N_2(v). The rotational distribution within each vibrational state was assumed in equilibrium with the translational degrees of freedom at temperature T. The instantaneous pumping of each level was calculated from Nighan's theory using measured values of voltage and current. Rate equations for the energy stored in each mode were written and numerically solved. The model predicts the already described pulseshape of TEA lasers for different gas compositions (as well as instantaneous population inversion, gain, delay, etc.) and confirms earlier qualitative predictions. Output pulse energy is about 18 Joules/litre and may rise up to 35 Joules/litre in well UV preionised, seeded atmospheric pressure mixtures [105]. With FWHM pulses of typically 50 nsec, and assuming half the energy is in the N_2 associated pulse, peak powers of the order of 200 MW can be obtained for a relatively modest 0.5 litre discharge volume. Moreover, these lasers are volume scalable.

The high bandwidth of TEA lasers makes them particularly appropriate for short pulse generation by mode locking. At atmospheric pressure eq. (10) gives over 3 GHZ bandwidth and short pulses (around 0.5 - 1 nsec) can be generated. [56], [106].

11.4. TEA configurations

In this section we will give a brief description of the main features of some representative TEA configurations (fig. 29).

The scheme of Dumanchin and Rocca-Serra (a) [16] was only capable of sub-atmospheric operation, but had some primitive way of preionisation due to the external electrodes. In Beaulieu's scheme(b) [17] the current was externally limited and shared by the use of external 1 kΩ resistors. Laurie and Hale(b) [107] were able to produce uniform discharges in pin-rod systems without external resistors using faster electric pulses.

Johnson [108] used independent capacitors for each pin, thus increasing the efficiency (no Ri^2), but at a higher initial cost. A different category of lasers is due to the work of Dumanchin et al.(e) [94], Laflamme (c) [18] and Lamberton and Pearson (d) [95]. These

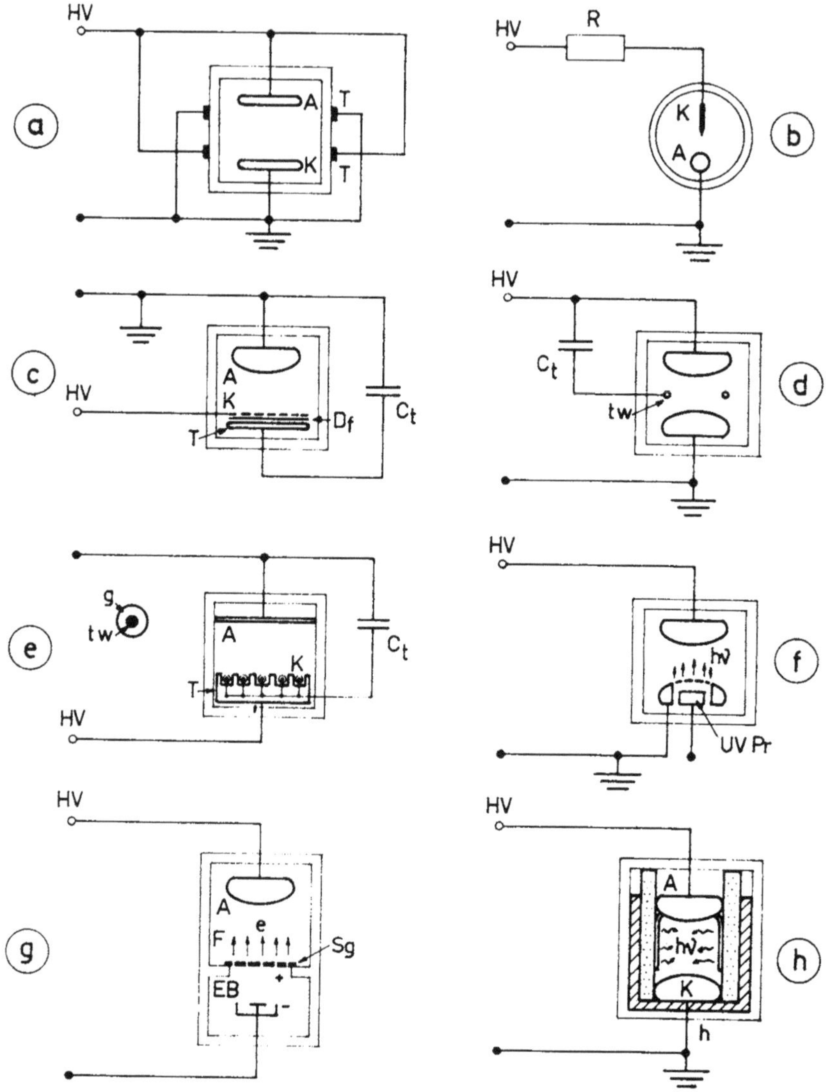

Fig. 29. Schemes of high pressure CO_2 lasers. A: anode, K: cathode, T: trigger, C_t: trigger capacitor, D_f: dielectric foil, t_W: trigger wire, F: Titanium or mylar foil, S_g: supporting rid, EB: electron gene, SG: spark gap and R: charging resistors.

lasers are inherently more efficient because some degree of preionisation is provided by means of energy limited discharges, allowing operation at lower values of E/N. In the first two cases the source of electrons is the corona discharge appearing all over the cathode due to the extremely high electric fields (10^5 V/cm) between cathode and trigger. In the third case, a weak sheet discharge is produced between anode and the wire. This discharge, which is limited by the capacitor C_t, generates UV radiation providing the required preionisation. Specially contoured electrodes (see later) are used to avoid field concentration. A third category of strongly preionised lasers is due to the work of Richardson et al. (f) [96], Basov et al. (g) [98] and Fenstermacher et al. (g) [99]. In the first case an externally pulsed and carefully timed UV source (rows of arcs) provides the necessary preionisation, whereas in the second case an electron current generated by an external electron gun is injected into the laser volume through a thin foil. Both systems are very efficient because they allow operation at low value of E/N, thus producing more effective electron pumping.

Later [109], [110] a simpler technique involving corona discharges along dielectric plates placed sidewise along the laser channel has been successfully used for UV preionization (h).

The very uniform electric field distributions necessary to excite arc-free discharges require specially contoured electrodes. As proposed by Chang [111] electrodes can be designed following a pair of symmetric equipotentials can be obtained from conformal transformations. From this Chang obtained the equations of electrode contours which give the more uniform field distributions in the central part of the interelectrode region.

Another electrode contour also used is the Rogowski type [59], derived from a couple of semiinfinite plane electrodes (fig. 30 b). As each electrode will be composed of two interrupted Rogowski curves and a flat central part some irregularities or discontinuities appear at the joining points. Ernst [112] proposed another profile obtained adding extra terms to Chang's conformal transformation. This reduces the electrode width for the same active region as compared with Chang's profile.

Fig. 31 shows a pair of Chang electrodes in an UV prionized design 8 J/pulse TEA laser (design and construction at CITEFA-CEILAP) [113].

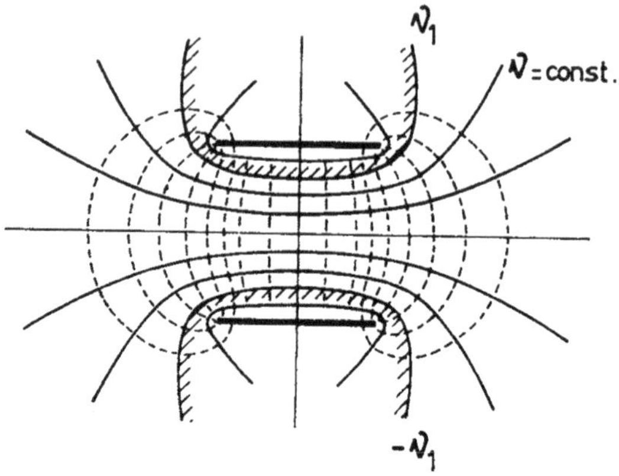

Fig. 30a. Chang profile electrodes.

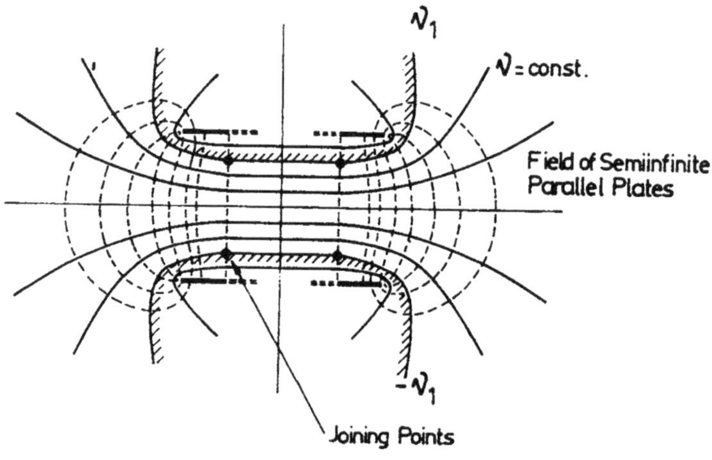

Fig. 30b. Rogowski profile electrodes.

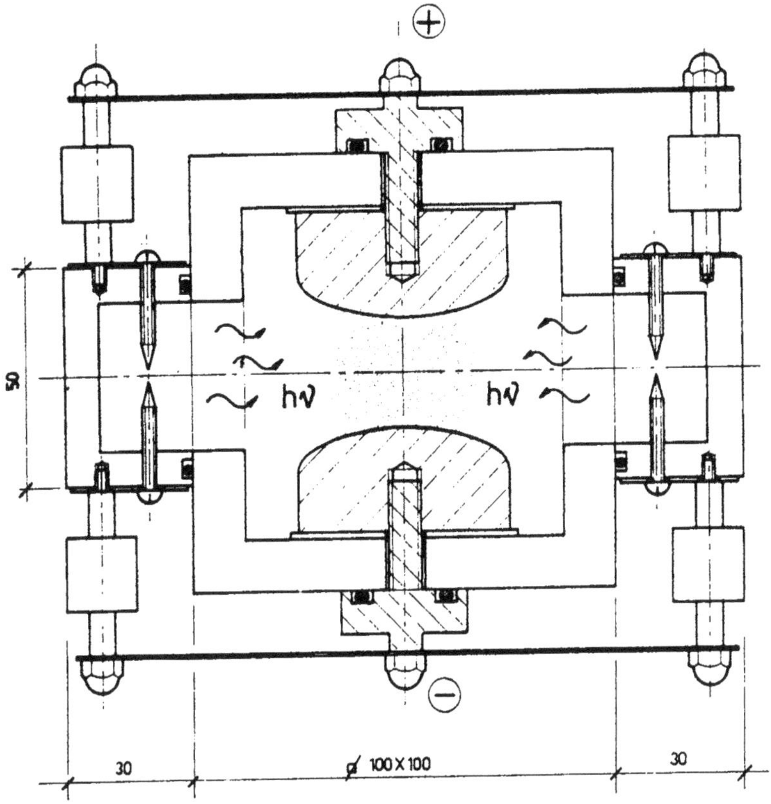

Fig. 31. Chang electrodes in a UV preionized design 8 J/pulse TEA-laser (CEILAP).

12. Outline of CO_2 lasers applications

12.1. Introduction

CO_2 lasers are widely used because of high output power available, high efficiency and low cost per Watt. As the generated wavelengths around 10 μm are strongly absorbed by most non metals these lasers are useful in cutting fabrics, plastics, leather, ply wood and in surgery. The increase of the normally low 10.6 μm absorption by metals associated with the high power densities available makes these lasers also adequate for cutting, welding and heat treating metals.

The tunability of CO_2 lasers in the 9-11 μm range allows the resonant strong interaction with polyatomic molecules leading to research in laser chemistry and its applications.

12.2. CW CO_2 laser applications

High reflectivity at 10.6 μm and high thermal diffusivity are adverse properties for CO_2 laser processing of metals. However this can be overcome because absorption increases at the high power densities available with these lasers, particularly when reaching the melting temperature and above. Therefore CO_2 lasers are widely used for metal processing. Power density levels are of the order of 10^3-10^5 W/cm^2 for heat treating, 10^5-10^7 W/cm^2 for welding and 10^6-10^7 W/cm^2 for drilling and cutting [114].

Laser hardening can be precisely localized with little thermal distortion of the surrounding region.

Heat treating requires reaching a critical temperature in a given zone, below its melting point. This allows the use of a defocused beam of relatively low power density, but in order to achieve sufficient energy absorption, the surface of the component must be covered with an absorbing layer, such as manganese phosphate.

Moreover, the cooling is normally fast enough to be self quenching.

The advantages of laser welding over other methods as electrom beams or arcs are narrow welds, narrow Heat Affected Zone (HAZ), easy automation, independence of operator skill, and access to otherwise inaccessible areas.

Tensile strength tests show that in general breaking takes place outside the weld and HAZ.

A coaxial or sometimes parallel to the surface inert gas jet is used to avoid oxidation. Argon is used in the low power range and is replaced by Helium at high power levels in order to avoid ionization and the consequent generation of an absorbing plasma.

Table 2 shows some representative figures for welding of metals. Comparison of values of different workers is difficult because of the differences in beam mode quality, focal length, surface state and gas flow rates.

TABLE 2
WELDING PERFORMANCE FOR METALS

MAT.	THICKNESS mm	POWER kW	CUTTING SPEED cm/min	REFERENCE
Steel	2-2.5	3.5	130	114
Steel	5	1	965	115
Titanium alloy	3-3.5	3.2	100	114
Titanium alloy	2	2	295	115
Stainless Steel	0.125	0.25	126	122
Stainless Steel	6.5	2	35	115

In cutting and drilling a coaxial oxygen jet is generally used particularly in ferrous materials because the exothermicity of the reaction accelerates the cutting speed. Another important factor that influences the efficiency of energy absorption is the polarization of the laser output. When the output is linearly polarized extracavity optical components are usually required to produce good cut quality, since otherwise the effect will be different depending on the direction of cutting.

Table 3 shows some results for cutting metals and non-metals.

Low power (20-30 W) CO_2 lasers are used for marking plastics, paper, glass and ceramics and microsoldering PC boards in non-contact, computer-controlled systems. They are also practical in removing coatings from metals without significant heating the substrate and in evaporating non metallic materials for deposition of thin films in high-vacuum systems.

CO_2 lasers with power outputs in the range 1 to 100 W are useful in surgery as the 10.6 μm light vaporizes the cells without heating the surrounding tissue. The beam directed to the healing area by either an articulated arm or a joystick mounted mirror associated with a surgical microscope can be used to cut tissue and debulk tumors, cauterizing at the same time as cutting. The main advantages attributed to the laser technique are precision, absence of bleeding, less spreading of malignant cells and, in some cases, considerable shorter period of convalescence.

TABLE 3
CO_2 CUTTING PERFORMANCE FOR METALS

MATERIAL	THICKNESS mm	POWER Kw	SPEED cm/min	GAS	REF.
Aluminium	12.7	3	81	O_2	[115]
Aluminium	12.7	15	228	none	[116]
Aluminium	1.0	0.4	202	O_2	[115]
Aluminium	3.17	2.0	165	O_2	[115]
Stainless Steel	9.5	3.0	187	O_2	[115]
Stainless Steel	3.0	3.0	252	O_2	[117]
Stainless Steel	3.17	0.5	152	O_2	115]
Stainless Steel	0.1	0.2	457	O_2	[118]
Steel	54	6.0	33	O_2	[117]
Steel	3	4.0	406	O_2	[117]
Titanium	25.4	10	508	O_2	[119]
Titanium	50.6	3.8	51	O_2	[117]
Titanium	3	4.3	127.2	O_2	[117]
Acrylic	3.1	0.3	183	Air	[120]
Acrylic	6.35	0.5	432	Air	[115]
Acrylic	5	0.2			[121]
Mylar	$25x10^{-3}$	0.5	$3.5x10^4$	Air	[115]
Mylar	0.3	0.1		N_2	[121]
Plywood	5	0.225	110	Air	[120]
Hard wood	25.4	2	114	Air	115]
Asbestos board	20	2	240	Air	[115]
Asbestos board	9	0.21		N_2	[121]
Cardboard	0.1	0.5	10^4	Air	[115]
Leather	3	0.225	305	Air	[120]

In Gynecology, the CO_2 laser has been used in various premalignant and malignant conditions, and in surgery for fertility and treatment of endometriosis.

In Dermatology, it has been shown to be very useful to remove tatoos and other cutaneous formations. In Neurosurgery, Ascher and Heppner demonstrated that it was possible to vaporize large tumours of the brain stem and spinal cord. [123].

The CO_2 laser is also used currently in other specialties, as otorhinolaryngology, dentistry and general surgery. The dramatic increase in the number of books, papers

[124], laser meetings and other medical activities, evidences the importance of laser surgery. It is used also in research related with Cardiology [125], [126] and many other branches of Medicine. The CO_2 laser instruments for surgery are currently used in hospitals all around the world. In reference [127] it is possible to find details about the design and construction at the CEILAP of a surgical device.

Low power (10 W) CO_2 lasers show a good performance for range finding for distances below a few kilometers. CO_2 lidar transmitters are also useful in measuring wind speed and concentration of pollutants in the atmosphere.

Radiation at 10.6 μm can also be useful to initiate and control chemical reactions of polyatomic molecules, leading in some cases to an increase in the cost advantage of the process. A great number of molecules absorb energy at wavelengths emitted by the CO_2 laser, such as NH_3, SF_6, C_2H_4, BCl_3, CF_2Cl_2, etc. Particularly C_2H_4 is converted into acetylene [128] and CF_2Cl_2 into $CClF_3$ [129].

These lasers are also used for in-situ monitoring exhaust gas emissions and for remote sensing of physical characteristics of a process, as temperature or particle size.

12.3. TEA CO_2 laser applications

The availability of simple tunable pulsed TE CO_2 lasers led to the study of the interaction of powerful infrared radiation with polyatomic molecules having vibrational absorption bands in the 9-11 um region. Collisionless multiphoton dissociation was obtained, that means excitation through vibrational states in the ground electronic state up to dissociation. The main goal of this result, as known, is isotope separation. Many reviews have been published on this subject [130], [131].

Isotopes such as D, Cl^{37}, C^{12} are of great interest and their obtention by the laser method is a subject of intensive research, in order to study the economic feasibility at industrial scale. Measurements of the dependence of energy absorption and dissociation of molecules containing these isotopes on laser wavelength, fluence, focusing of radiation, pressure of the sample, buffer gas and other parameters permit a better understanding of the phenomenon of multiphoton dissociation. Some studies on this subject at the CEILAP have been published in [132], [133].

Furthermore the laser isotope separation effectiveness is influenced by the relaxation phenomena occurring in natural samples due to energy transfer through collisions between the excited molecules and the surrounding.

The infrared fluorescence emitted by the molecules resonant with the CO_2 laser radiation decays with a time constant characteristic of the process of energy transfer, which can be vibration-vibration between molecules, vibration-translation between molecules or molecules and a buffer, and diffusion. The relaxation rates of these processes have been also extensively studied at the CEILAP [134]-[137].

References

[1]. C.K.N. Patel, W.L. Faust and R.A. McFarlane, Bull. Am. Phys. Soc. 9, 500 (1964).

[2]. C.K.N. Patel, Phys.Rev. 136A, 1187 (1964).

[3]. C.K.N. Patel, Phys.Rev.Lett. 12, 588 (1964).

[4]. F. Legay and N.Legay-Sommaire, Comp. Ren. Ac. Sciences 259, 99 (1964).

[5] C.K.N. Patel, Phys.Rev.Lett. 13, 617 (1964).

[6] C.K.N. Patel, Appl.Phys.Lett. 7, 15 (1965).

[7] T. Bridges and C.K.N. Patel, Appl. Phys. Lett. 7, 244 (1965).

[8] G. Moeller and J. Ridgen, Appl. Phys. Lett. 7, 274 (1965).

[9] C.Patel, P. Tieu, J.H. Mcfee, Appl. Phys. Lett. 7, 290 (1965).

[10] W.Witteman, Appl.Phys.Lett. 11, 337 (1967).

[11] C.Freed, IEEE J. Quantum Electron. QE-3, 203 (1967).

[12] M. Kovacs, G. Flynn and A. Javan, Appl. Phys. Lett. 8, 62 (1966).

[13] A. Hill, Appl.Phys.Lett. 12, 324 (1968).

[14] A. Hill, Appl.Phys.Lett. 16, 423 (1970).

[15] C. Brown and J. Davis, Appl.Phys.Lett. 21, 480 (1972).

[16] R. Dumanchin and J. Rocca-Serra, Comp. Ren. Ac. Sciences 269,916 (1969).

[17] A. Beaulieu, Appl. Phys. Lett. 16, 504 (1970).

[18] A.K. Laflamme, Rev. of Sc. Inst. 41, 1578 (1970).

[19] N. Basov and A. Oraevskii, Sov. Phys. JETP 17, 1171 (1973).

[20] E.T. Gerry, IEEE Spectrum 7, 51 (1970).

[21] A.E. Siegman, Proc. IEEE 53, 277 (1965).

[22] T. Cool, R. Stephens and T. Falk, Int. J.Chem.Kin. 1, 495 (1969).

[23] T. Chang and O.Wood III, Appl.Phys.Lett. 21, 19 (1972)

[24] T. Bridges, E. Burkhardt and P. Smith, Appl. Phys. Lett. 20, 403 (1972).

[25] V.P. Tychinskii, Sov. Phys. Uspekhi. 10, 131 (1967).

[26] N.N. Sobolev and V.V. Sokovikov, Sov. Phys.Uspekhi, 10, 153 (1967)

[27] P.K. Cheo, "CO_2 Lasers", Chapter 2 of "Lasers,a Series of Advances", Marcel Dekker, N.Y., 1971.

[28] A.J. DeMaria, Proc. IEEE, 61, 731 (1973).

[29] C.S. Willett, "Introduction to gas Lasers: Population inversion mechanisms", Pergamon Press, (1974).

[30] G.A. Abilsiitov et al., Sov. J. Quantum Electron., 11, 1535, (1981).

[31] O.R. Wood, Proc. IEEE, 62, 355 (1974).

[32] W.W. Duley "CO_2 Lasers", Academic Press, N.Y., (1976)

[33] W.J. Witteman, "The CO_2 Laser", Springer-Verlag, Berlin, (1987).

[34] G. Herzberg, Molecular spectra and molecular structure II: Infrared and Raman spectra of polyatomic molecules, Van Nostrand, Princeton, NJ (1964).

[35] J.A. Howe and R.A. Mc Farlane, J. Mol. Spectr. 19, 224 (1966).

[36] B. Hartman and B.Kleman, Can. J. Phys. 44,1609 (1966).

[37] John D. Anderson Jr.: "Gasdynamic Lasers: An Introduction", Academic Press, (1976).

[38] S.A. Losev, "Gasdynamic Lasers", Springer-Verlag, (1981).

[39] E.T. Gerry, Gas Dynamic Lasers, paper N 71-23, A.I.A.A. 9th Aerospace Sciences Meeting. N.Y. Jan.(1971).

[40] R.W. Gross and J.F. Bott, "Handbook of Chemical Lasers", J. Wiley & Sons, (1976).

[41] A. Ben-Shaul, Y. Haas, K.L. Kompa and R.D. Levine, "Lasers and chemical change", Springer-Verlag (1981).

[42] J.M. Sirota and W.H. Christiansen, IEEE J. Quantum Electron. QE-21, 1777 (1985).

[43] R.J. Insuik and W. H. Christiansen, IEEE J. Quantum Electron. QE-20, 622 (1984).

[44] O.P. Judd, unpublished, as cited by O.R. Wood, Proc. IEEE 62, 355 (1974).

[45] J.S. Ostrem, H. Gamo: Proc. IEEE 61, 1656 (1973).

[46] E.T. Gerry, D.A. Leonard: Appl. Phys. Lett. 8, 227 (1966).

[47] C.K.N. Patel: Phys. Rev. Lett 12, 588 (1964).

[48] D. Rapp and P. Englander-Golden, J. Chem. Phys. 40, 573 (1964).

[49] W.L. Nighan, Appl. Phys. Lett. 15, 355 (1969).

[50] W.L. Nighan, Phys. Rev. A2, 1989 (1970).

[51] A. Burgos, H. Mancini, E. Quel and V. Slezak, Jap. Journal of Appl. Phys. 15, 2387 (1976).

[52] G.J. Schulz, Phys. Rev. 116, 1141 (1959).

[53] G.J. Schulz, Phys. Rev. 135, A988 (1964).

[54] W.W. Rigrod: J. Appl. Phys. 36, 2487 (1965)

[55] B. Etcheverry, C.A. Rosito: Com. Cient_fica 82, CITEFA, (1979).

[56] A. F. Gibson, M.F. Kimmitt, C.A. Rosito: Appl. Phys. Lett. 18, 546 (1971b).

[57] A.M. Robinson, J.S. Weiss: Can. J. Phys. 60, 1656 (1982).

[58] D.C. Smith and J.H. McCoy: Appl. Phys. Lett. 15, 282 (1969).

[59] J.C. Cobine: "Gaseous Conductors", Dover, N.W. (1958).

[60] E. Nasser: "Fundamentals of gaseous ionization and plasma electronics", Wiley-Interscience, N.Y. 1971.

[61] P.K. Cheo, IEEE J. Quantum Electron. QE-3, 683 (1967).

[62] J.L. Lachambre, J.L. Ladrambre, J. Macfarlane, G. Otis and P. Lavigne, Appl. Phys. Lett. 32, 652 (1978).

[63] C.P. Christensen, Ch. Freed and H.A. Haus, IEEE J. Quantum Electron. QE-5, 276 (1969).

[64] O. Ersoy, Optical and Quantum Electronics. 7, 247 (1975).

[65] W.J. Witteman, IEEE J. Quantum Electron. QE-3, 373 (1967).

[66] F. Strumia and N. Joli, Physics of new laser sources, NATO Series B, Phys. 132 (1985).

[67] K.D. Laakmann, U.S. Patent N 4, 169, 251 (1979).

[68] G.D. Griffith, U.S. Patent N 4, 352, 188 (1982).

[69] D. He and D.R. Hall, J. Appl. Phys. 54, 4367 (1983).

[70] D. He and D.R. Hall, Appl. Phys. Lett. 43, 726 (1983).

[71] J.P. Crancon and B. Lavarini, French Patent No. 2108912 (1970).

[72] G.I. Kozlov and V.A. Kuznetsov, Sov. J. Quantum Electron. 15, 362 (1985).

[73] D.G. Youmans, Appl. Phys. Lett. 44, 365 (1984).

[74] A.F. Glova et al., Sov. Tech. Phys. Lett. 11, 102 (1985).

[75] L.A. Newman et al., Appl. Phys. Lett. 48, 1707 (1986).

[76] J.G. Xin and D.R. Hall, Appl.Phys.Lett. 51, 469 (1987).

[77] C.J. Buczek, R.J. Freiberg, P. Chenausky and R.J. Wayne, Proc. IEEE. 59, 659 (1971).

[78] R. Beck, Appl. Phys. B. B42, 233 (1987).

[79] D. Campbell, F. Gonzalez, C.A. Rosito, G.D. Santiago, Third Latinamerican Symposium on Laser and Applications, Mar del Plata, 1-3 Sept., 1988.

[80] S.L. Ream, Laser Focus, p. 43, (Dec. 1982).

[81] R. Targ and W.B. Tiffany, Appl.Phys.Lett. 15, 302 (1969).

[82] N. Tabata et al. 5th International Symposium on Gas Flow and Chemical Lasers, Proceedings, Oxford 20-24th August, (1984).

[83] Li Zaiguang et al., ibid.

[84] T. Akiba, H. Nagai and M. Hishii, IEEE J. Quantum Electron. QE-15, 162 (1979).

[85] M. Kasamatsu et al., IEEE J. Quantum Electron. QE-18, 173, (1982).

[86] T.W. Johns and J.A. Nation, Rev. Sci. Instrum., 44, 169 (1973).

[87] A.V. Artamonov et al., Sov.J. Quantum Electron. 8, 528 (1978).

[88] A. Echbreth, and J.W. Davis, IEEE J. Quantum Electron. QE-8, 139 (1972).

[89] A. Seguin et al., Appl. Phys. B 43, 127 (1987).

[90] H. Kyong, J.J. Seguin and J. Tulip, IEEE J. Quantum Electron. QE-15, 44 (1979).

[91] H. Seguin, K. Nam and J. Tulip, J. Quantum Electron, QE-15, 50 (1979).

[92] A. Nath, H. Seguin and V. Seguin, J. Quantum Electron. QE-2, 268 (1986).

[93] V.E. Marchant, L. Focus. 21, 162 (1985).

[94] R. Dumanchin, M. Michon, J.C. Farcy and J. Rocca-Serra, IEEE J. Quantum Electron. QE-8, 163 (1972).

[95] H.M. Lamberton and P.R. Pearson, Electron. Lett. 7,141 (1971).

[96] M.C. Richardson, A.J. Alcock, K. Leopold and P. Burtyn, IEEE J. Quantum Electron. QE-9, 236 (1973).

[97] A.J. Alcock, K. Leopold and M.C. Richardson, IEEE/OSA Conf. Laser Engineering and Applications, Washington D.C., May 30-June 1, 1973.

[98] N.Q. Basov et al., JETP Lett. 14, 285 (1971).

[99] C.A. Fenstermacher, M.J. Nutter, W.T. Leland and K. Boyer, Appl. Phys. Lett. 20, 56 (1972).

[100] A.J. Beaulieu, Proc. IEEE 59, 667 (1971).

[101] E. Morikawa, J. of Appl. Phys. 48, 1229 (1977); and references therein.

[102] R.L. Taylor and S. Bitterman, Rev. Mod. Phys. 41, 26 (1969).
[103] K.R. Manes and H.J. Seguin, J. Appl. Phys. 43, 5073 (1972).
[104] W.J. Witteman, Phillips Res. Rep. 21, 73 (1966).
[105] B.J. Reits and H.M. Olbertz, Appl. Phys. Lett. 27, 24 (1975).
[106] A.F. Gibson, C.A. Rosito, C.A. Raffo and M.F. Kimmitt, Appl.Phys.Lett. 21, 356 (1972a).
[107] K.A. Laurie and M.M. Hale, IEEE J. Quantum Electron. QE-7, 530 (1971).
[108] D.J. Johnson, IEEE J. Quantum Electron. QE-7, 185 (1971).
[109] G.J. Ernst, Rev. Sci. Instrum. 48, 1281 (1977).
[110] D.J. Brink and V. Hasson, J. Appl. Phys. 49, 2250 (1978).
[111] T.Y. Chang, Rev. Sc. Instrum. 44, 405 (1973).
[112] G.J. Ernst, Opt. Comm. 49, 275 (1978).
[113] C.A. Rosito, R. D'Angelo, F. Gonzalez, D. Krysa, O. Neguembor, 1er. Simposio Laseres y Aplicaciones, CITEFA-CEILAP, 15-16 October 1981.
[114] G.A. Abilsiitov, E.P. Velikhov, Opt. and Laser Tech. 30, (1984).
[115] Control Laser Corporation, Pub CO_2, CL 30, June 1983.
[116] E.V. Locke, E.D. Hoag and R.A. Hella, IEEE J. Quantum Electron. QE-8, 132 (1972).
[117] C.M. Banas and R. Webb, Proceedings of the IEEE, 70, 556 (1982).
[118] W.W. Duley, J.N. Gonsalves, Opt. Laser Tech. 6, 78 (1974).
[119] E.V. Locke and R.A. Hella, IEEE J. Quantum Electron. QE-10, 179 (1974).
[120] J.E. Harry and F.W. Lunau, IEEE Trans. Ind. Appl., IA-8, 418 (1972).
[121] D. Krysa and C. Alzola, private communication.
[122] J.M. Webster, Metals Progr. 98, 59 (1970).
[123] P. Ascher and F. Heppner, Acta Chir. Aust. 2, 32 (1977).
[124] J.A. Dixon, Proceedings of the IEEE. 70, 579 (1982) and references therein.
[125] R. Rabinovich, E. Quel, E. Valero, J. Milei, R. Pesce, R. Beigelman, C. Rosito, R. Favaloro, Cong. of Cardiology, Rio (1986).
[126] E. Quel, E. Valero, R. Rabinovich, J. Milei, R. Pesce, R. Beigelman, C. Rosito, R. Favaloro, Internat. Cardiology Meeting, Boston (1986).
[127] C. Rosito, D. Krysa, L. de Pablo Pardo, M. Ferrari, F. Gonzalez y G. Santiago, Commun. presented at "III Encuentro Latinoamericano sobre Laser y Aplicaciones", Mar del Plata, Sep. (1988).
[128] E. Quel, X. de Hemptime, Ann. Soc. Sci. Bruxelles 83, 262 (1969).

[129] V. Slezak, J. Caballero, A. Burgos and E. Quel, Chem. Phys. Lett. 54, 105 (1978).

[130] V.S. Letokhov, Nonlinear laser chemistry, Springer-Verlag-Berlin (1983).

[131] V.N. Bagratashvili, V.S. Letokhov, A.A. Makarov, E.A. Ryabov, Multiple photon infrared laser photophysics and photochemistry, Harwood Academic Publishers (1985).

[132] M.L. Azcarate, E.J. Quel, B. Toselli, J.C. Ferrero, E.H. Staricco, J. Phys. Chem. 92, 403 (1988).

[133] M.L. Azcarate, E.J. Quel, J. Phys. Chem. (to be published).

[134] M.L. Azcarate, E.J. Quel, Appl. Phys. B (to be published).

[135] E.M. Alonso, R.J. D'Angelo, A.L. Peuriot, E.J. Quel, V. Slezak, Optics Comm. 68, 265 (1988).

[136] E.M. Alonso, R.J. D'Angelo, E.J. Quel, Appl.Phys. B. 47, 233 (1988).

[137] E.M. Alonso, R.J. D'Angelo, E.J. Quel, Appl. Phys. B, (to be published).

INFRARED MULTIPLE-PHOTON EXCITATION OF $CDCl_3$ INDUCED BY A TEA CO_2 LASER

M. L. Azcarate* and E. J. Quel
CEILAP (CITEFA-CONICET). Zufriategui 4380. (1603) Villa Martelli. Argentina

B. Toselli, J. C. Ferrero, and E. H. Staricco
INFIQC, Departamento de Fisicoquimica, Facultad de Ciencias Quimicas, Universidad Nacional de Cordoba, Sucursal 16, C.C. 61, 5016 Cordoba, Argentina

I. INTRODUCTION

The multiple-photon excitation processes (MPE) induced by an intense IR source in polyatomic molecules have been the subject of intense research since they provide information on the dynamics of the absorption and reaction mechanisms [1-4], and, on energy-transfer mechanisms in collisional processes of vibrationally excited molecules [5-12].

The widely accepted model of IRMPE [1,4,13] assumes that the vibrational structure of polyatomic molecules can be described by an initial set of discrete anharmonic states followed by a dense manifold of states referred to as the quasi-continuum. The discrete region is characterized by coherent excitation, while in the quasi-continuum vibrational relaxation gives rise to incoherent transitions. Further excitation through the quasi-continuum leads finally to unimolecular dissociation.

For small molecules the onset of the quasi-continuum occurs after the absorption of ~ 5 photons. Initially these molecules are distributed among the rotational levels of the ground vibrational state and only a small fraction will be in resonance with the laser radiation. For pumping rates higher than the rotational relaxation rate, a "rotational bottleneck" to absorption occurs [1]. Therefore, the IRMPE of small polyatomic molecules is restricted by rotational and anharmonic bottlenecks at the first stage of excitation, and consequently, the dissociation process becomes inefficient. Collisions between excited molecules and a non-reactive gas induce different relaxation processes. Rotational relaxation, which does not change the total energy content of the molecule, increases the number of molecules in resonance with the laser radiation. Thus, the reaction yield increases, and the dissociation process becomes more efficient. V-V and V-T/R energy-transfer processes induced by collisions decrease the total vibrational energy content of the excited molecule, diminishing therefore, both the dissociation yield and the isotopic selectivity in the case of a gas mixture of different isotopic composition.

All these features become relevant when dealing with the IR multiple-photon dissociation (IRMPD) of small deuteriated polyatomics in laser isotope separation (LIS) of deuterium experiments. Since it has proved to be isotopically selective, IRMPD of many deuteriated halocarbons has been extensively studied in view of attaining high isotopic enrichment [1,14-19].

CDF_3 appears to be the most promising working molecule for deuterium LIS [20,21]. Even so, the favorable results to date still cannot compete, on a larger scale, with the widely developed Girdler-Sulfide process based on the chemical exchange between water and SH_2. One of the major drawbacks met by the heavy-water process based on IRMPD of fluoroform is the low rate of redeuteriation in water or natural gas. Chloroform has attractive features to be considered a potential candidate for deuterium LIS:

a. It has a higher rate of H/D exchange in base-catalysed aqueous solutions than fluoroform ($\geq 10^4$ times) and a low hydrolisis rate [22].

b. It should undergo IRMPD more easily than fluoroform since its density of states rises more quickly with excitation energy [21].

c. The ν_4 mode of $CDCl_3$ (914 cm^{-1}) is resonant with the 10P(50) CO_2 laser line while $CHCl_3$ is transparent in this spectrum zone.

d. Molecular beam IRMPD of $CDCl_3$ indicates that molecular elimination [23], eq. 1,

$$CDCl_3 \rightarrow CCl_2 . + DCl \quad (1)$$

with deuterium localized in a stable molecule is the dominant dissociation channel (>99.1%).

We have studied the IR multiple-photon absorption (IRMPA) spectrum of $CDCl_3$ at the initial steps of the excitation as well as the IRMPD spectrum at different sample pressures and laser fluences. Experimental results were fitted by the solution of an appropriate master equation to give the energy dependence of the laser absorption cross section, $\sigma_L(\Phi)$, the IRMPD probability, $P(\Phi)$, and the average energy transferred in a deactivating collision, $<\Delta E>_d$.

We have studied the effect of collisions in the IRMPD of $CDCl_3$ in equimolar mixtures of $CDCl_3/CHCl_3$. We were interested in the relaxation processes induced by collisions of the non-absorbing isotopic species and added inert gases with the excited $CDCl_3$. Experimental results were compared with those obtained by using an energy grained master equation model valid for small molecules. Average energy transferred in deactivating collisions values, $<\Delta E>_d$, were determined as well as their dependence with the excitation energy. In addition lower limit values for the isotopic selectivity were calculated from $CDCl_3$ reaction volumes.

II. EXPERIMENTAL PROCEDURE

A homemade TEA CO_2 multimode line-tunable laser was used as radiation source throughout this work. The laser frequency was verified with a CO_2 spectrum analyzer (Optical Engineering). The laser pulse consisted of a 90 ns FWHM spike followed by a 200 ns tail as measured with a photon-drag detector. The laser output energy at P(48), 10.911 μm, was typically 1.2 J/pulse at a repetition rate of 0.3 Hz. Pulse-to-pulse variations in the laser output was within 10%. The laser energy was measured with a Scientech 362 disc calorimeter and attenuated with CaF_2 flats.

Different experimental set-ups were used to study the $CDCl_3$ IRMPA and IRMPD [24]. The energy absorbed by a sample was determined by direct calorimetric measurement. Incident and transmitted energy were monitored with a dual piroelectric detector arrangement (GENTEC ED 200 and ED 500), amplified and stored on an oscilloscope Tektronix 7633. The beam diameter was determined from the pattern burnt on heat sensitive paper. Cells of different lengths were used at different $CDCl_3$ pressures so as to keep absorption below 20%. Typically the signal was averaged for several pulses both for empty and filled cell.

In the IRMPD experiments, the laser pulses were focused into the center of a pyrex reaction cell (10 cm long, 200 cm^3) sealed with NaCl windows at both ends, through a 13 cm focal length GaAs lens. The spot size resulting from the $1/e^2$ decrease in the transmission through a regulable diafragm was 0.17 cm. The focal fluence was calculated as the incident laser energy divided by the spot area. The number of irradiation pulses varied between 500-1000 depending on the pulse energy and sample pressure so as to keep $CDCl_3$ conversion below 10%. In the IRMPD of pure $CDCl_3$ experiments, sample concentrations were determined before and after irradiation, by measuring the $CDCl_3$ ν_4 absorption with a 225 Perkin-Elmer IR spectrophotometer.

Reaction yields of the IRMPD of $CDCl_3$ in equimolar mixtures of $CDCl_3/CHCl_3$ were determined from IR spectrometry, too, following depletion of the $CDCl_3$ and $CHCl_3$ ν_4 absorption band at 914 and 1221 cm^{-1}, respectively. Mass spectrometric determination of the reaction yield was also made comparing the mass peaks of $CD^{37}Cl_2^+$ and $CH^{37}Cl_2^+$ at 88^+ and 87^+, respectively, in the photolyzed sample with those in an unphotolyzed reference sample with an EMBA II Extranuclear quadrupole mass spectrometer.

$CDCl_3$ (99.8% D) was purchased from Merck while the $CHCl_3$ used, stabilized with 0.5% of absolute C_2H_5OH, was from Carlo Erba. Both species were used without purification except outgasing of non-condensable gases at 77^0K. Ar (99.9995%) and Xe (99.995%) were Matheson Research Purity and were used directly from cylinders. Sample pressures below 10 Torr were measured with a MKS Baratron capacitance manometer while sample pressures between 10 and 50 Torr were measured with an oil manometer.

III. RESULTS AND DATA ANALYSIS

III.1- Absorption Cross Sections Measurements

a). Effect of Laser Fluence, Wavelength and Pressure

Absorption cross sections values for $CDCl_3$ were measured for different fluences and sample pressures, (0.5-5 Torr), in the wavelength range 929.07-916.76 cm^{-1} corresponding to the P(36)-P(48) lines of the 10.6 μm emission band of the CO_2 laser [25]. The $CDCl_3$ v_4 mode, (914 cm^{-1}), is near resonant with the P(48) line. The $CDCl_3$ IRMPA does not follow Beer's law and the absorption cross sections, σ_L, depend on the $CDCl_3$ pressure. The $\sigma_L(\Phi)$ was calculated at each pressure from $\ln T = N\, l\, \sigma_L(\Phi)$; T is the laser fluence transmittance, N is the concentration in molecules/cm^3, l is the cell length in cm and σ_L is the laser absorption cross section.

Figure 1 illustrates the behaviour of σ_L as a function of the laser fluence for the laser lines ranging from P(48) near the single-photon absorption maximum to the P(36) at the "blue" wing of the absorption band. For all the laser lines considered, σ_L decreases at higher fluences. For the laser fluence range considered, 0.01-1.4 J/cm^2, the average number of photons absorbed per molecule, $\langle n \rangle = \sigma_L \Phi / hc\omega$, were σ_L is the laser absorption cross section, Φ is the laser fluence, h is Planck's constant, c is the speed of light and ω is the laser photon wavelength in cm^{-1}, extends to $\langle n \rangle \leq 6$ for all laser lines.

The laser absorption cross sections measured with the line 10P(48) in the fluence range 0.01-0.8 J/cm^2, for five different pressures from 0.5 to 5 Torr of $CDCl_3$ are shown in Figure 2. At low sample pressures, $\sigma_L(\Phi)$ values are much smaller than the linear absorption cross section values, σ_{10}.

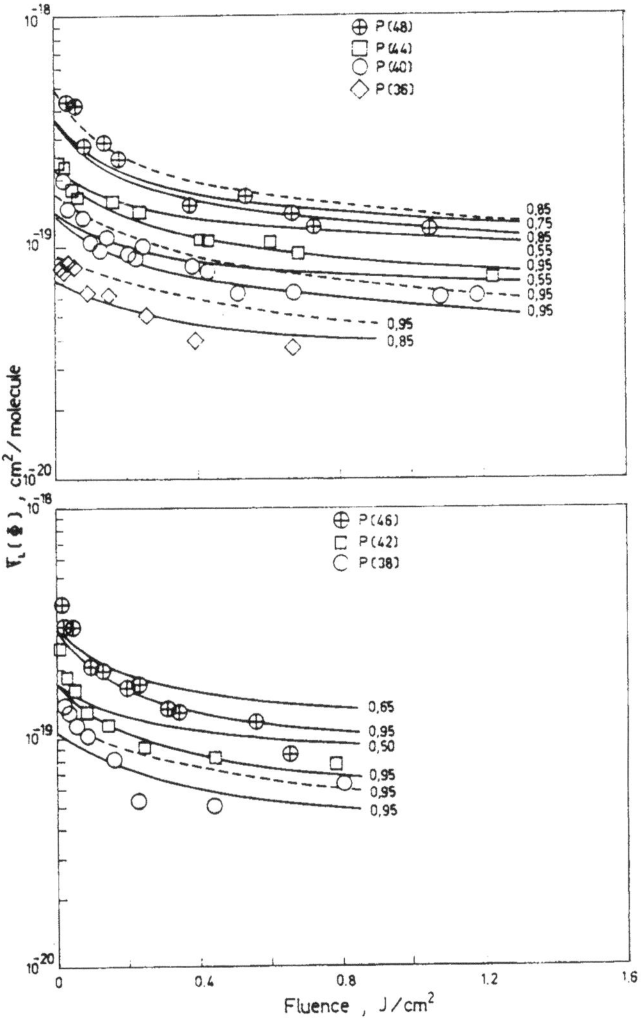

Figure 1: Fluence dependence of the laser absorption cross section, $\sigma_L(\Phi)$, for different lines of the 10.6 μm emission band of the CO_2 laser. The lines are the results from master equation calculations using the level-to-level cross sections, $\sigma_{i+1,i} = \sigma_{10}/(i+1)^m$, with the σ_{10} values from Table III and the m values shown. The solid lines correspond to calculations with $k_{rot} = 1.5 \times 10^{-3}$ ns^{-1} Torr^{-1} and the broken lines to $k_{rot} = 6 \times 10^{-3}$ ns^{-1} Torr^{-1}.

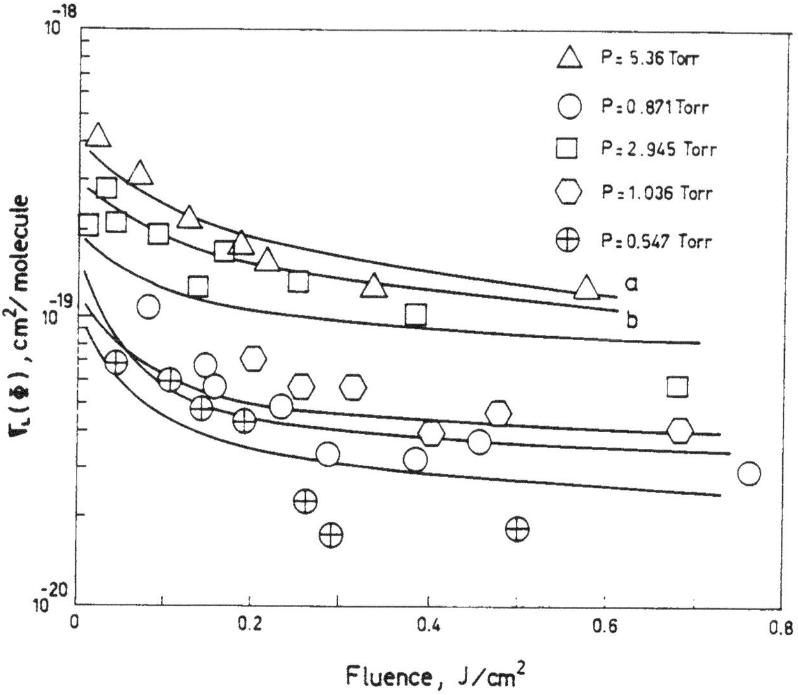

Figure 2: Fluence dependence of the laser absorption cross section, $\sigma_L(\Phi)$, for different pressures of $CDCl_3$. The lines are the results from master equation calculations with $\sigma_{i+1,i} = \sigma_{10}/(i+1)^{0.9}$ and $k_{rot} = 1.0 \times 10^{-3}$ ns^{-1} $Torr^{-1}$. The numbers indicate the $CDCl_3$ pressure in Torrs.

b). Effect of Bath Gas

The effect on the IRMPA of $CDCl_3$ of the addition of a non-reactant bath gas was studied for different laser fluences [26]. Absortion cross sections of 0.5 Torr of $CDCl_3$

were measured varying the pressure of Ar or $CHCl_3$ as shown in Figure 3 for three different laser fluences. As for pure $CDCl_3$, an effect of total pressure is evidenced resulting in the increase of σ_L with the increase of bath gas pressure. For a given $CHCl_3$ pressure the increase in the $\sigma_L(\Phi)$ value is more pronounced than with Ar and is similar to that with $CDCl_3$. This effect shows the contribution of collisions to remove the rotational bottleneck to absorption at the lower stages of excitation. Also, the dependence of $\sigma_L(\Phi)$ on fluence in the presence of a bath gas (Figure 3) is similar to that obtained for pure $CDCl_3$ (Figure 2).

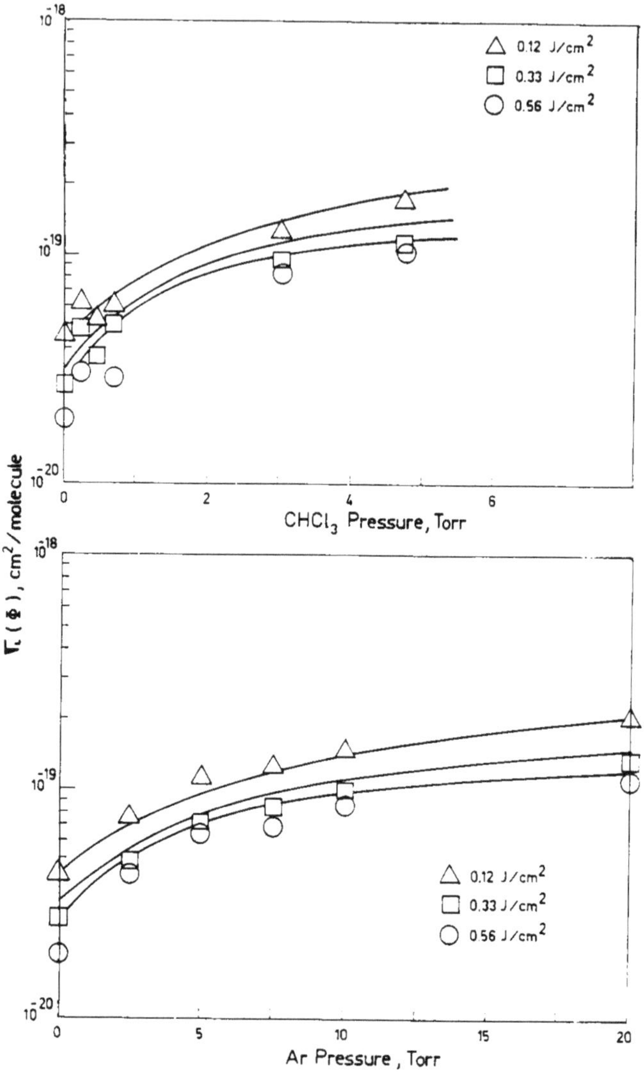

Figure 3: Laser absorption cross section, $\sigma_L(\Phi)$, for 0.57 Torr of $CDCl_3$ as a function of added $CHCl_3$ and Ar pressure for different laser fluences. The lines are the results from master equation calculations with $\sigma_{i+1,i} = \sigma_{10}/(i+1)^{0.9}$ and with $k_{rot}[CDCl_3\text{-}CHCl_3] = 1.0 \times 10^{-3}$ ns^{-1} $Torr^{-1}$ and $k_{rot}[CDCl_3\text{-}Ar] = 3.0 \times 10^{-4}$ ns^{-1} $Torr^{-1}$.

III.2.- Dissociation Probability Measurements

a). Pure $CDCl_3$

i). Effect of Number of Pulses

The reaction volume $V_R = V_c/n \ln ([CDCl_3]_0/[CDCl_3]_F)$ cm^3/pulse, where V_c is the cell volume, and $[CDCl_3]_0$ and $[CDCl_3]_F$ are the initial and final concentrations of $CDCl_3$ respectively, vs. the number of irradiation pulses, n, is plotted in Figure 4. The linearity of this plot indicates that the reaction volume V_R is independent of the extent of reaction. For a given laser energy, the effective volume of reaction is constant, and therefore, it is a good measure of the reaction yield. These results show too, that depletion of 2 Torr $CDCl_3$ is first order in $CDCl_3$ in agreement with the kinetics found by Magnotta et al. [27] and Yokoyama et al. [28]. Mc Alpine et al. [21] have found that $CDCl_3$ depletion is second order in $CDCl_3$.

ii). Effect of Irradiation Geometry

Reaction yields were measured as reaction volumes, V_R, determined from IR and mass spectra.

The fluence dependence of the reaction volume of the IRMPD of $CDCl_3$ was studied for irradiation with different rotational lines of the 10.6 μm laser emission band.

A linear least-squares fit of the log-log plot of these curves yields a slope m = 1.57 $\pm$ 0.8, as is expected for irradiations in higly focused geometries [29-32]. The reaction volume then can be approximated by the double-cone shape geometry.

These reaction volumes obtained by irradiation in a focused geometry were deconvoluted according to a general method which assumes an homogeneous spatial profile for the laser beam [33] to obtain the corresponding $CDCl_3$ dissociation probabilities per pulse, $P(\Phi)$.

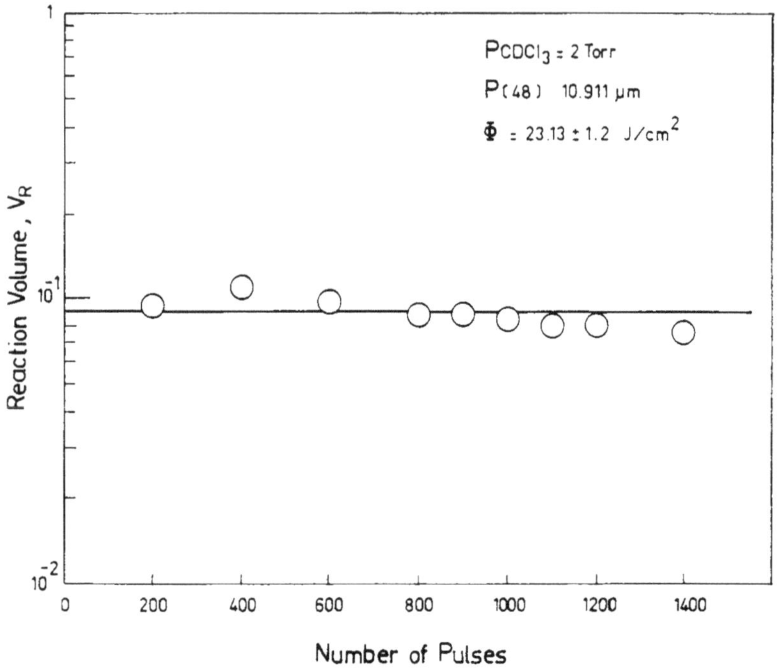

Figure 4: Reaction volume V_R vs. number of pulses for irradiation with P(48), 10.91μm, of 2 Torr $CDCl_3$ at 23 J/cm^2 focal fluence.

iii). Effect of Laser Fluence and Frequency

The fluence dependence of the $CDCl_3$ reaction volumes was measured in the range 7-45 J/cm^2, for $CDCl_3$ pressures in the range 0.3-5 Torr and irradiation wavelengths ranging from 907.77 to 929.07 cm^{-1} [27,28,34].

The reaction probability of 0.3 Torr of $CDCl_3$ as a function of laser fluence, in the range 7-60 J/cm^2, for irradiation with the 10P(48) line is plotted in Figure 5, while the fluence dependence of the dissociation probability of 2 Torr $CDCl_3$ at a number of different laser frequencies is illustrated in Figure 6. For a given laser fluence, there is a large variation with laser frequency in the dissociation probability vs. fluence plot. At 20 J/cm^2, reaction probabilities of 1, 0.66, 0.40, 0.24 and 0.124 were obtained for irradiation of 2 Torr $CDCl_3$ with P(23) 907.77 cm^{-1}, P(48) 916.76 cm^{-1}, P(42) 923.02 cm^{-1}, P(40) 925.06 cm^{-1} and P(38) 927.08 cm^{-1}, respectively.

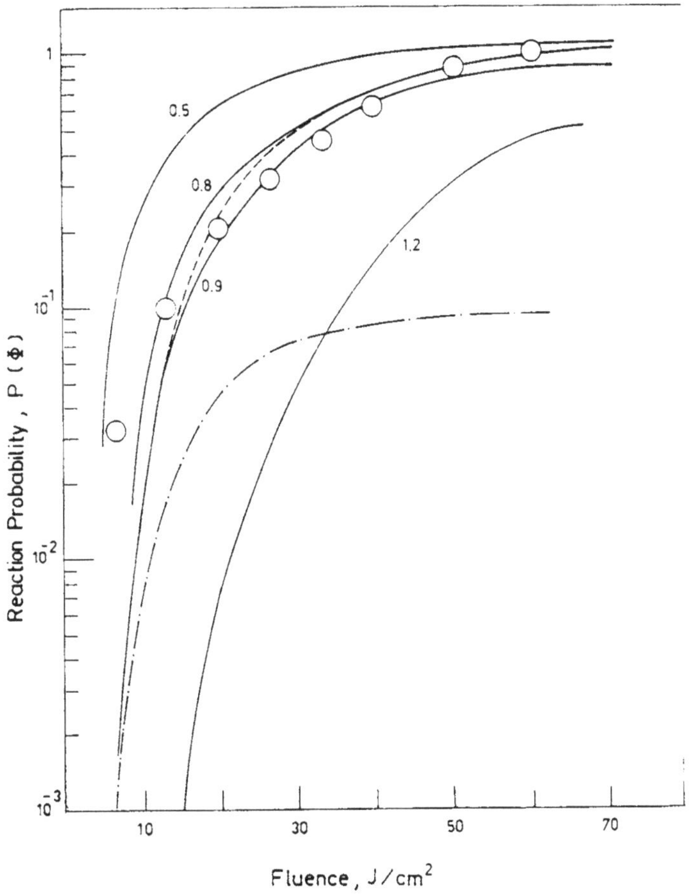

Figure 5: Dissociation probability of 0.3 Torr of $CDCl_3$ as a function of fluence for irradiation with 10P(48) laser line. The lines are the results from master equation calculations with $\sigma_{i+1,i} = \sigma_{10}/(i+1)^m$ and : (—)k_{rot} = 1.0 x 10^{-3} ns^{-1} $Torr^{-1}$, (---)k_{rot} = 1.5 x 10^{-3} ns^{-1} $Torr^{-1}$, and (– · –) k_{rot} = 1.5 x 10^{-3} ns^{-1} $Torr^{-1}$. The values of m are shown on the plot.

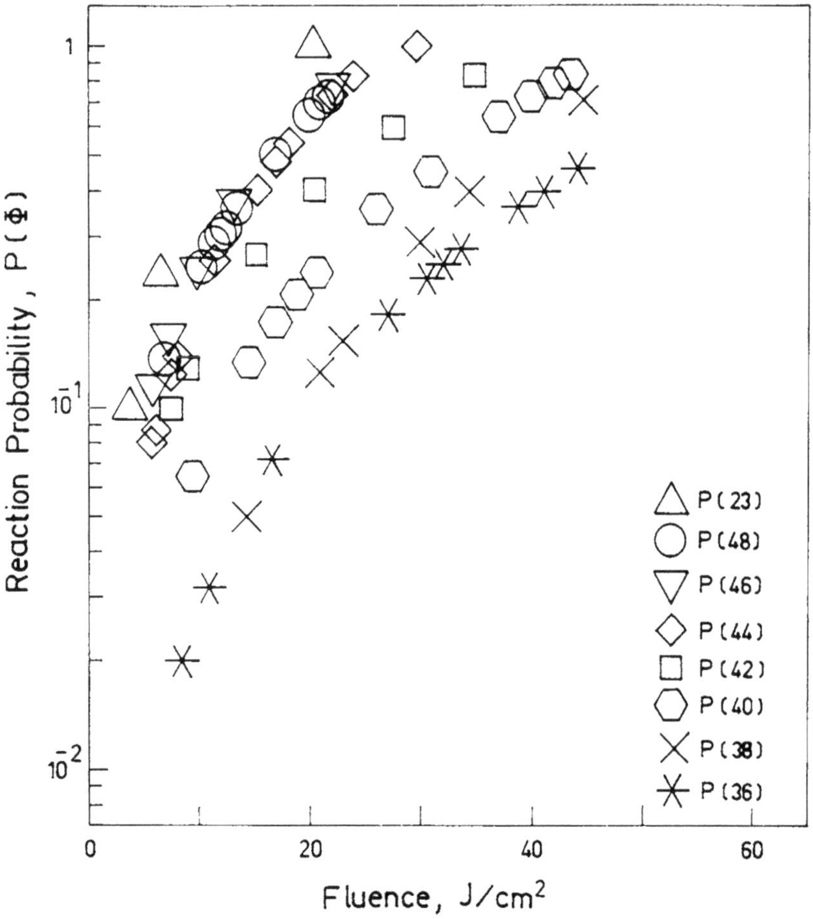

Figure 6: Experimental dissociation probability of 2 Torr $CDCl_3$ as a function of fluence for irradiation with different CO_2 laser lines.

The dissociation fluence threshold increases with increasing laser frequency detuning from the band center.

The wavelength dependence of $CDCl_3$ IRMPD when the ν_4 mode is excited is plotted in Figure 7. The small signal transmission curve is shown too for comparison. Laser wavelengths were scanned beginning at the P(36), 10.765 μm at the blue wing of the ν_4 band and up to the P(23), 11.016 μm to the red of the band center. Further scanning to the red wing could not be accomplished since no laser emission could be obtained in that zone. The maximum of the IRMPD spectrum is red-shifted $\geq 6\ cm^{-1}$ as compared to the linear absorption spectrum maximum. Mc Alpine et al. [21] found that the IR multiple photon absorption spectrum maximum is red-shifted $\geq 5\ cm^{-1}$. Similarly, the peak of the $CTCl_3$ MPD spectrum is shifted ~ $10\ cm^{-1}$ to the red of the band center [27]. The anharmonic bottleneck in the IRMPD should not be unexpected in view of the large anharmonic constant of the $CDCl_3$ ν_4 mode ($X_{44} = -10.6\ cm^{-1}$ [35]). Calculation of the Rabi frequency $\omega_R = (3.55 \times 10^{-8})I^{1/2}\ cm^{-1}$ at the maximum intensity for a 20 J/cm^2 pulse yields $1.8\ cm^{-1}$. Taking the TEA CO_2 laser linewidth to be 3 GHz, the resulting power-broadened linewidth, $2\ cm^{-1}$, is not large enough to overcome the anharmonic bottleneck. Therefore, collisional processes and irradiation source detuning should play an essential role in overcoming anharmonicity.

iv). Effect of Pressure

$CDCl_3$ dissociation volume increases with pressure in the range 0.3 - 5 Torr, as shown in Figure 8 for different laser lines in agreement with the results obtained by Yokoyama et al. [28]. For $CDCl_3$ the rotational hole is filled at about 1.5 Torr [26], so that initial increase of V_R with pressure is expected. However, an increase in the reaction volume with pressure has been observed in our experiments as well as in those of Yokoyama et al. [28] for pressures higher than that corresponding to the rotational hole filling. A least-squares fit of the experimental results yields different slopes for curves corresponding to different irradiation frequencies as shown in Figure 8. In all cases, the slopes are < 1. This effect evidences a deactivating mechanism which must be related to V-V energy transfer. Faster V-V than V-T energy transfer takes place among excited $CDCl_3$ molecules and promotes reaction prior to V-T deactivation. The fact that the curves corresponding to different laser lines have different slopes indicates that the V-V energy transfer mechanism depends on the irradiation frequency.

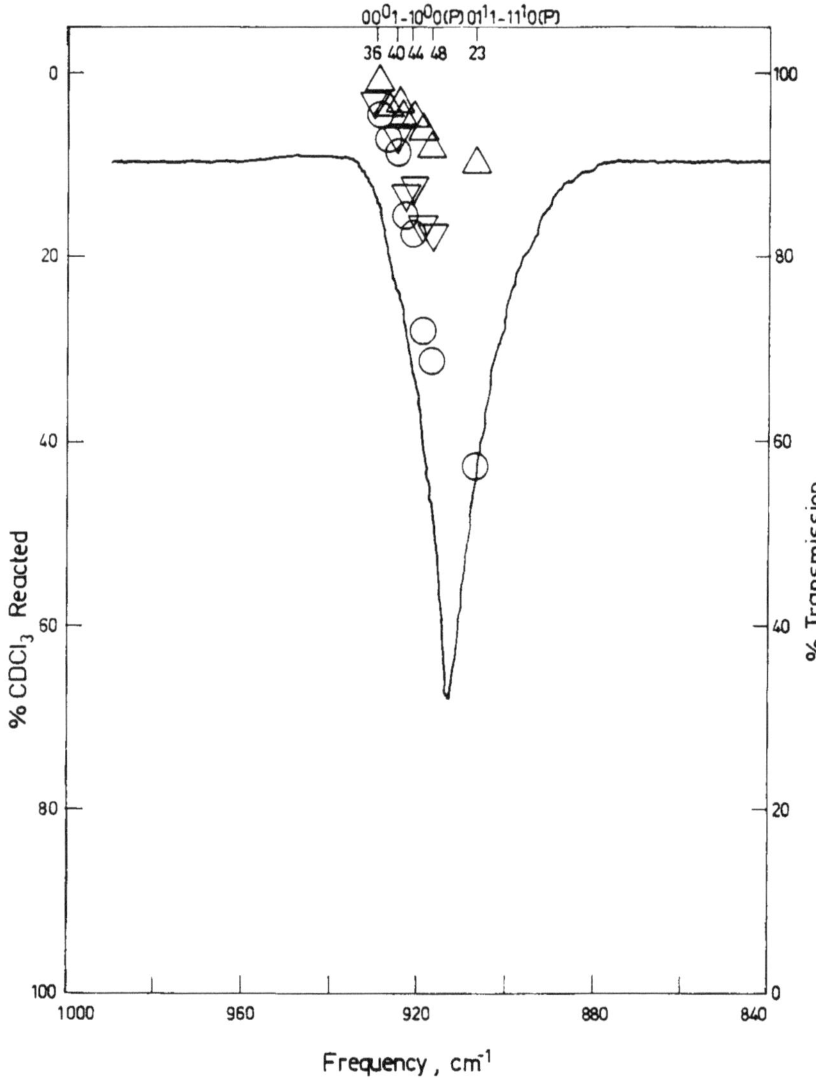

Figure 7: Wavelength dependence of the IRMPD of 2 Torr $CDCl_3$ for three different laser fluences (left axis): (Δ) 7 J/cm^2, (∇) 15 J/cm^2 and (O) 20 J/cm^2. Linear transmission spectrum for 2 Torr $CDCl_3$ is shown for comparison (right axis).

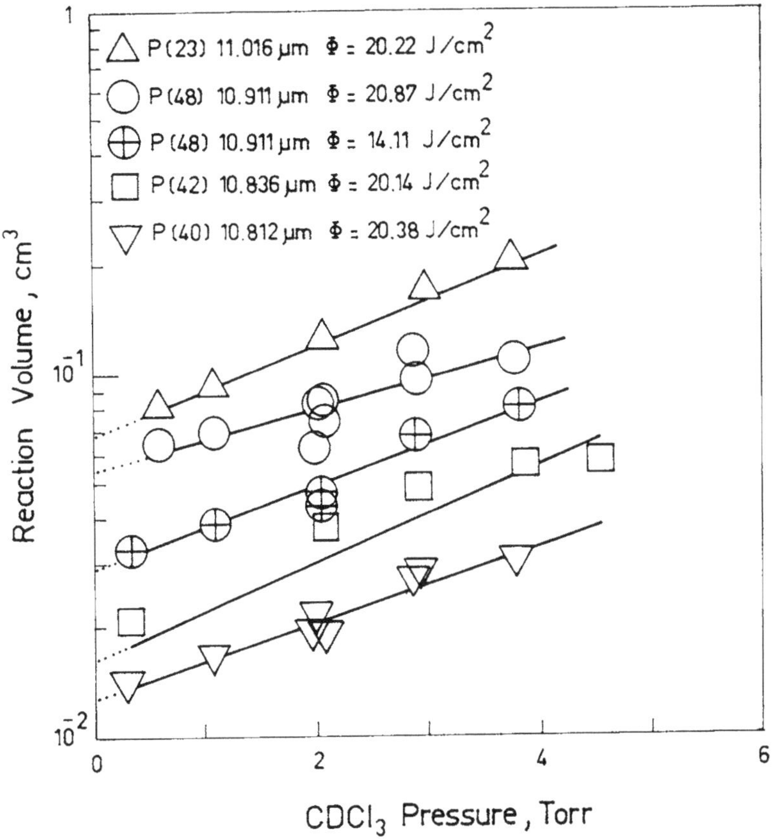

Figure 8: Reaction volume vs. $CDCl_3$ pressure for different laser emission lines. Full lines represent linear least-squares fit to experimental data. Broken lines are extrapolations to collisionless conditions.

v). Effect of Bath Gas

The reaction probability of 0.3 Torr $CDCl_3$ was measured as a function of $CHCl_3$ and Ar pressure at 20 and 33 J/cm^2 [26]. The increase in the bath gas pressure initially results in an increase in the fraction of resonant molecules with the laser radiation, and consequently, in the reaction yield. Once rotational relaxation is completed, further increase in $CHCl_3$ or Ar pressure results in vibrational deactivation of the excited molecules by collisions and therefore, in a decrease of the dissociation yield (Figures 9 and 10). Competition between rotational relaxation and vibrational deactivation results in a maximum for the reaction probability as a function of bath gas pressure. At 33 J/cm^2, the maximum obtained is about 0.7 for 1 Torr of added $CHCl_3$ or 20 Torr of Ar, representing an enhancement factor of 1.5 with respect to the value for 0.3 Torr of pure $CDCl_3$. At 20 J/cm^2 the maximum is about 0.5 for both bath gases, but the reaction probability value with respect to the value of pure $CDCl_3$ is a factor 2.5 larger. This effect evidences a more efficient remotion of rotational bottleneck at the lowest fluence. The deactivating effect is also greater at the lowest fluence.

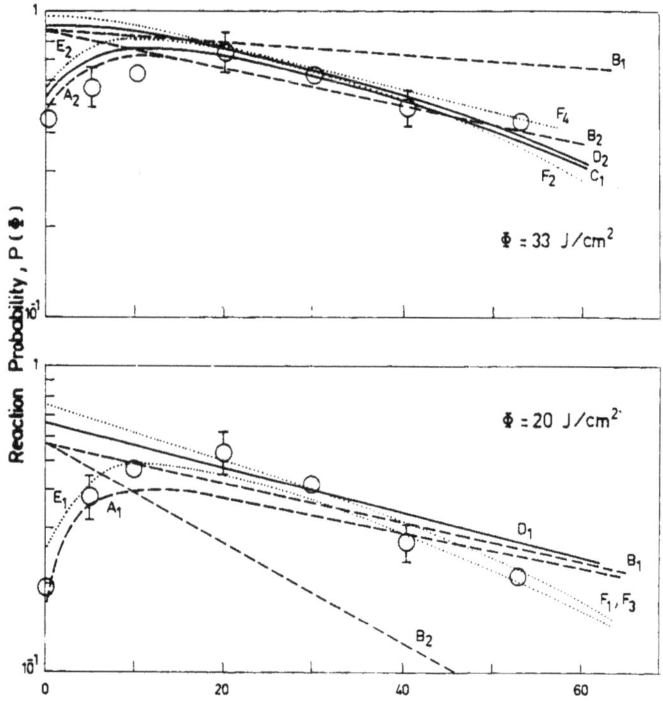

Figure 9: Reaction probability of 0.3 Torr of $CDCl_3$ vs. added Ar pressure at 20 and 33 J/cm^2. The points represent experimental data while the lines represent calculated results with different deactivation models as described in detail in Table IV.

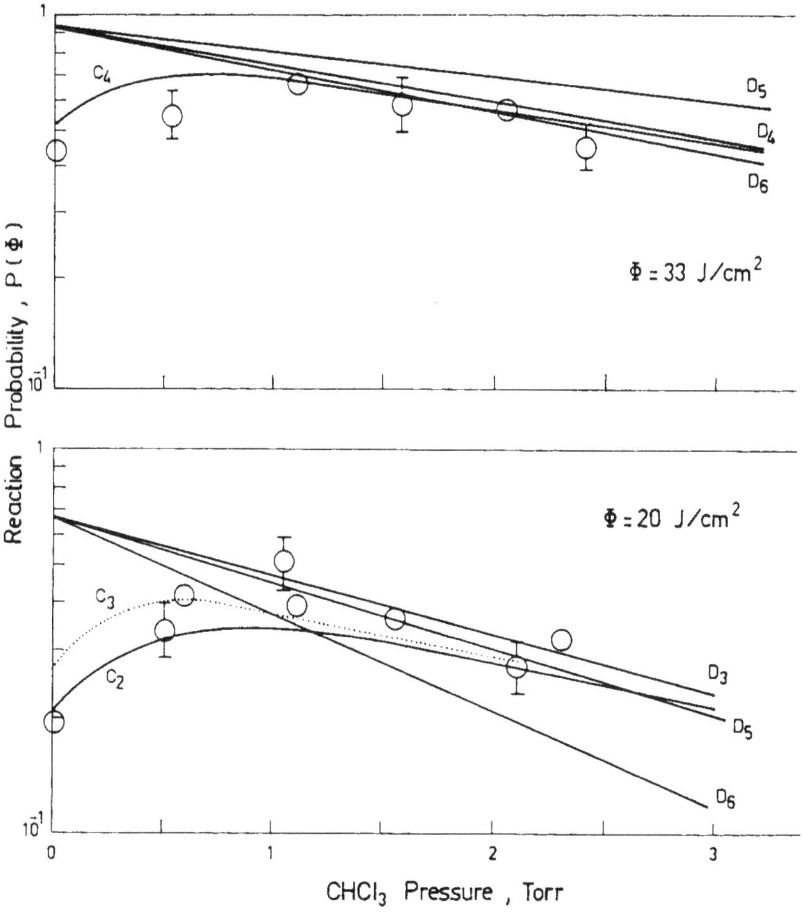

Figure 10: Reaction probability of 0.3 Torr of $CDCl_3$ vs. added $CHCl_3$ pressure at 20 and 33 J/cm^2. The points represent experimental data while the lines represent calculated results with different deactivation models as described in detail in Table IV .

b). Equimolar $CDCl_3/CHCl_3$ Mixtures

i). Effect of Laser Fluence and Pressure

Figures 11 and 12 show plots of the $CDCl_3$ dissociation probability as a function of $CDCl_3$ pressure in $CDCl_3/CHCl_3$ equimolar mixtures for different laser fluences [24,36].

The initial increase with pressure in the reaction probability at 20 J/cm^2 evidences the existence of a rotational bottleneck which is overcome by collisions with the non-absorbing species. As pressure continues to increase, collisional deactivation of excited $CDCl_3$ by V-V and V-T/R transfer to $CHCl_3$ competes with dissociation leading to a linear decrease in the $CDCl_3$ dissociation yield.

Reaction probabilities increase with fluence since the excitation and dissociation rates are much higher and more molecules are allowed to be pumped up and dissociated before they are deactivated as can be seen by the curve at 44 J/cm^2.

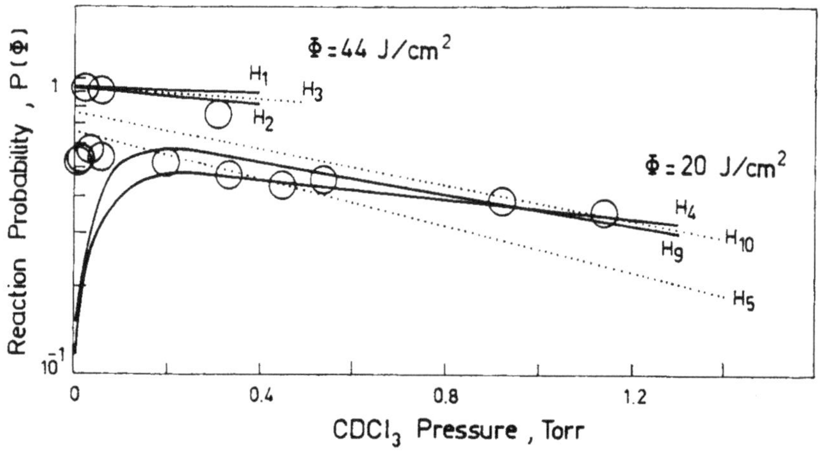

Figure 11: Reaction probability of $CDCl_3$ in equimolar $CDCl_3/CHCl_3$ mixtures vs. $CDCl_3$ pressure at 44 and 20 J/cm^2. The points are the experimental results. The lines are calculated results described in detail in Table VI.

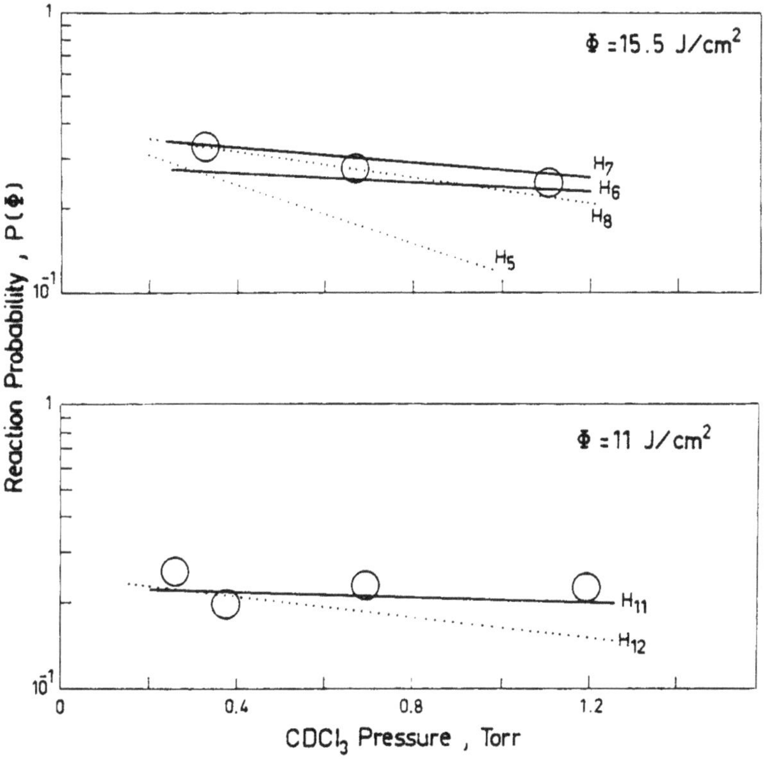

Figure 12: Reaction probability of $CDCl_3$ in equimolar $CDCl_3/CHCl_3$ mixtures vs. $CDCl_3$ pressure at 15.5 and 11 J/cm^2. The points are the experimental results. The lines are calculated results described in detail in Table VI.

ii). Effect of Bath Gas

Figure 13 shows the $CDCl_3$ dissociation probability in $CDCl_3/CHCl_3$ equimolar mixtures as a function of added Ar pressure. Initially, the reaction probability increases with increasing Ar pressure due to the rotational relaxation induced by collisions which help to overcome the rotational bottleneck. Once the maximum is attained, corresponding to the filling of the rotational hole, V-T deactivation of excited $CDCl_3$ causes the reaction probability to decrease with increasing Ar pressure.

The slope in the deactivation region of the curve in presence of Ar is smaller in the case of $CDCl_3/CHCl_3$ equimolar mixtures than in pure $CDCl_3$ [24,26] showing again that $CHCl_3$ is a more efficient vibrational relaxation partner. The maximum of the $CDCl_3$ reaction probability was obtained for 20 Torr of Ar in pure $CDCl_3$ while in the equimolar mixtures of $CDCl_3/CHCl_3$, it is obtained for 10 Torr of Ar, thus showing the contribution of collisions with $CHCl_3$ to rotational hole filling, too.

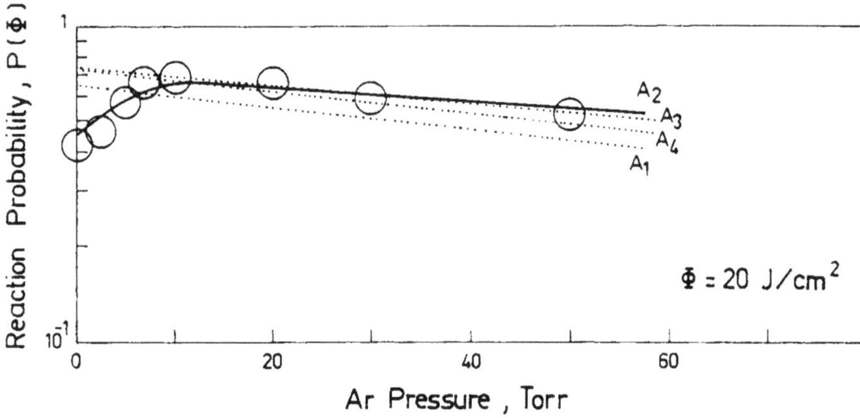

Figure 13: Reaction probability of $CDCl_3$ in $CDCl_3/CHCl_3$ equimolar mixtures vs. added Ar pressure at 20 J/cm^2. The points represent experimental data while the lines represent calculated results described in detail in Table VI.

Figure 14 shows the dependence of the reaction probability of the $CDCl_3$ pressure in $CDCl_3/CHCl_3$ equimolar mixtures with addition of 10 Torr of Ar or Xe. An enhancement of the reaction probability of 20% with respect to the values obtained without inert gas addition is evidenced. This result shows again, the important role of collisions in removing the rotational bottleneck of the IRMPA process. Within experimental error no difference is evidenced in the effect of Ar or Xe addition.

iii). Isotopic Selectivity

The isotopic selectivity, s, in a mixture of $CDCl_3/CHCl_3$ is given by the general formula [1]

$$s = V_D/V_H \qquad (2)$$

where V_D and V_H are the $CDCl_3$ and $CHCl_3$ reaction volumes respectively.

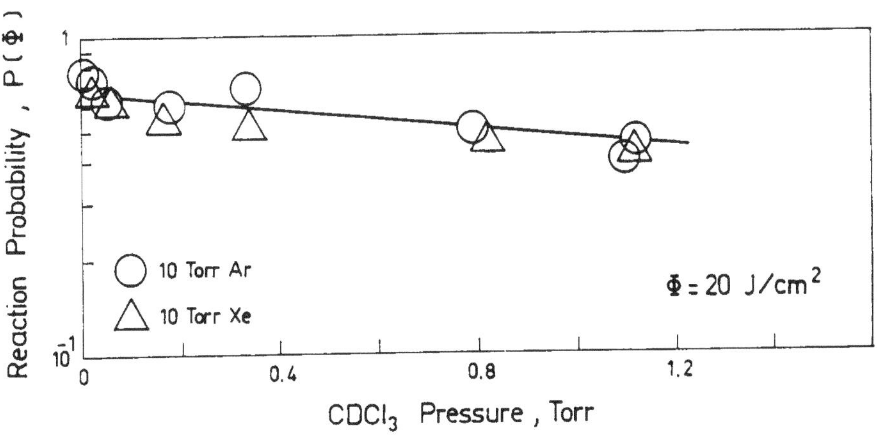

Figure 14: Reaction probability of $CDCl_3$ in equimolar mixtures of $CDCl_3/CHCl_3$ vs. $CDCl_3$ pressure with addition of 10 Torr Xe and 10 Torr Ar. Points represent experimental data while the line represent calculated results with $k_{rot} = [1.5 \times 10^{-3} \times P_{chloroform} + 6 \times 10^{-3}]^{-1}$ ns, m = 0.85 and $\langle \Delta E \rangle_d = 914\ cm^{-1}$.

In none of the IRMPD of $CDCl_3$ in $CDCl_3/CHCl_3$ equimolar mixtures was $CHCl_3$ dissociation detected. Moreover, 1 Torr of pure $CHCl_3$ was irradiated with 10,000 pulses (P(48), 10.91 μm) and no dissociation was observed within experimental error.

The reproducibility of the detection system, about 3%, was taken as an upper limit of the $CHCl_3$ reaction volume for calculating the isotopic selectivity. The values thus obtained represent a lower limit to the selectivity factor and are shown in Table I. These values are consistent with those obtained by Yokoyama et al. [28].

At a given fluence, the isotopic selectivity decreases with the increase in total sample pressure due to the vibrational deactivation of excited $CDCl_3$ by $CHCl_3$. An increase in the isotopic selectivity with increasing fluence is observed in contrast to the results obtained in the isotopic enrichment in ^{13}C, in $CHClF_3$ [19] and CDF_3 [37], and in D, in CDF_3 [38]. This effect is related to anharmonic restrictions, much more severe for $CDCl_3$ ($X_{44} = -10.6\ cm^{-1}$) [35] than for CDF_3 ($X_{55} = -0.25\ cm^{-1}$) [39]. At higher fluences, power broadening of the discrete levels due to interaction with the intense electric field of the laser partially removes the anharmonic bottleneck. In consequence, more molecules are allowed to be excited and then dissociated, and the isotopic selectivity increases as there is no $CHCl_3$ dissociation.

Table I shows that the addition of inert gases increases considerably the dissociation selectivity by the removal of the rotational and anharmonic bottlenecks through collisions.

TABLE I : Isotopic Selectivity Values.

FLUENCE (J/cm^2)	$P[CDCl_3+$ $+CHCl_3]$ (Torr)	P[Ar] (Torr)	s
44	0.046	0	149
	0.116		145
	0.614		120
20	0.066	0	83
	0.120		77
	0.668		68
	1.082		65
	2.290		50
15.5	0.654	0	53
	1.344		44
	2.208		39
11	0.516	0	41
	1.386		34
	2.208		34
20	0.052	10	106
	0.104		99
	0.684		83
	2.238		68

III.3. Master Equation Calculations

The experimental results of the $CDCl_3$ IRMPA and IRMPD were modeled using an energy-grained master equation formulation which includes rotational and anharmonic restrictions to absorption and collisional effects which has been described in detail in references 24, 26 and 40.

The time evolution of the population of the vibrational level i, $N_i(t)$, is given by the rate equation 3:

$$\frac{dN_i(t)}{dt} = I(t)\,[\sigma_{i,i-1}N_{i-1} + \sigma_{i,i+1}N_{i+1} - (\sigma_{i+1,i} + \sigma_{i-1,i})N_i] + \\ + w\,\Sigma\,P_{ij}\,N_j - w\,N_i - k_i\,N_i \qquad (3)$$

I(t) is the pulse intensity and $\sigma_{i+1,i}$ is the cross section for transfer from level i to level i+1 by absorption of laser energy which is related to the stimulated emission cross section $\sigma_{i,i+1}$ by detailed balance: $\sigma_{i,i+1} = (g_i/g_{i+1})\,\sigma_{i+1,i}$, where g_i is the vibrational density of states of level i. k_i is the RRKM rate for unimolecular decay fron level i, w is the collisional frequency and P_{ij} is the collisional transition probability from level j to level i.

Rotational partition of the first excitation step was included to account for the fact that only a small fraction $f_i(t)$ of the vibrational ground state population in certain rotational states is able to absorb laser photons, so that in eq. 3 the vibrational ground state population was considered to be: $N'_j = f_1(t)\,N_1$. The fraction of resonant molecules, $f_i(t)$, for the different frequencies of the laser emission lines, was calculated for different band widths, from the simulated spectrum of $CDCl_3$ using the spectroscopic data of reference 29 as

$$f_1 = a + bw + cw^2 + dw^3 + ew^4 \qquad (4)$$

with $w = w_R + w_L$, where w_R is the Rabi frequency, $w_R = 1.514 \times 10^{-14}\, I^{1/2}\ cm^{-1}$ and w_L is the laser linewidth, which was 0.1 cm^{-1} for our laser. The values for the polynomial coefficients for the different laser lines are given in Table II.

The anharmonicity of the pumped mode was introduced considering that only that fraction of the population coming from the ground level through absorption of one laser photon for which power broadening is greater than the anharmonic shift, can further absorb. This fraction is given by eq. 5

$$f_2 = \frac{wR + wL - 2\ |X44|}{wR + wL} \qquad (5)$$

with X_{44} = -10.6 cm^{-1} [35]. Equations 6 and 7 introduce the rates of rotational hole filling and anharmonic restriction removal by collision, respectively:

$$dN'_1/dt = k_{rot}(N_1^* - N_1^{eq})\,N_T \qquad (6)$$

$$dN'_2/dt = k_{rot}\,(N_2^* - N_2^R)\,N_T \qquad (7)$$

where N_T represents the total density of molecules. N_1^* and N_2^* are the non-resonant instantaneous populations of the ground and first excited levels, respectively, N_1^{eq} is the same population as if a Boltzmann distribution would have been reestablished and N_2^R is the resonant population of level 2 in collisionless conditions. k_{rot} is a rotational constant for the rotational hole filling.

Table II : Fraction of Resonant Molecules for Different Laser Emission Lines, $f_1 = a + bw + cw^2 + dw^3 + ew^4$

w range (cm^{-1})	Frequency (cm^{-1})	a	b	c	d	e
0.01-0.1	916.76	-0.00172	0.7429	-16.696	187.67	751.64
0.1 -10		-0.004609	0.25194	-0.01622	0.000089	0
0.01-10	918.86	0.000376	0.21075	-0.01637	0.000528	0
	920.95	0.001006	0.08159	0.16084	-1.019	0
	923.02	-0.000059	0.14255	0.00046	-0.000050	0
	925.06	0.000230	0.21775	0.05485	-6.21190	0
	927.08	0.001863	0.08493	-0.00054	0	0
	929.07	-0.000195	0.06269	0.00014	0.00003	0

Unimolecular dissociation rate constants, k_i, were calculated with RRKM theory considering that the only reaction channel is molecular elimination of DCl. Arrhenius parameters used in the calculation were those given in reference 41, log A = 14.3 and E_a = 62 kcal/mol, and the vibrational frequencies used were those from reference 24 and 26. The density of vibrational states was calculated by direct counting up to an energy equal to five laser photons and by Haarhoff approximation [42] thereafter. Collisional frequencies were calculated from Lennard-Jones collision parameters [24,26].

Absorption cross sections were calculated assuming a model $\sigma_{i+1,i} = \sigma_{10} / (i+1)^m$, where σ_{10} is the $1 \leftarrow 0$ absorption cross section determined from conventional IR spectroscopy, and m is a parameter to be determined. The σ_{10} values corresponding to the different laser lines are given in Table III.

Table III: Small Signal Absorption Cross Sections for Different Laser Emission Lines

Laserline	Frequency	σ_{10} x 10^{19}
(cm^{-1})	(cm^2/molecule)	
P(48)	916.76	5.500
P(46)	918.86	4.439
P(44)	920.95	3.259
P(42)	923.02	2.303
P(40)	925.06	1.878
P(38)	927.08	1.457
P(36)	929.07	0.997

In order to integrate numerically the master equation, parameters m and k_{rot} related to the level-to-level absorption cross sections and rotational hole filling, respectively, must be supplied as well as a model for the collisional deactivation transition probability, P_{ij}. Parameters m and k_{rot} are best obtained from modeling $CDCl_3$ absorption, with and without the addition of bath gases (Figures 1-3), and low pressure dissociation measurements (Figure 5).

IRMPA was simulated and the calculated $\sigma_L(F)$ values from the master equation formulation for the different laser lines and pressures are presented in Figures 1 and 2.

The level-to-level cross sections $\sigma_{i+1,i}$, with m = 0.9 ± 0.5, together with averaged rotational constant values for rotational hole filling of k_{rot} = (1.5 - 6) x 10^{-3} ns^{-1} $Torr^{-1}$, adequately reproduce the $\sigma_L(\Phi)$ vs. Φ data for different sample pressures and laser lines. The dependence of the absorption cross section on the bath gas pressure was calculated and good fits to the experimental curves for different fluences were obtained for m = 0.9, and $k_{rot}[CDCl_3\text{-}CHCl_3]$ = 1.0 x 10^{-3} ns^{-1} $Torr^{-1}$ and $k_{rot}[CDCl_3\text{-}Ar]$ = 3.0 x 10^{-4} ns^{-1} $Torr^{-1}$, for $CHCl_3$ and Ar, respectively (Figure 3). The influence of m and k_{rot} was also tested modeling the reaction probability at low pressure of $CDCl_3$. The results are shown in Figure 5. Values of m in the range 0.8-0.9 fit well the experimental curve for k_{rot} = 1.0 x 10^{-3} ns^{-1} $Torr^{-1}$, while calculated curves with m = 0.5 and 1.2 deviate from the experimental points. Changing k_{rot} to 1.5 x 10^{-3} ns^{-1} $Torr^{-1}$ with m = 0.9 produces little effect in the calculated results.

The microscopic absorption cross sections and rotational relaxation constants were therefore determined by simulation of the low energy IRMPA and low pressure IRMPD. The values of the parameters which produced the best fits to the experimental results were m = 0.85 ± 0.5, $k_{rot}[CDCl_3\text{-}CHCl_3]$ = (1.5-6) x 10^{-3} ns^{-1} $Torr^{-1}$ and $k_{rot}[CDCl_3\text{-}Ar]$ = (3-6) x 10^{-4} ns^{-1} $Torr^{-1}$. Vibrational deactivation processes did not affect these results and were therefore not included in these calculations.

V-T energy transfer processes become relevant in the IRMPD of $CDCl_3$ at high pressures. Therefore, these experiments were used to select an appropriate model for the vibrational transition probabilities, P_{ij}. The values of m and $k_{rot}[CDCl_3\text{-}M]$ used in these calculations were those determined above.

Different functional forms for P_{ij} have been proposed for efficient and inefficient colliders [2,5,9]. The exponential model [2,5,40] with P_{ij} given by $P_{ij} = C \exp[-(E_j-E_i)/<\Delta E>_d]$ seems to be best suited for inefficient colliders such as inert gases. The exponential reverse model [5,6] given by $P_{ij} = C\,[3 - 2(E_j-E_i)/<\Delta E>_d]$ or the step-ladder model [2,6,11] $P_{ij} = 1 - P_{ij}$ for $E_j-E_i = <\Delta E>_d$ and zero otherwise, seem to be a better choice for more efficient colliders such as polyatomic molecules. Up transitions were calculated from detailed balance and completeness. Introducing these energy transfer probability models in the master equation calculations to fit experimental data allows average energy transferred in a deactivating collision values, $<\Delta E>_d$, to be determined.

The step-ladder model was used with $<\Delta E>_d$ values independent of the excitation energy of the molecule being deactivated by collisions. In the exponential and exponential reverse models, $<\Delta E>_d$ was considered to depend upon the energy of the level from which deactivation took place through a linear equation given by: $<\Delta E>_d = a + bE_j$, so that a and b are the fitting parameters.

IRMPD of 0.3 Torr of $CDCl_3$ in the presence of a bath gas was modeled with an exponential model for P_{ij} for the effect of Ar and with the step-ladder model for the effect of $CHCl_3$. In both cases, calculations were made with an energy-independent $<\Delta E>_d$. The most illustrative results of the calculations are shown in Figures 9 and 10, and the conditions are listed in Table IV . The most remarkable result is that, for a given value of m, larger $<\Delta E>_d$ are obtained at the highest fluence for both bath gases. This result is independent of the value of m and of the deactivation probability model. For added Ar, $<\Delta E>_d$ values of 456 and 304 cm^{-1} were obtained for fluences of 33 and 20 J/cm^2, respectively, with m = 0.8. For m = 0.9 , the corresponding values of $<\Delta E>_d$ were 304 and 183 cm^{-1}. Calculations with m = 0.9 and the same value of $<\Delta E>_d$ cannot fit the curves at both fluences, simultaneously. For added $CHCl_3$, good fits were obtained for $<\Delta E>_d$ values of 3656 and 1828 cm^{-1} for fluences of 33 and 20 J/cm^2, respectively, with m = 0.85. For m = 0.9 , the corresponding values of $<\Delta E>_d$ were 2280 and 914 cm^{-1}. In the deactivation zone of the curves the data set obtained from these calculations coincide with those obtained from considering only vibrational deactivation showing that rotational partition is relevant only in the low pressure region. Calculations with an energy-dependent $<\Delta E>_d$ were made only in the deactivation zone of the curves, so that, rotational partition was not considered. For Ar as bath gas, good fits to the experimental curves were obtained with m = 0.8, an exponential model and, $<\Delta E>_d = (152 + 0.020E)$ cm^{-1} at 33 J/cm^2 and $<\Delta E>_d = (152 + 0.015E)$ cm^{-1} at 20 J/cm^2 (Figure 9). For $CHCl_3$ as bath gas, good fits to the experimental curves were obtained with m = 0.85, an exponential reverse model and, $<\Delta E>_d =(304 + 0.20E)$ cm^{-1} at 33 J/cm^2 and $<\Delta E>_d =$ $(304 + 0.1E)$ cm^{-1} at 20 J/cm^2 (Figure 10). Therefore, the energy-dependent model of $<\Delta E>_d$ which best describes the vibrational deactivation process of $CDCl_3$ at different fluences is $<\Delta E>_d = 152 + (0.013 \pm 0.007)E$ cm^{-1} for added Ar and $<\Delta E>_d = 304 +$ $(0.13 \pm 0.07)E$ cm^{-1} for added $CHCl_3$.

Dissociation probability data of 2 Torr of $CDCl_3$ with P(48) line could be adequately reproduced by master equation calculations with the m and k_{rot} values determined from IRMPA and low pressure IRMPD simulation, only for energy dependent $<\Delta E>_d$ values as shown in Figure 15 and Table V. Calculations with a fixed $<\Delta E>_d$ value could not reproduce the curve shape.

When modeling the dissociation probabilities vs. Φ for the laser lines in the range P(46)-P(36) with the m and k_{rot} values determined from IRMPA and low pressure IRMPD simulation, the calculated $P(\Phi)$ values obtained were orders of magnitude lower than the experimental ones, even for $<\Delta E>_d = 0$. Little improvement was obtained for variations of k_{rot}, even for calculations without rotational partition.

Dissociation probabilities vs. Φ for non-resonant lines could only be fitted by solving the master equation with the same k_{rot} and energy dependent $<\Delta E>_d$ values used in the modeling of the P(48) data, but with different m values. Although many calculations were made, only the most relevant results are shown in Figures 15 and 16 and Table V for clarity.

Excellent agreement between experimental and calculated data was obtained at higher and medium fluences. At the lower fluences, the calculated curves fall more sharply than the experimental. At the dissociation probability threshold energy, $<\Delta E>_d = 0$ was considered for the different laser lines.

Reaction probability vs. Φ at P(23), 907.77 cm^{-1}, could not be calculated since the master equation formulation used does not include the anhamonicity of the vibrational levels higher than v=1.

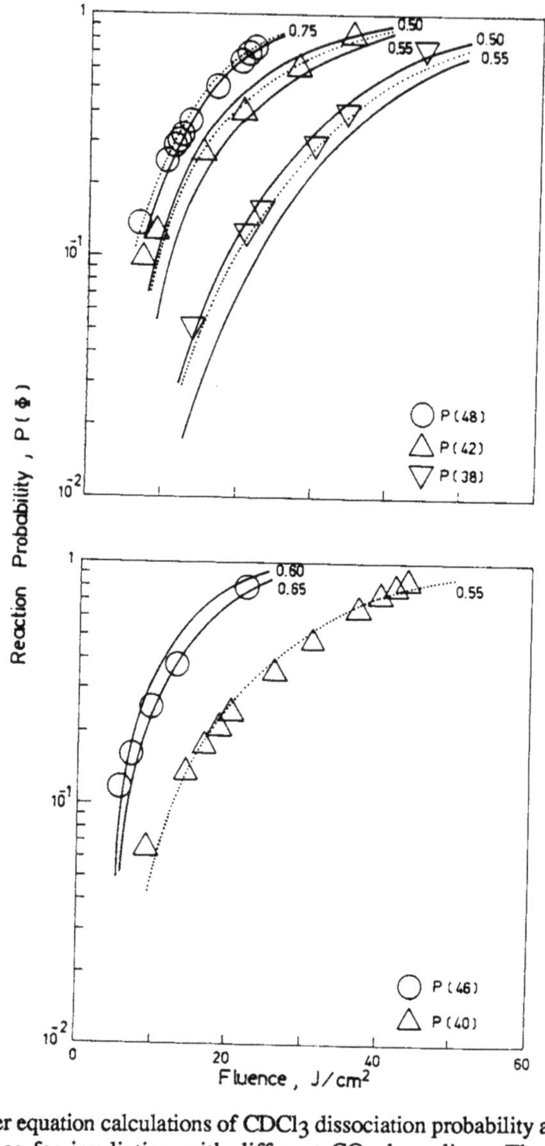

Figure 15: Master equation calculations of $CDCl_3$ dissociation probability as a function of fluence for irradiation with different CO_2 laser lines. The level-to-level absorption cross sections, $\sigma_{i+1,i} = \sigma_{10}/(i+1)^m$, were used with the σ_{10} values given in Table III and the m values shown. The experimental P(Φ) data are shown for comparison. Model conditions for master equation calculations are described in Table V.

TableIV:Model Conditions for the Master Equation Calculations

	m	Rotational partition	M	k_{rot}x10^4	$\langle\Delta E\rangle_d$= a+bE		P_{ij} model
				ns^{-1} $Torr^{-1}$	cm^{-1} a	b	
A_1	0.90	yes	Ar	3.0	183	0	SL
A_2	0.90	yes	Ar	1.5	304	0	SL
B_1	0.90	no	Ar		183	0	SL
B_2	0.90	no	Ar		304	0	SL
B_3	0.90	no	Ar		152	0.007	E
B_4	0.90	no	Ar		152	0.010	E
B_5	0.90	no	$CHCl_3$		304	0.1	ER
B_6	0.90	no	$CHCl_3$		304	0.06	ER
C_1	0.85	yes	Ar	1.5	366	0	SL
C_2	0.85	yes	$CHCl_3$	10	1828	0	SL
C_3	0.85	yes	$CHCl_3$	15	1828	0	SL
C_4	0.85	yes	$CHCl_3$	10	3656	0	SL
D_1	0.85	no	Ar		228	0	SL
D_2	0.85	no	Ar		366	0	SL
D_3	0.85	no	$CHCl_3$		1828	0	SL
D_4	0.85	no	$CHCl_3$		3656	0	SL
D_5	0.85	no	$CHCl_3$		304	0.1	E
D_6	0.85	no	$CHCl_3$		304	0.2	E
D_7	0.90	no	$CHCl_3$		2280	0	SL
D_8	0.90	no	$CHCl_3$		914	0	SL
E_1	0.80	yes	Ar	1.5	304	0	SL
E_2	0.80	yes	Ar	1.5	456	0	SL
F_1	0.80	no	Ar		304	0	SL
F_2	0.80	no	Ar		456	0	SL
F_3	0.80	no	Ar		152	0.015	E
F_4	0.80	no	Ar		152	0.020	E

SL = step-ladder, ER = exponential reverse, E = exponential.

TABLE V : Model Conditions for the Master Equation Calculations

Laser line	m	Φ (J/cm^2)	$<\Delta E>_d$ (cm^{-1})
10P(48)	0.75	30	1828
		25	1828
		20	914
		15	914
		10	304
		6	0
10P(46)	0.65	30	1828
		25	1828
		20	914
		15	914
		10	304
		5	0
10P(44)	0.55	35	1828
		30	1828
		20	914
		15	914
		10	304
		5	0
10P(42)	0.55	40	1828
		30	1828
		20	914
		15	456
		10	0
10P(40)	0.55	45	1828
		40	1828
		30	1828
		20	914
		15	456

		10	0
10P(38)	0.45	50	1828
		40	1828
		30	1828
		20	914
		15	456
		10	0
10P(36)	0.45	50	1828
		45	1828
		40	1828
		30	914
		20	456
		14	0

IRMPA of non-resonant lines was simulated again with these new m values, but the calculated $\sigma_L(\Phi)$ values did not reproduce well the experimental results as depicted in Figure 1.

IRMPD of $CDCl_3$ in the $CDCL_3/CHCl_3$ equimolar isotopic mixture for different laser fluences was modeled using the step-ladder and exponential reverse models for vibrational deactivation. Typical calculated data sets are shown in Figures 11 and 12 and described in Table VI.

For dissociation at fluences of 44 J/cm^2 no rotational bottleneck was found so that modeling included only vibrational deactivation. It was found that at high levelsof excitation, the dissociation probability is less sensitive to changes of m and $\langle\Delta E\rangle_d$ values than at lower excitation levels. At 20 J/cm^2 a rotational bottleneck is observed so that rotational partition as well as vibrational deactivation is considered in calculations using k_{rot} values of (1.5-6) x 10^{-3} ns^{-1} $Torr^{-1}$. Analogous results are obtained for calculations at 11 and 15.5 J/cm^2.

Table VI : Model Conditions for the Master Equation Calculations

	m	Rotational partition	M	$k_{rot} x 10^3$ $ns^{-1}Torr^{-1}$	$\langle \Delta E \rangle_d = a + bE$ cm^{-1}	P_{ij} model
					a b	
H_1	0.90	no	$CHCl_3$		3656 0	SL
H_2	0.90	no	$CHCl_3$		7296 0	SL
H_3	0.90	no	$CHCl_3$		304 0.2	ER
H_4	0.90	yes	$CHCl_3$	6	914 0	SL
H_5	0.90	no	$CHCl_3$		304 0.1	ER
H_6	0.90	yes	$CHCl_3$	6	456 0	SL
H_7	0.90	no	$CHCl_3$		456 0	SL
H_8	0.90	no	$CHCl_3$		304 0.025	ER
H_9	0.85	yes	$CHCl_3$	6	1824 0	SL
H_{10}	0.85	no	$CHCl_3$	304 0.1	ER	
H_{11}	0.85	no	$CHCl_3$	152 0	SL	
H_{12}	0.85	no	$CHCl_3$	304 0.01	ER	
A_1	0.85	no	Ar		152 0.001	E
A_2	0.80	yes	Ar	0.6	152 0	SL
A_3	0.80	no	Ar	152 0.003	E	
A_4	0.80	no	Ar	152 0.004	E	

SL = step-ladder, ER = exponential reverse, E = exponential.

Again, the most relevant feature of this modeling is the fact that in order to fit the dissociation vs. $CDCl_3$ pressure curves, different $\langle \Delta E \rangle_d$ values for different fluences must be used. For m values of 0.85, the dissociation curves were fitted with $\langle \Delta E \rangle_d$ values of 7296, 1824, 914 and 152 cm^{-1} for fluences of 44, 20, 15.5 and 11 J/cm^2, respectively. For $m = 0.9$, $\langle\Delta E\rangle_d$ values of 5484, 914 and 456 cm^{-1} fit the dissociation curves at 44, 20 and 15.5 J/cm^2, respectively.

When using an exponential reverse model for the deactivation probability, best fits to the experimental data were found for $m = 0.85$, with $\langle\Delta E\rangle_d = (304 + 0.3E)\ cm^{-1}$ at $\Phi = 44\ J/cm^2$, $\langle\Delta E\rangle_d = (304 + 0.1E)\ cm^{-1}$ at $\Phi = 20\ J/cm^2$ and $\langle\Delta E\rangle_d = (304 + 0.05E)\ cm^{-1}$ at $\Phi = 15.5\ J/cm^2$. With $m = 0.9$, $\langle\Delta E\rangle_d = (304 + 0.2E)\ cm^{-1}$ at $\Phi = 44\ J/cm^2$, $\langle\Delta E\rangle_d = (304 + 0.06E)\ cm^{-1}$ at $\Phi = 20\ J/cm^2$ and $\langle\Delta E\rangle_d = (304 + 0.025E)\ cm^{-1}$ at $\Phi = 15.5\ J/cm^2$. Within experimental and calculation uncertainty, the vibrational deactivation process of $CDCl_3$ by collisions with $CHCl_3$ in equimolar mixtures of $CDCl_3/CHCl_3$ at different fluences is described by an energy dependent $\langle\Delta E\rangle_d$ given by $\langle\Delta E\rangle_d = 304 + (0.2 \pm 0.09)\ E\ \ cm^{-1}$.

The effect of Ar addition to the $CDCl_3$ IRMPD in equimolar mixtures of $CDCl_3/CHCl_3$ at 20 J/cm^2 was modeled using the step-ladder model for the P_{ij}. Best fits to the curve were found for $\langle\Delta E\rangle_d = 102 - 131\ cm^{-1}$ with m = 0,85 and $\langle\Delta E\rangle_d = 152\ cm^{-1}$ with m = 0,8 (Figure 13). When using the exponential model for P_{ij} the curve was fitted with $\langle\Delta E\rangle_d = 152 + 0.02\ E\ \ cm^{-1}$.

Modeling of the IRMPD of $CDCl_3$ in equimolar isotopic mixtures at 20 J/cm^2 with the addition of 10 Torr of Xe or Ar gave satisfactory results for $m = 0.85$ and $k_{rot} = (1.5\text{-}6) \times 10^{-4}\ ns^{-1}\ Torr^{-1}$ and $\langle\Delta E\rangle_d = 914\ cm^{-1}$ as shown in Figure 14.

In all calculations, the master equation was truncated at 76 kcal/mole corresponding to 29 photons of the 10P(48) CO_2 laser line. Since the unimolecular reaction constant is $5.5 \times 10^9\ s^{-1}$ at $E_{vib} = 76$ kcal/mole, molecules with this energy will rapidly react without being significantly excited over 76 kcal/mole.

The energy grain size used in the calculations with energy independent $\langle\Delta E\rangle_d$ was equal to $\langle\Delta E\rangle_d$ or $\langle\Delta E\rangle_d/2$. Anyhow convergence was always checked up to $h\nu/12$. When an energy dependent $\langle\Delta E\rangle_d$ was used to model the dissociation results, an energy grain of 152 cm^{-1}, ($h\nu/6$), was used for the exponential reverse model while an energy grain of 76 cm^{-1} was used for the exponential model when modeling Ar addition effects.

Temperature effects on the reaction probability were not considered explicitly in these calculations since they were found to be negligible in our experimental conditions [26].

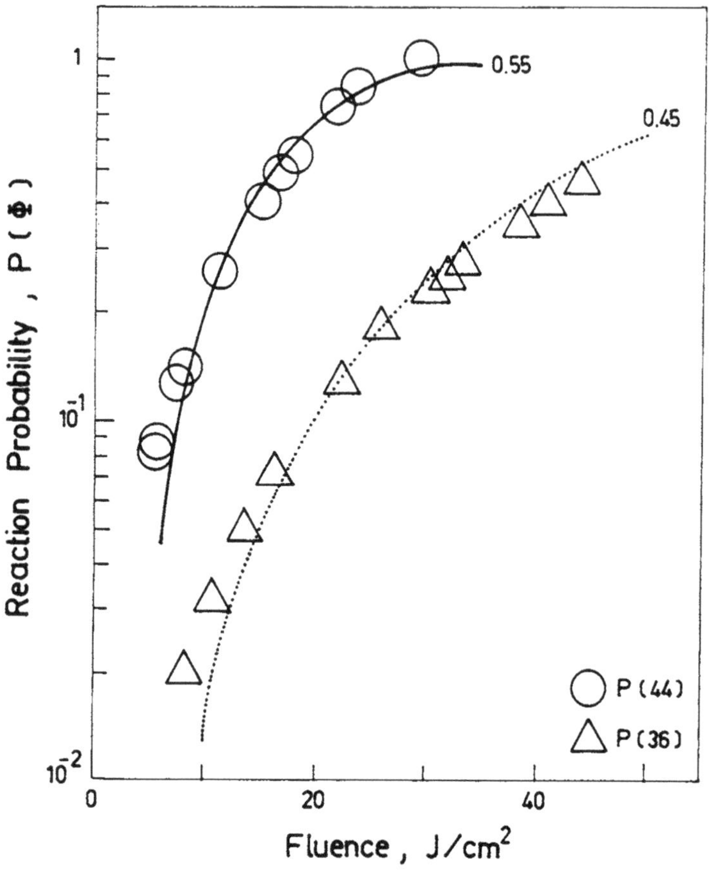

Figure 16: Master equation calculations of $CDCl_3$ dissociation probability as a function of fluence for irradiation with different CO_2 laser lines. The level-to-level absorption cross sections, $\sigma_{i+1,i} = \sigma_{10}/(i+1)^m$, were used with the σ_{10} values given in Table III and the m values shown. The experimental P(Φ) data are shown for comparison. Model conditions for master equation calculations are described in Table V.

III.-DISCUSSION

IRMPE forms a broad population distribution. The IRMPA and IRMPD experiments performed at different laser fluences allow information about the population distribution at different molecular stages of excitation to be obtained. The IRMPA experiments probe the discrete, low-energy levels of $CDCl_3$ while the IRMPD measurements in different experimental conditions give information about the quasi-continuum and threshold energy regions. Master equation simulations of the low pressure IRMPA and IRMPD experiments provide an adequate model for the level-to-level absorption cross section $\sigma_{i+1,1}$. Rotational and vibrational relaxations occur at different time scales, and affect in different ways the different levels of excitation. An increase in the sample pressure at the low stages of excitation increases the fraction of resonant molecules and consequently, the absorption cross section. At high laser fluences and low sample pressures, the rate of up pumping is sufficiently fast so that the population distribution is formed above the reaction threshold where the dissociating constants, k_i, are very fast, and therefore, most molecules will rapidly react before being deactivated by collisions. As anharmonicity is not overcome by power broadening even at the highest fluences used in the IRMPD experiments [34], collisions must play an important role, too, in overcoming the anharmonic bottleneck. Therefore, simulation of the low energy IRMPA of $CDCl_3$ in the presence of a bath gas and of the low pressure IRMPD experiments yields the values of k_{rot}, an averaged rate constant which accounts for collisional relaxation of the rotational and anharmonic bottlenecks.

At high laser fluences and high sample pressures, all molecules of the absorbing species interact with the laser radiation, and, deactivating collisions with the bath gas compete with up pumping and unimolecular reaction. Since the energy transfer processes proceed from different excitation levels, collisions with the buffer gas will influence the population distributions by shifting them towards lower energies and changing their shape. The high-energy wing of the population distribution will be less affected by the deactivating collisions than the low energy wing, since it will react more rapidly. Consequently, lower excitation rates are observed at high bath gas pressures. Master equation simulation of these experiments allow, therefore, the vibrational deactivation transition probabilities P_{ij} to be determined. The energy transfer processes depend on the nature of the colliding partners. Recent experiments support the dependence of $<\Delta E>_d$ on the level of excitation of the colliding molecule [5-12]. Results from the modeling calculations of the IRMPD of $CDCl_3$ at high sample pressures show a dependence of the $<\Delta E>_d$ values on molecular vibrational excitation energy.

The red shift of the IRMPD spectrum includes the anharmonicity of the first excitation steps as well as the interactive coupling between the pumped mode and the various vibrational modes of the molecule in the quasi-continuum.

An important feature of the IRMPD spectrum is that it does not follow the small signal absorption spectrum, specially at the "blue" wing. Master equation calculations with different m values than those found in modeling the IRMPA should be made in order to reproduce the fluence dependence of the dissociation probability at different laser lines. Moreover, for all laser lines considered, the low fluence zone of the calculated curve falls more sharply than the experimental, even when supposing no rotational partition. These results indicate that a larger fraction of molecules than that expected from power broadened laser linewidth and rotational relaxation effects is interacting with the laser radiation. Recent studies of the IR absorption spectroscopy of highly vibrationally excited molecules have shown that, at high levels of excitation, the line shape deforms due to anharmonic interactions of close-lying vibrational modes, and the wings of the spectrum fall more slowly than Lorentzian wings [44-46]. The broadening of the blue wing would allow more molecules to interact with the laser radiation, and therefore, larger dissociation probability values than expected would be obtained on detuning the laser radiation from the band center. The only way in which the master equation formulation used to model the experimental results can account for increasing absorption at high levels of excitation at the wings of the spectrum is by changing the m value in the level-to-level absorption cross section model. Lowering of the m value results in an increase of the absorption cross section with excitation energy. The apparent mismatch between IRMPA and IRMPD simulation results can thus be explained. The IRMPA spectrum was measured up to a level of excitation corresponding to $<n> \sim 5$. This level of excitation is still in the discrete region of the spectrum and thus IRMPA takes into account only rotational distribution of molecules and anharmonicity of the pumped mode. No quasi-continuum effects are evidenced. Since IRMPD takes place at the quasi-continuum levels of excitation, intermode anharmonic coupling effects should be considered when calculating the dissociation probabilities at different excitation frequencies. These quasi-continuum effects are accounted for by increasing the level-to-level absorption cross section at the quasi-continuum energy region.

Large rotational and anharmonic bottlenecks to absorption are undesirable features in a working molecule for deuterium isotopic separation since they cause a large waste of photons. The high density of vibrational levels of $CDCl_3$ and the addition of a bath gas to induce rotational and anharmonic relaxation, partially overcome these restrictions so that higher dissociation probabilities are obtained at lower fluences than for CDF_3. The main drawback encountered is that the $CDCl_3$ v_4 mode absorption peak is resonant with the P(50) line in the low gain region of 10.6 μm emission band of the CO_2 laser. The IRMPD spectrum obtained shows that the use of higher gain, less resonant lines to dissociate $CDCl_3$ results in $P(\Phi)$ values much higher than those expected from laser and molecular spectroscopic parameter considerations. This is a great advantage when working with TEA CO_2 lasers delivering poor energy values at the low gain region of the 10.6 μm band. When this is not the case, red- shifting the laser frequency to the 11 μm emission band allows much higher dissociation probabilities than in resonant irradiation to be obtained. Higher isotopic selectivity values than those reported by Yokoyama et al. [28] were obtained with the addition of 10 Torr of Ar or Xe to the IRMPD of $CDCl_3$ in isotopic mixtures of $CDCl_3/CHCl_3$. The values given in Table I represent lower limits to the isotopic selectivity imposed by the sensitivity of the detection system.

AKNOWLEDGEMENTS

Aknowledgement is made to the Consejo Nacional de Investigaciones Cientificas y Tecnicas (CONICET), CITEFA and the Organization of American States (O.A.S.) for finantial support. Late Dr. A. H. Burgos participated in the early phase of this work. Authors thank O. Vilar, M. Seifert and J. Luque for expert technical assistance. Authors are grateful to Lic. M. Alonso for the careful reading of this manuscript and for helpful suggestions.

References

* CONICET Scientific Carreer Member.

[1] V.S. Letokhov. **"Non-Linear Laser Chemistry"**, ed. F.P. Schafer. Springer-Verlag. Berlin. (1983)

[2] W.C. Danen, and J.C. Jang. **"Laser-Induced Chemical Processes"**, ed. J.I. Steinfeld. Plenum Press. New York, 45 (1981).

[3] R.L. Woodin, and A. Kaldor. **"Photoselective Chemistry", Part 2**, ed. J. Jortner, R.D. Levine y S.A. Rice. John Wiley & Sons. New York, 3 (1981).

[4] N. Bloembergen, and E. Yablonovitch. Physics Today **31**, 23 (1978).

[5] M.J. Rossi, J.R. Pladziewicz, and J.R. Barker. J. Chem. Phys. **78(11)**, 6695-6708 (1983).

[6] J.R. Barker, and R.E. Golden. J. Phys. Chem. **88**,1012- 1017 (1984).

[7] H. Hippler, L. Lindemann, and J. Troe. J. Chem. Phys. **83(8)**, 3906-3912 (1985).

[8] B. Herzog, H. Hippler, L. Kang, and J. Troe. Chem. Phys. Lett. **120(2)**, 124-127 (1985).

[9] J.M. Zellweger, T.C. Brown, and J.R. Barker. J. Phys. Chem. **90**, 461-465 (1986).

[10] J.E. Baggot, and D.W. Law.J. Chem. Phys. **85(11)**, 6475- 6488 (1986).

[11] G.Y. Chung, and R.W. Carr. J. Phys. Chem .**91**, 2831-2837 (1987).

[12] D.L. Snavely, R. N. Zare, J.A. Miller, D.W. Chandler. J. Phys. Chem. **90**, 3544 (1986).

[13] T.B. Simpson, J.G. Black, I. Burak, E. Yablonovitch, and N. Bloembergen. J. Chem. Phys. **83(2)**, 628-640 (1985).

[14] J.B. Marling, I.P. Herman, S.J. Thomas. J. Chem. Phys. **72**, 5603 (1980).

[15] D.S. King, J.C. Stephenson. J. Am. Chem. Soc. **100:23**, 715 (1978).

[16] D.K. Evans, R.D. Mc Alpine, H.M. Adams. J. Chem. Phys. **77**, 3551 (1982).

[17] F. Magnotta, I.P. Herman, F.T. Aldridge. Chem Phys. Lett. **92**, 600 (1982).

[18] J. Moser, P. Morand, R. Duperrex, H. Van den Bergh. Chem. Phys. **79**, 277 (1983).

[19] M. Gauthier, A. Outhouse, Y. Ishikawa, K.O. Kutschke, P.A. Hackett. Appl. Phys. B **35**, 173 (1984).

[20] V. Parthasarathy, S.K. Sarkar, K.V.S. Rama Rao, J.P. Mittal. Appl. Phys. B. **39**, 187 (1986).

[21] R. Mc Alpine, J.W. Goodale, D.K. Evans. Can. J. Chem. **63**, 2995 (1985).

[22] F.T. Aldridge, I.P. Herman, J.B. Marling. L.L. Wood. Lawrence Livermore Laboratory, UCID-17736-78-1.

[23] I.P. Herman, F. Magnotta, R.J. Buss, Y.T. Lee. J. Chem. Phys. **79**, 1789 (1983).

[24] M.L. Azcarate. Doctoral Thesis. University of Buenos Aires (1986).

[25] M.L. Azcarate, E.J. Quel. Appl. Phys. B. **47**, 223-228 (1988).

[26] M.L. Azcarate, E.J. Quel, B. Toselli, J.C. Ferrero, E.H. Staricco. J. Phys. Chem. **92**, 403-408 (1988).

[27] F. Magnotta, I.P. Herman. J. Chem. Phys. **81**(5), 2363-2374 (1978).

[28] A. Yokoyama, K. Suzuki, G. Fujisawa, N. Ishikawa, M. Iwasaki. J. Nucl. Sci. Technol. **18**(9), 737-738 (1981).

[29] A. Ruoff, H. Burger. Spectrochim. Acta **26A**, 989-997 (1970).

[30] S. Speiser, J. Jortner. Chem. Phys. Lett. **44**(3), 399-403 (1976).

[31] W. Fuss, T.P. Cotter. Appl. Phys.B **12**, 265-276 (1977).

[32] J.L. Lyman, S.D. Rockwood, S.M. Freund. J. Chem. Phys. **67**(10), 4545-4556 (1977).

[33] A. Yokoyama, K. Suzuki, G. Fujisawa, and N. Ishikawa. Appl. Phys. B **38**, 99-105 (1985).

[34] M.L. Azcarate, E.J. Quel. Appl. Phys. B. **47**, 239-242 (1988).

[35] M.V. Garcia, R. Escribano, G. del Rio, J.M. Orza. Anal. Quim. **71**, 745-754 (1975).

[36] M.L. Azcarate, E.J. Quel. J. Phys. Chem. **93**, 697-702 (1989).

[37] K. Sugita, Y. Ishikawa, S. Arai. Appl. Phys. B **36**, 111 (1985).

[38] I.P. Herman, J.B. Marling. J. Chem. Phys. **72**, 516 (1980).

[39] R.D. Mc. Alpine, D.K. Evans. **"Photodissociation and Photoionization"**, ed. K.P. Lawley. John Wiley & Sons. New York, 31 (1985).

[40] a) B. Toselli, J.C. Ferrero, E.H. Staricco. J. Phys. Chem. **89**, 1492-1499 (1985). b).B. Toselli, J.C. Ferrero, E.H. Staricco. J. Phys. Chem.**90**, 4562 (1986).

[41] I.P. Herman, F. Magnotta, R.J. Buss, Y.T. Lee. J. Chem. Phys. **79**(4), 1789-1794 (1983).

[42] P.J. Robinson, K.A. Holbrook. **"Unimolecular Reactions"**. Wiley-Interscience: London. (1972).

[43] J.C. Jang-Wren, D.W. Setser, J.C. Ferrero. J. Phys. Chem. **89**, 414-426 (1985).

[44] A.A. Stuchebrukhov, I.E. Khromov. Chem. Phys. Lett. **134(3)**, 251-254 (1987).

[45] V.N. Bagratashvili, V.N. Burimov, S.I. Ionov, A.P. Sviridov, A.A. Stuchebrukhov, I.M. Turovetz. Chem. Phys. Lett. **137(1)**, 45-50 (1987).

[46] D.W. Lupo, M. Quack. Chem. Phys. Lett. **130(5)**, 371-376 (1986).

EXCIMER LASER FLASH PHOTOLISIS STUDIES OF FSO_3 RADICAL REACTIONS

C.J. Cobos, A.E. Croce de Cobos and E. Castellano
Instituto de Investigaciones Fisicoquimicas Teoricas y Aplicadas
(INIFTA), Facultad de Ciencias Exactas, Universidad Nacional de La Plata, Argentina.

Abstract

The kinetics of the recombination reactions $F + FSO_3 + M \rightarrow FSO_2OF + M$ (M=He, N_2, CF_4 and SF_6) and $FSO_3 + FSO_3 + M \leftrightarrow F_2S_2O_6 + M$ (M=He and CF_4), have been studied using a 193-nm excimer laser flash photolysis/visible absorption tecnique between 293-381 K, in the pressure range 5-760 Torr. Time dependent FSO_3 concentration profiles were monitored by absorption spectroscopy at 450 nm. The first reaction was found to be in the fall-off region. The following room temperature limiting high and low pressure rate constants: $K_{rec,\infty} = (7.6 + 1.0) \times 10^{-11}$ cm^3 molecule^{-1} s^{-1}, $K_{rec,0}$ = [He] $(4.11 + 0.5) \times 10^{-28}$, [$N_2$] (5.2 + cm^3 molecule^{-1} s^{-1}, were determined. Temperature dependences of $T^{0.23}$ and $T^{-3.1}$, were obtained for $K_{rec,\infty}$ and $K_{rec,0}$ respectively. The second reaction was first-order with $K_{rec,\infty} = 4.64 \times 10^{-14} (T/300)^{(0.72 \pm 0.25)}$ cm^3 molecule^{-1} s^{-1} and $K_\infty = 2.09 \times 10^{14} \exp(-(11090 + 290)/T)$ s^{-1} over all the experimental conditions. Equilibrium constants and thermochemical parameters for both reactions systems are reported. A theoretical analysis in terms of Troe's unimolecular rate theories is also presented.

I. Introduction

The recombinations of radicals play a relevant role in numerous photochemical and thermal systems such as those which occur in atmospheric and combustion chemistry. From the theoretical point of view, these reactions provide a reliable testing ground for unimolecular rate theory. In spite of the fact that many reactions of the fluorosulfate radicals have been studied under stationary conditions [1] , very little has been investigated in direct measurements [2]. The excimer laser flash photolysis technique here employed allows the study of fast reactions in real time-resolved employed allows the study of fast reactions in real time-resolved experiments. The rate constants of the recombination reactions (1) [3] and (2) [4],

$$F + FSO_3 + M \rightarrow FSO_2OF + M \quad (1)$$

$$FSO_3 + FSO_3 + M \leftrightarrow F_2S_2O_6 + M \quad (2,-2)$$

together with the reverse of reaction (2), have been here directly determined. To follow the kinetics, transient FSO_3 radical absorption at 450 nm was analysed. In both cases, the absorption cross section of the absorbing species must be known accurately if the recombination rate constant is to be determined. Moreover, both systems provide clean radical sources, rendering them appropriate for FSO_3 absorption cross section measurements.

In the case of reaction (2, -2), the analysis of the data enables rate constants for both the forward and the backward reactions to be evaluated under identical conditions by a direct, time-resolved technique. The characteristics of the experiment, whereby the equilibrium is perturbed by partially flash photolysing the $F_2S_2O_6$, correpond to what is essentially a relaxation method.

By and large, the analysis of our experiments provides an insight on theoretical aspects of radical-radical recombination reactions.

II. Experimental

The apparatus is described in detail elsewhere [3,4] so that just and outline will be given here. The FSO_3 radicals were generated by flash photolysing FSO_2OF or $F_2S_2O_6$ molecules with 17ns pulses from an ArF excimer laser (Lambda Physik EMG 101 MSC). The chemical relaxation of the system back to reform FSO_2OF or $F_2S_2O_6$ according to reactions (1) and (2) was observed by measuring the temporal change in the FSO_3 absorption at 450 nm. A 150 W xenon arc lamp (Hanovia 901 C-1) was employed when reaction (1) was investigated, while a 250 W tungsten lamp powered by a dc supply was used for the experiments corresponding to reaction (2). The reaction vessel was a cylindrical quartz cell contained into an electrically heated aluminium housing with a cross-beam geometry for the photolysis and analysis beams. A calibrated chromel-alumel thermocouple placed in contact with the wall of the reaction vessel was used to measure the temperature of the reaction mixtures (293-381°K). The analysis light passed through the cell to a prism monochromator (Zeiss MM12) with a spectral resolution of 3.3 nm. The emerging ratiation was displayed on a digital oscilloscope (Nicolet 2090) electronically triggered by the laser. The single-shot signals were stored on the oscilloscope floppy disks and analysed on a measured using a pyroelectric detector (Gentec). The pressure measurements of the reactants, FSO_2OF (up to 5 Torr) and $F_2S_2O_6$ (3-6 Torr) and of the bath gases M = He, N_2, CF_4 and SF_6 (5-760 Torr) were done with a sensitive quartz spiral gauge. FSO_2OF and $F_2S_2O_6$ were prepared by photolysing F_2 in the presence of SO_3 and separated by distillation [3, 4]. The purity of these gases was checked by IR spectrophotometry (Perkin Elemer 325). The absorption spectra were recordedd in a Cary 14 spectrophotometer. High purity bath gases were employed in all the experiments.

III. Results and discussion

III.1 Reaction $F + FSO_3 + M \rightarrow FSO_2OF + M$.

The FSO_2OF presents a strong absorption at the photolysis wavelength, being its absorption cross- section σ (FSO_2OF, 193 nm) = (1.57 ± 0.07) x 10^{-18} cm^2 $molecule^{-1}$[5]. On the other hand, FSO_3 radicals are the only absorbing species at the monitoring wavelength of 450 nm [6]. From the temperature dependence of $F_2S_2O_6 \leftrightarrow 2\ FSO_3$ equilibrium mixtures (70-90 Torr) between 310 and 334 °K, the FSO_3 absorption cross section was determined. The slope of an absorbance vs FSO_3 concentration (= $(K_c\ [F_2S_2O_6])^{1/2}$, were K_c is the equilibrium constant [6]) plot yields directly the absorption crosss section σ(FSO_3, 450 nm) = (3.64 ± 0.32) x 10^{-18} cm^2 $molecule^{-1}$. At this wavelength, the FSO_2OF absorption cross section is about a factor of 10^5 smaller [5]. The primary quantum yield for $FS0_3$ formation was measured from the FSO_3 concentration formed inmediately after the laser pulse and the number of 193 nm photons absorbed, namely Φ (FSO_3) = $[FSO_3]_{t=0/hv=193nm}$ = 1.09 ± 0.15 FSO_3 radicals/quantum. On the time scale of our experiments, the microsecond scale, the photoproducts generated by the laser radiation react exclusively through reaction (1). This process was investigated over the temperature range 298-378°K in the presence of the diluent gases M = He, N_2, CF_4 and SF_6. The decay curves were analysed by a second order kinetics

$$\frac{1}{\ln(Io/I)} = \frac{Krec\ .\ t}{(FSO3,\ 450\ nm).\ l} + \frac{1}{\ln(Io/I')} \qquad (3)$$

Here l is the absorption path length, I_o the incident ligth intensity and I and I' are, respectively, the transmited ligth intensities at time t and inmediately after the pulse. Figure 1 shows a typical $(\ln(I_o/I))^{-1}$ vs time representation. The pressure dependence of the derived rate constants, Krec, sohws that our results are in the by the factors of 4.2 (M = He), 3.7 (N_2), 2.7 (CF_4) and 2.3 (SF_6) over the whole pressure range. Therefore, the experimental data were fitted and low pressure rate constants $k_{rec,\infty}$ and $K_{rec,0}$, respectively. The deteils of this procedure are given elsewhere [3]. From the pressure and temperature dependences of the experiments with M = CF_4, the following equations were derived:

$$K_{rec,\ \infty} = 7.2 \times 10^{-11}\ (T/300)^{0.23}\ cm^3\ molecule^{-1}\ s^{-1} \qquad (4)$$

$$K_{rec,0} = [CF_4]\ 1.1 \times 10^{-27}\ (T/300)^{-3.1}\ cm^3\ molecule^{-1}\ s^{-1} \qquad (5)$$

The $K_{rec,0}$ values [He] $(4.1 \pm 0.5) \times 10^{-28}$, $[N_2]$ $(5.2 \pm 0.7) \times 10^{-28}$ and $[SF_6]$ $(13.0 \pm 2.0) \times 10^{-28}$ $cm^3\ molecule^{-1}\ s^{-1}$ were extrapolated from the room temperature fall-off curves. The absolute values of $K_{rec,\infty}$ and $K_{rec,0}$ and their temperature dependences are typical for this kind of reactions [7]. The $K_{rec,\infty}$ values were combined with the high pressure dissociation rate constants K_∞ of the reverse of reaction (1) [9] to evaluate the equilibrium constant of the system $K_c = K_\infty/K_{rec,\infty}$. In this way, the following expression was obtained in the range 298-394 °K

$$K_c = 9.0 \times 10^{24} \exp(-16408/T) \quad molecule\ cm^{-3} \qquad (6)$$

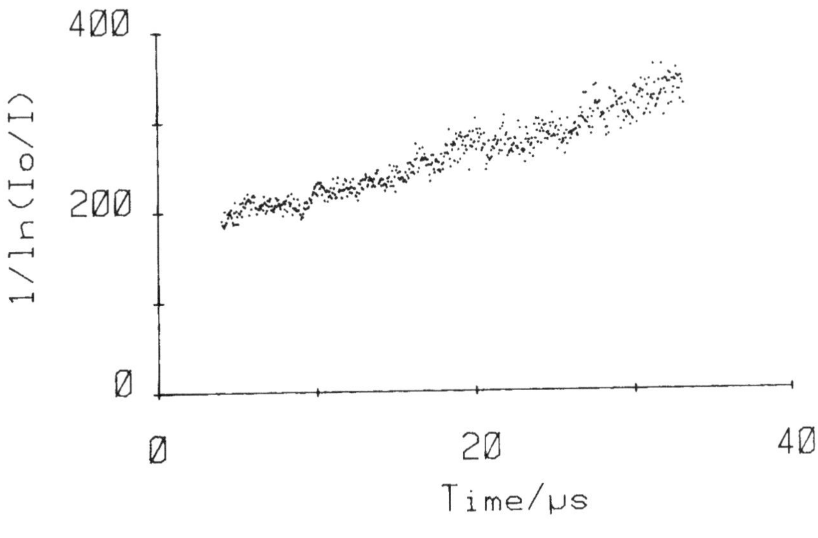

Figure 1. Typical $1/\ln(I_o/I)$ vs. time representation. Average of 7 experiments performed with 1.5 Torr of FSO_2OF and 153.3 Torr of CF_4 at 294 K. The slope leads to $k_{rec} = 4.46 \times 10^{-11}$ cm^3 $molecule^{-1}\ s^{-1}$.

The standard enthalpy and entropy changes for reaction (1), $\Delta H°_{298} = 33.1 \pm 0.9$ kcal mol^{-1} and $\Delta S°_{298} = 27.2 + 1.2$ cal mol^{-1} K^{-1}, were determined from the temperature dependence of $K_p = K_c RT$.

The $K_{rec,\infty}$ values were interpreted in terms of the canonical version of the statistical adiabatic channel model, SACM [10]. In this model, the Morse parameters, α/β, of the interaction potential. α takes into account the attenuation of the transitional modes and β is the range parameter of the radial potential. The room temperature $K_{rec,\infty}$ value is reproduced when an $\alpha/\beta = 0.49$ is employed in the calculations. This value is in very good agreement with the average value that fits a large number of recombination-dissociation reactions, $< \alpha/\beta > = 0.46 \pm 0.09$ [11]. The SACM also predicts a $T^{0.15}$ temperature dependence which agrees very well with the experimental (equation (4)).

The collision efficiencies for the different bath gases can be derived as $\beta_c = K_{rec,0}/K_{rec}{}^{SC}{}_{,0}$ from the comparison of the experimental $K_{rec,0}$ values and the theoretically evaluated strong collision limiting low pressure rate contants, $K_{rec,0}$. In this way, the values $\beta_c = 0.11$, 0.19, 0.46 and 0.50 were computed at 298 K for the gases He, N_2, CF_4 and SF_6 respectively. The β_c values allow the calculation of the average energy transferred in both activating and deactivating collisions between the energized FSO_2OF adduct at energies close to the FSO_2O-F bond dissociation, and the diluent gases, $< \Delta E>$.

$-<\Delta E> = 40$, 82, 346 and 414 cm^{-1} were calculated from the above β_c values for He, N_2, CF_4 and SF_6 respectively. These values compare well with directly measured $<\Delta E>$ values for other molecules [13]. Finally, the above $K_{rec,\infty}$, $K_{rec.0}$ and β_c values can be easily combined with computed Lennard-Jones collision frequencies, Z_{LJ}, to evaluate specific rate constants thermally averaged over a Boltzmann distribution of total energies, E, and total angular momenta, J, $k(\langle E\rangle,\langle J\rangle) = \beta_c Z_{LJ} K_{rec,\infty}/K_{rec,0}$. The values 8.7×10^6, 1.3×10^7 and 1.8×10^7 s^{-1} were computed at 298, 339 and 378 K respectively. Due to the fact that k(E,J) normally decreases with increasing J [14] at low excitations, the observed positive temperature dependence can be attributed to the increasing thermal energy.

III.2 Reaction $FSO_3 + FSO_3 \leftrightarrow F_2S_2O_6$.

The $F_2S_2O_6$ and the FSO_3 radicals coexist in chemical equilibrium at ordinary temperatures[6] but fortunately their absorption spectra do not overlap [6,15]. The $F_2S_2O_6$ absorbs below 270 nm, being its absorption cross section at the photolysis wavelength $\sigma(F_2S_2O_6$, 193 nm) = 4.14×10^{-18} cm^2 $molecule^{-1}$. UV flash spectroscopy studies show that only FSO_3 radicals are formedby $F_2S_2O_6$ photolysis [2]. The FSO_3 radicals present several electronically excited states [15] ($^2A_1$, $^2E(1)$ $^2E(2)$ and ^4E) wich can be populated after the $F_2S_2O_6$ photodissociation. Therefore, the relaxation of the excited photoproducts by collisional energy transfer will lead to a heating of the reaction volumen and an appreciable temperature jump (1-5 K). At 293K, such increase of temperature does not affect the decay profiles and a second order rate law accounts for the reaction kinetics. However, at the higher temperatures, the temperature jump induces measurable changes in the FSO_3 equilibrium concentration. In this case, after the chemical relaxation a new concentration. In this case, after the chemical relaxation a new equilibrium at the higher temperature is established (see Figure 2). Over 321.5 K, the temporal evolution of the FSO_3 concentrations is given the equation,

$$\{ d[FSO_3]/dt \} = -2 k_2[FSO_3]^2 + 2 k_{-2}[F_2S_2O_6] \qquad (7)$$

where k_2 and k_{-2} are the recombination and dissociation rate constants of reaction (2). A non-linear least squares fit of the FSO_3 concentration resulting of the integration of equation (7) to the experimental decay profiles, delivered a set of k_2 and k_{-2} values. We found that over the whole pressure range investigated (250-760 Torr of He and CF_4) these rate constants can be certainly assimilated to the high pressure values, $k_2 = k_{rec,\infty}$ and $k_{-2} = k_{\infty}$. The following equation fits very well all the recombination rate constants measured between 293 and 381 K,

$$K_{rec,\infty} = 4.64 \times 10^{-14} (T/300)^{(0.72 + 0.25)} \text{ cm}^3 \text{ molecule}^{-1} \text{ s}^{-1} \qquad (8)$$

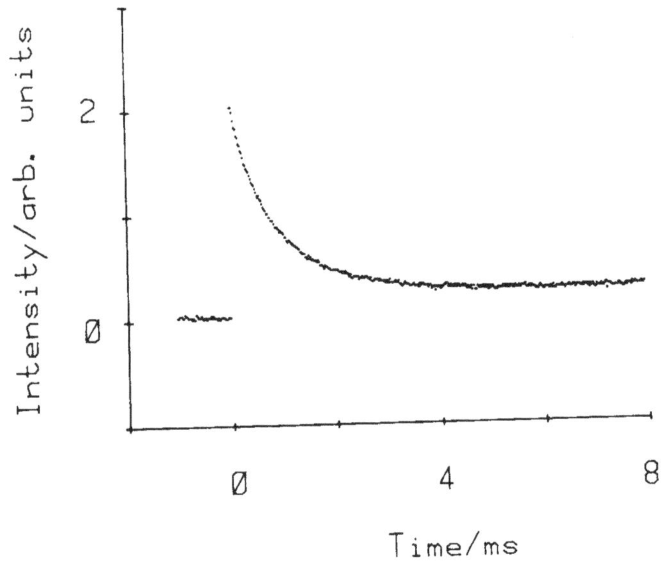

Figure 2. Time dependence of the FSO_3 radical absorption at 450 nm, resulting from the 193-nm photolysis of 4.8 Torr of $F_2S_2O_6$ in the presence of 402.3 Torr of CF_4 at 356 K.

Our results confirm the very small $K_{rec,\infty}$ values reported in reference 2. These values are notably smaller than the accepted as normal values for this type of reactions, i.e. of about 10^{-12} to 5×10^{-11} cm^3 $molecule^{-1}$ s^{-1}[7,11]. The experimental dissociatin rate constants are here represented over the range 321.5 and 381 K by the Arrhenius expression

$$k_\infty = 2.09 \times 10^{14} \exp(-(11090 \pm 290)/T)\ s^{-1} \qquad (9)$$

The individual k $/k_{rec,\infty}$ values lead to the equation

$$k_c = 1.45 \times 10^{27} \exp(-(10730 \pm 380)/T) \qquad molecule\ cm^{-3} \qquad (10)$$

The thermochemical parameters $\Delta H°_{298} = 22.1 \pm 0.7$ kcal mol^{-1} and $\Delta S°_{298} = 38.2 \pm 1.0$ cal mol^{-1} K^{-1} were derived as in the previous section. These values are in very good agreement with those reported in reference 6.

As above mentioned, the small temperature jump after the laser pulse shifted the position of the strongly temperature-dependent $FSO_3 + FSO_3 \leftrightarrow F_2S_2O_6$ equilibrium in the experiments performed at the higher temperatures. Therefore, the system relaxes to a new euqilibrium. Relaxation theory [16] gives for the first-order approach to equilibrium

$$k_2 = 1 / [\tau (4 [FSO_3]_e + K_{c,e})] \qquad [11]$$

where $[FSO_3]_e$ and $K_{c,e}$ are values taken at the new established equilibrium and

$1/\tau = - [d(\ln \Delta[FSO_3]) / dt]$ is the inverse of the relaxation time. When the ratio $\Delta[FSO_3]/[FSO_3]_e$ is small, changes in FSO_3 concentration are proportional to changes in the transmitted light intensities. In Figure 3 a plot of -ln (I/I') vs. time for the experiments performed at 381 K is depicted. Its slope yields $k_2 = 6.1 \times 10^{-14}$ cm^3 $molecule^{-1}$ s^{-1} in very good agreement with the value consigned above (equation 8). Relaxation experiments of this system at higher temperatures are underway in our laboratory [17].

Due to the anomalously low $k_{rec,\infty}$ values obtained for this reaction, it seems to be interesting to compare them with the theoretical predictions of the SACM. The experimental rate constants were exactly reproduced when the average value $<\alpha/\beta> = 0.46$ [11] was employed.

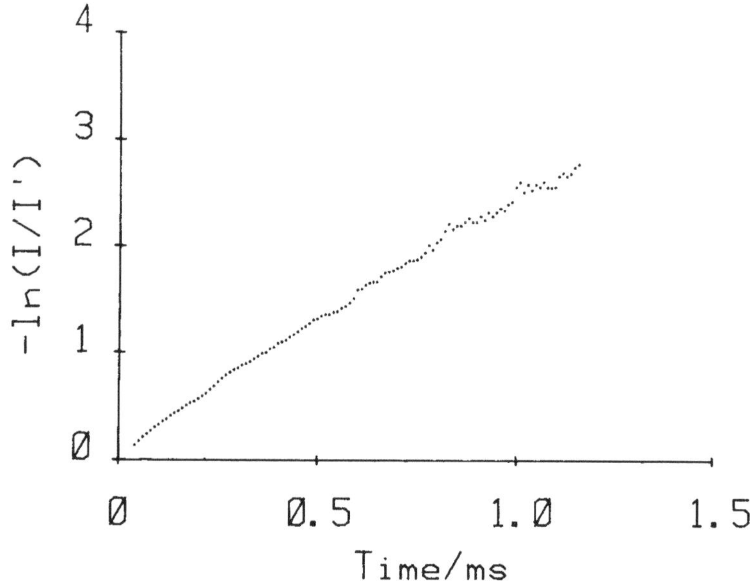

Figure 3. Relaxation plot of -ln(I/I') vs. time. Average of 6 experiments carried out with 4.8 Torr of $F_2S_2O_6$ and 394.6 Torr of CF_4 at 381 K. From the slope the value $k_2 = 6.1 \times 10^{-14}$ cm^3 $molecule^{-1}$ s^{-1} is obtained.

Therefore, the present experimental observables provide a valuable test of the SACM, which appears to be able to predict high pressure rate constants over a wide range of values (over about four orders of magnitude [11]). Factorization of $k_{rec,\infty}$ into a loose rate constant $k_{rec,\infty}$ (PSL) corresponding to the phase space theory and into a rigidity (or steric) factor $f_{rigid} = k_{rec,\infty}$

$/k_{rec,\infty}$ (PSL) [11], indicates that the extremely small f_{rigid} of 3.2×10^{-4} is the responsible of the anomalous $k_{rec,\infty}$ value. The factor f_{rigid} provides an indication of the tightness of the transition states. The low f_{rigid} value found here shows that, at the critical configuration, the fragments are strongly restricted by angular forces arising from the transitional modes.

Achnowledgements

This research project is supported by the Consejo Nacional de Investigaciones Cientificas y Tecnicas and the Comision de Investigaciones Cientificas de la Provincia de Buenos Aires. Part of the equipment used in the present work was provided thorough the Cooperation Agreement between the University of Mainz (W. Germany) and the University of La Plata (Argentina). We also thank the Stiftung Volkswagenwerk for a partnership project with Prof. J. Troe.

References

[1] H. J. Schumacher, J. Photochem. Photobiol., 7, 755 (1968).

[2] E. San Roman, P.F. Aramendia and H.J. Schumacher, Anales Asoc. Quim. Argentina, 70, 887 (1982). P.F. Aramendia, Doctoral Thesis, University of Buenos Aires.

[3] A.E. Croce de Cobos, C.J. Cobos and E. Castellano, J. Phys. Chem., in press.

[4] C.J. Cobos, A.E. Croce de Cobos, H. Hippler and E. Castellano, J. Phys. Chem., in press.

[5] A.E. Croce de Cobos, to be published.

[6] F.B. Dudley and G.H. Cady, J. Amer. Chem. Soc., 65, 3375 (1963). E. Castellano, R. Gatti, J.E. Sicre and H.J. Schumacher, Z. Phys. Chem. NF, 42, 174 (1964).

[7] D.L. Baulch, R.A. Cox, R.F. Hampson, J.A. Kerr, J. Troe and R.T. Watson, J. Phys. Chem. Ref. Data, 13, 1259 (1984).

[8] J. Troe, J. Phys. Chem., 83, 114 (1979).

[9] J. Czarnowski, E. castellano and H.J. Schumacher, Z. Phys. Chem. NF, 57, 249 (1968).

[10] J. Troe, J. Chem. Phys., 75, 226 (1981).

[11] C.J. Cobos and J. Troe, J. Chem. Phys., 83, 1010 (1985).

[12] J. Troe, J. Chem. Phys. , 66, 4745, 4758 (1977).

[13] J.E. Dove, H. Hippler and J. Troe, J. Chem. Phys., 82, 1907 (1985).

[14] L. Brower, C.J. Cobos, J. Troe, H.R. Dubal and F.F. Crim, J. Chem. Phys., 86, 6171 (1987), and references therein.

[15] G.W. King, D.P. Santry and C.H. Warren, J. Mol. Spectrosc., 32, 108 (1969).

[16] M. Eigen and L. De Maeyer, in Techniques of Chemistry, Vol. VI. Investigation of Rates and Mechanism of Reactions, Part II , Third Edition, J. Wiley & Sons, 1974, p. 63.

[17] C.J. Cobos, A.E. Croce de Cobos and E. Castellano, to be published

Electrodynamics of Single Atoms in a Cavity

S. HAROCHE and J.M. RAIMOND

Laboratoire de Spectroscopie Hertzienne de l'Ecole Normale Supérieure
(Université Pierre et Marie Curie, associé au CNRS UA 18)
24 rue LHOMOND 75231 PARIS CEDEX 05

1.Introduction

The radiative properties of simple atomic systems have been recently studied in various experimental situations which have become accessible due to the development of new laser techniques. It is now possible to prepare and detect single atoms or ions, to select their internal state with high precision, to confine them to well defined position in space and to control their velocities. Coupling such atoms with electromagnetic cavities leads to the study of very interesting effects such as the modification of the spontaneous emission rates, the shift of the atomic energy levels and more generally the alteration of the atom spectrum structure induced by the cavity. These effects have wide ranging applications in Quantum Electrodynamics. Among these, a special mention should be made of a new variety of Quantum oscillators, the micromasers: in these devices, atoms injected one by one in a high Q resonant cavity couple to the cavity field and produce a microwave with very small photon numbers, whose statistical properties are quite different from those of ordinary masers or lasers. This new field of atom-radiation interaction studies is now called "Cavity Quantum Electrodynamics". The lectures presented at the Mar del Plata school were intended to be an overview of this field, as it was by the end of 1988. In these notes, we discuss a few selected topics of Cavity QED, some of which have been investigated at Ecole Normale Supérieure in Paris and at Yale University over the last six or seven years. The discussion remains qualitative since all theoretical derivations and detailed experimental descriptions can be found in previously published articles given in references. These notes are thus intended as an introduction to the field of Cavity QED and a guide to further studies. Other reviews covering the field in general or some selected topics can be found in [1-6]. In Section 2, we discuss spontaneous emission of a single atom in a cavity, both in the low Q and high Q regimes and describe such effects as spontaneous emission inhibition and enhancement and reversible spontaneous emission (vacuum Rabi oscillation). In Section 3, we briefly analyze dispersive effects of Cavity QED which include shifts and splittings of atomic transition frequencies induced by the coupling with the cavity. Section 4 is devoted to a discussion of micromaser physics. We present a general description of these devices and show how they can be used to test various aspects of the theory of measurement in Quantum Mechanics or to generate non classical states of the electromagnetic field. The last Section examines a special example of micromaser, the two-photon Rydberg atom maser. This is the first quantum oscillator successfully operated on an intrinsically non linear effect, without any injected field. It provides a very good illustration of the main concepts of Cavity QED. Finally, in brief concluding remarks, we try to point out a few promising directions for this very active field in atomic and laser physics.

2 . Spontaneous emission in a cavity

Spontaneous emission can be viewed as resulting from the coupling of the atomic radiator with the continuum of field modes in the space surrounding it. Each mode is a quantum oscillator presenting zero point fluctuations which perturb the atom and induce its emission. If the atom is confined in a metallic box, the density of modes is modified by the box boundary conditions and becomes different from free space. This correspondingly modifies the spectrum of the vacuum fluctuations seen by the atom and may result in an important alteration of the spontaneous emission rates.

2.1 Inhibition of Spontaneous Emission in a simple geometry.

The simplest geometry to observe these effects corresponds to a cavity made of two plane parallel mirrors. The mode density in such a cavity is plotted as a function of frequency on Fig 1a. We distinguish the modes with their electric field parallel to the mirrors (solid line) from the ones with an electric field normal to the mirrors (dashed line). $\omega_0=\pi c/d$ is the angular frequency of the lowest cavity mode with its electric field parallel to the mirrors (d: mirror separation). All the modes polarized along the mirror surfaces with a wavelength larger than twice the mirror separation cannot accomodate an antinode between the mirrors and are thus suppressed in the cavity. As a result, an atomic excited state which in free space would radiate at such a wavelength with its electric field polarized along the metal surface

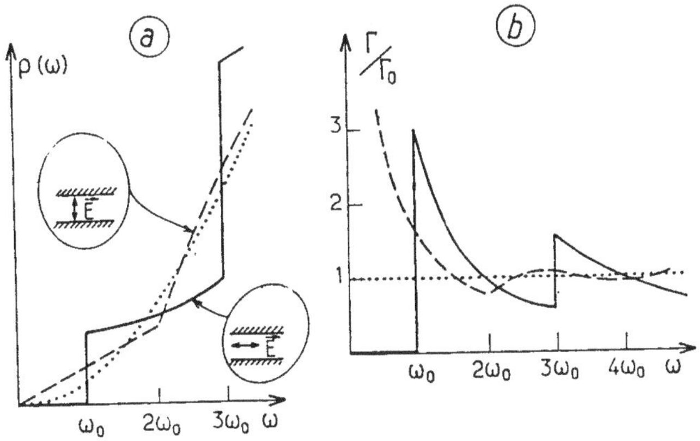

Figure 1:a) Mode density $\rho(\omega)$ versus ω in a cavity made of two plane parallel mirrors separated by a gap d ($\omega_0=\pi c/d$). Density evaluated at midgap position z=d/2. Solid line: electric field of the mode parallel to mirrors; dashed line: electric field normal to the plates. For comparison, the free space mode density is represented in dotted lines.

b) Ratio Γ/Γ_0 of the spontaneous emission rates in the cavity (Γ) and in free space (Γ_0) versus frequency ω. Solid line: transition with photon electric field parallel to plates; dashed lines: transition with photon electric field normal to plates.

does not radiate at all. On the other hand, the modes polarized normal to the surface, far from being suppressed have an enhanced density at low frequencies. As a result, excited states radiating with such a polarization decay faster than in free space. The radiation rates for these two different polarizations, normalized to the free space rate, are plotted on Fig1b versus the frequency of the transition. In order to inhibit spontaneous emission, it is necessary not only to confine the atom in a "microscopic cavity", but also to conveniently polarize the excited atomic state so as to decouple it from the modes which are not suppressed by the cavity geometry. The inhibition effect corresponding to this simple geometry has been first noticed in [7] and observed at MIT on transitions between Rydberg levels falling in the millimeter range of the spectrum [8]. The mirror separation was then of a fraction of a millimeter. More recently this effect has been observed at Yale on near infrared transitions, the atom being confined between mirrors only one micron apart [9]. Other experiments have been performed on visible transitions of dye molecules flowing between very closely spaced mirrors[10]. Although particularly simple to discuss in the plane parallel geometry, the inhibition effect is not restricted to it: mode alterations due to the boundary conditions also exist in cavities with different sizes and shapes and spontaneous emission rates alterations have been observed in various kinds of cavities in the microwave [11] as well as in the optical domain [12].

2.2 Enhancement of spontaneous emission .

Fig 1 shows that the mode density may for some frequencies or polarizations be larger in the cavity than in free space. This leads to an enhancement of the spontaneous emission rate for an atom radiating resonantly in the corresponding modes. The effect is particularly dramatic when the cavity is tuned to resonance with the atomic transition. The increase in mode density occuring at the cavity resonant frequencies has the effect of speeding up the spontaneous emission process by several orders of magnitude. We have observed this effect on a transition between Rydberg states of Na [13]. The free space spontaneous emission rate is then very small (150 s^{-1}). It is increased to $8 10^4 s^{-1}$ when the atom is placed in a Fabry-Perot resonator with $Q=10^6$. Similar spontaneous emission enhancement effects have been observed since with other geometries of cavities, in the microwave as well as optical domains [10-12].

2.3 Reversible Spontaneous emission in a very high Q cavity: the regime of vacuum Rabi nutation.

Consider the effect of spontaneous emission enhancement in a cavity whose quality factor Q is progressively increased. The mode density being correspondingly enhanced, the rate of spontaneous emission becomes larger and larger. At some point this rate becomes larger than the cavity damping rate itself. In other words, the photon is emitted faster than it can be absorbed in the cavity walls. The atom can then reabsorb this photon and be reexcited: spontaneous emission in the high Q cavity has become reversible. The transition between the irreversible and reversible regimes of spontaneous emission is simple to discuss by introducing two important parameters: the vacuum Rabi frequency Ω which is equal to the electric dipole matrix element of the transition times the rms vacuum electric field in the cavity (divided by $\hbar$), and the cavity bandwidth $\Delta\omega$ which is equal to the the resonance frequency divided by the cavity Q factor. The cavity acts as as a "damping reservoir" for the emitted photon if the "broad band" cavity condition $\Delta\omega >> \Omega$ is fulfilled. The emitted photon is then immediately lost in the cavity walls and the situation corresponds to the enhanced spontaneous emission case discussed in subsection 2.2. If on the other hand the cavity is

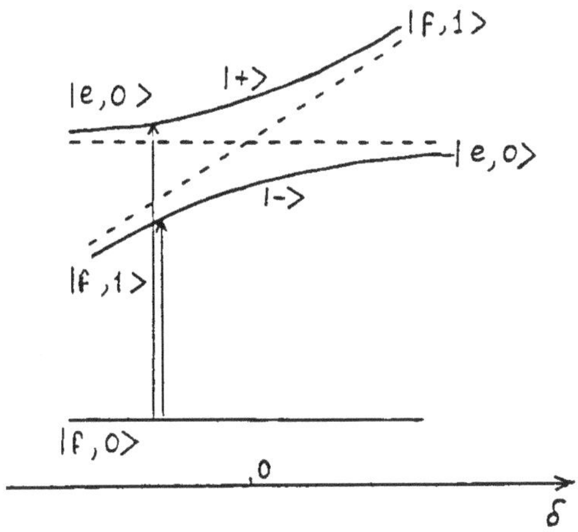

Figure 2: Lowest energy dressed states of the atom-cavity system, plotted against the atom-cavity detuning δ. The uncoupled states |e,0> and |f,1> are represented by dashed lines, crossing for δ=0. The coupled states |+> and |->, represented by solid lines, avoid crossing. The ground state |f,0> is uncoupled to any other level.

"narrow band" so that $\Delta\omega \ll \Omega$, the coherent Rabi precession is faster than the damping and the spontaneous emission process is reversible. A very simple and illuminating way of describing atom-cavity coupling effects in the case of a "narrow band" cavity is the so called "dressed atom" approach [1]. We call |e> and |f >the upper and lower atomic states pertinent to the transition of interest, |n> the photon states of the cavity modes. The spontaneous emission process corresponds to the coupling between the |e,0> and |f,1> states of the combined atom-cavity field system. We have represented on Fig2 the energies of these two uncoupled states as a function of the detuning δ between the atom transition and the cavity mode frequencies (dashed lines). We have also plotted on the figure the energy of the |f,0> state representing the atom in the lower level |f>, with the field in the vacuum (ground state of the atom field system). The two |e,0> and |f,1> states have the same energy at resonance and diverge from each other linearly as the atom field frequency difference is increased (level crossing for zero atom-cavity detuning). When atom field coupling is taken into account, the combined system eigenstates become linear combinations |+> and |–> of |e,0> and |f,1> which avoid crossing (solid lines on Fig2). The |f,0 >state on the other hand which is not coupled to any other state of the system remains unperturbed. The spontaneous emission experiment starts by preparing the system in the |e,0> state, which is no longer an eigenstate of the system, but in turn decomposes as a superposition of the two |+> and |-> eigenstates: the atom + cavity system then evolves at the Bohr frequency between the two |+> and |-> states, which, at resonance, is equal to 2Ω. The probability of finding the system excited at time t after the system preparation into state |e,0> is $P(t) = \cos^2(\Omega t)$. Spontaneous emission has become fully reversible. The above analysis completely neglects cavity damping. This damping has the effect of coupling the |+> and |-> states to the |f,0> one (field relaxation decreasing the photon number). When taken into account, this damping affects the probability P(t) which is multiplied by an exponentially decaying function. Strong damping corresponds to an overdamped regime (irreversible spontaneous emission, discussed in Section 2.2).

The reversible spontaneous emission regime has never been directly observed so far, due to the difficulty of realizing the $\Delta\omega \ll \Omega$ condition, at a temperature low enough to make sure that the cavity is initially empty (no thermal photons). At finite temperatures, more excited "dressed states" are also populated. Fig 3 shows, at resonance ($\delta=0$), the dressed atom

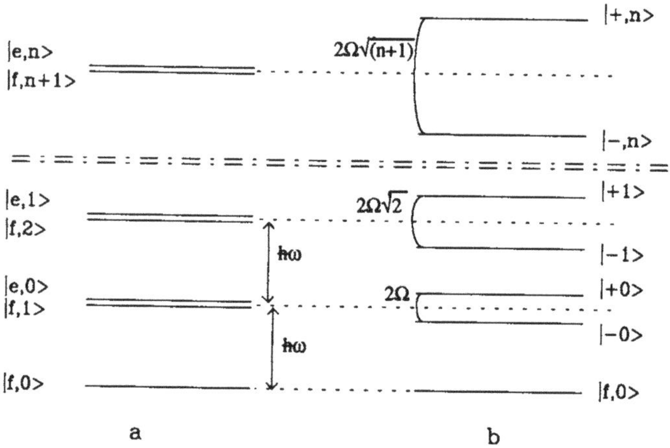

Figure 3: Dressed energy states of the atom+cavity field system. a)Energy diagram of the "uncoupled" hamiltonian near resonance ($\delta=0$). b)Energy diagram when the coupling is taken into account.

energy structure. We must now consider the $|e,n\rangle$ and $|f, n+1\rangle$ doublet states which are mixed by the atom-field coupling into eigenstates superpositions $|+,n\rangle$ and $|-, n\rangle$. The splitting between these states is $2\Omega\sqrt{(n+1)}$. The probability p(n) of having the system in the n manifold is given by the Planck distribution law. The probability P(t) of finding the atom in level e at time t after having prepared it at time t=0 in this state no longer varies as a pure sinusoidal function, but rather as a superposition of sinus functions with frequencies $2\Omega\sqrt{(n+1)}$, each corresponding to one populated manifold of Fig3. Such complex oscillations, whose exact time dependance reflects through the p(n) the actual temperature of the field, have been observed in Rydberg atom experiments [14]. As we will see below, the "dressed atom" diagrams of Fig 2 and 3 are also very useful to interpret other kinds of Cavity QED experiments.

2.3 Applications of spontaneous emission alterations

The possibility of adjusting the spontaneous emission rates of atomic transitions by turning off or on at will the vacuum fluctuations and by changing the spectral width of the field reservoir coupled to the atomic system opens the way to interesting experiments. We have to distinguish here the cases of "broad band " and "narrow band " cavities. In the first case, the cavity merely alters the rate of emission, and this alteration depends upon the polarization state, the position of the atom in the cavity and the frequency of the atomic transition. In the narrow band cavity case, the emission is qualitatively changed, since it produces a coherent quasi reversible atomic evolution. This situation is exploited in the micromasers, and we will come back to it in Section 4. We thus restrict the discussion here to the "broad band" cavity case. The most attractive application of spontaneous emission inhibition would be of course to take advantage of it in order to suppress the natural line width and thus get basically infinitely narrow spectral lines, the dream of a spectroscopist. Unfortunately, the cavity surrounding the atom not only changes the emission rates but also the emission frequencies. As a result, one ends up performing spectroscopy of an atom "dressed" by the cavity vacuum. This point is discussed in Section 3.

Interesting effects also arise from the fact that the cavity vacuum is, unlike the free space one, anisotropic and strongly frequency dependant. Magnetic or hyperfine sublevels belonging to an excited state manifold are coupled to different field polarizations which may be affected in different ways by the cavity (compare the mode densities with electric fields parallel and normal to the cavity mirrors on Fig 1a). As a result, these sublevels have different life times in the cavity, which is not the case in free space. Application of a magnetic field can produce a mixing of levels with different emission rates and thus modifies

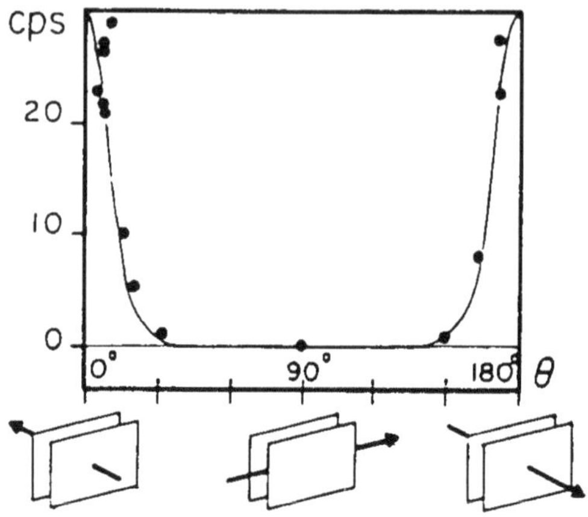

Figure 4 Excited state transmission through a narrow mirror gap, as a function of the angle θ between an applied magnetic field and the normal to the mirrors.

the life time of an atom prepared initially in one of these levels. This effect, which reflects the anisotropy of the vacuum fluctuations in the cavity has been observed [9]. Fig4 shows the transmission of a beam of excited atoms crossing the plane parallel mirror structure, as a function of the angle θ between an applied magnetic field and the normal to the mirrors. When the field has a component along the surface, it mixes the long lived states radiating with an electric field along the surfaces with short lived ones radiating with their electric field normal to the mirrors. This results in a strong decrease of the number of excited state atoms reaching a detector located behind the mirror gap.

The strong frequency dependance of the vacuum fluctuation spectrum near a cavity resonance can also result in different life times for closely spaced levels which in free space would radiate at the same rate. This is the case if the transitions corresponding to these levels fall on the wing of the cavity resonance, the splitting between the lines being of the order of the cavity resonance width. This effect has been investigated recently [15-16] by studying the radiation of an atom dressed by a strong resonant monochromatic laser field. It is well known that the coupling to such a field has the effect of splitting the atomic fluorescence into a symmetrical triplet, the spacing of the lines being the Rabi frequency in the laser field [17]. This triplet can be interpreted as a cascade spontaneous emission between dressed states of the combined atom-laser field (note that the dressing is here due to a laser field containing a very large number of photons, and not to an empty cavity field) . If this "dressed" atom is now placed in an empty cavity having a resonance slightly off the atomic line the vacuum mode density becomes different at the three frequencies of the triplet, resulting in an imbalance of the dressed atom spectrum line intensities. This also produces an imbalance in the spontaneous emission rates between the various dressed states and a modification of the steady state atomic populations in the two levels concerned by the atomic transition [16]. In some cases atomic inversion should be observed, whereas a resonant field acting on an atom in free space can only achieve population equality between these states. Note that in these experiments, the cavity is empty, the laser field being in a mode not subtended by the cavity mirrors. The role of the cavity is only to tamper with the vacuum surrounding the atom, which may have dramatic effect on the system dynamics and steady state.

In the above discussion, we have described the cavity QED effects as resulting from a perturbation of the atomic system by an altered field vacuum. An alternative interpretation is to view them as an effect of electric image perturbations: the atom in the cavity induces its images in the cavity walls and these images react on the atomic dipole to perturb its damping mechanism (radiation reaction point of view). These two pictures are obviously related since the image model is a simple method to take into account field boundary conditions. In fact the image model allows one to retrieve all the results of cavity QED discussed above. Whether to choose the vacuum field or the electric image point of view is often a matter of convenience. It is also possible to describe the cavity effects as a combination of vacuum fluctuation effects and radiation reaction effects, in a proportion which depends upon the ordering of field and atomic operators in the atom-radiation interaction hamiltonian [18-19].

3. Dispersive effects in Cavity QED.

The coupling of the atomic system to the cavity has the effect not only of altering the emission rates (resonant effects), but also of modifying the atomic energy levels. This may for example be seen as resulting from the change in the mode density of the vacuum field at frequencies off resonant with the atomic transitions (alteration of virtual transition processes). In the case of a resonant cavity, the coupling with the atom has the effect of splitting the transition frequency into a doublet, here again modifying the atomic transition frequencies with respect to their free space value. We restrict our discussion here to the simple case of a two-level atom coupled to a quasi resonant high Q cavity. This situation is quite reminiscent of the problem of two coupled oscillators, in which frequency splitting effects are well known. Here again, the concept of an atom dressed by the cavity vacuum is very convenient.

Consider the energy diagram of Fig2.with its three levels $|+\rangle$, $|-\rangle$ and $|f,0\rangle$. Assume now that we irradiate the atom by a "probing " field, not coupled to the cavity (the cavity is for example an open Fabry-Perot structure made of two mirrors and the probing field is propagating sideways). This field acts as a probe allowing us to determine by spectroscopy the transition frequencies of the dressed atom. Clearly we will observe the transitions shown by vertical arrows on Fig 2. If the cavity is not at resonance ($\delta \neq 0$), we get a line very close to the unperturbed atomic line e —> f and a line whose frequency is very close to the cavity frequency ($|f0\rangle \longrightarrow |f1\rangle$). The former one, which is nothing but the ordinary atomic absorption line is slightly pulled upward in frequency (if $\delta<0$). The latter one is a Rayleigh scattering line, the atom scattering preferentially in the cavity mode the laser light when the laser frequency corresponds to the cavity one. This Rayleigh line is pulled downward in frequency (when $\delta<0$). The direction of the frequency shifts is reversed when the sign of the atom-cavity detuning is changed. These "dressed atom" absorption lines far from the atom-cavity resonance have been observed [20]. When the two oscillator frequencies are getting close and if their damping is smaller than the coupling, a very interesting splitting effect is expected: instead of merging at the common atomic and cavity frequency, the two eigenfrequencies avoid each other and at exact resonance the spectrum consists of a doublet symmetrical with respect to the unperturbed atomic line [21]. This effect reflects in the frequency domain the existence of the anti crossing which in the time domain produced the coherent Rabi oscillation discussed in Section 2. This Rabi vacuum splitting effect can be interpreted as a dynamical Stark shift induced by the vacuum fluctuations in the cavity. It has not been observed yet because it requires very large parameters Ω and extremely high Q's difficult to realize in the optical domain. In the microwave domain, the high Ω and Q's are easier to achieve (Rydberg atoms in superconducting cavities) but the probing is difficult to perform since the cavities have a closed geometry.

Let us note that effects similar to the ones discussed here have been observed in systems made of N two level atoms in a cavity [2]. The atomic sample then behaves as a collective system, similar to a spin J=N/2 (the one atom case corresponds of course to J=1/2). The coupling of the collective atomic system to the cavity field is caracterized by a collective Rabi frequency $\Omega\sqrt{N}$ instead of Ω. This frequency being much larger than Ω as soon as N is larger than a few units, the coherent Rabi oscillations [22] and dynamical stark splitting effects [23] become much easier to observe than on one atom systems.

4.Micromasers

4.1 Principle of micromasers.

A two level atom coupled to a single mode of the electromagnetic field is the prototype model for a laser in Quantum Electronics (Jaynes-Cummings model [24]). The atom-field system actually studied in Cavity QED thus leads directly to the realization of very simple, fundamental maser systems called micromasers. These masers usually operate on Rydberg transitions, with atoms confined in small microwave cavities. This system ensures very large atom-field couplings Ω. In a maser , the field contains more than one photon and we have to consider the interaction with the field of an ensemble of atoms. This can be achieved in two different ways: (i) the atoms can be introduced in the cavity by bursts (N atom at a time). The system is then a pulsed maser, operating in a transient regime related to superradiance, whose dynamical behaviour rests on the collective properties of the N-atom sample coupled to the cavity mode [1,2,25]. Contrary to ordinary maser or laser systems, where N is enormous, the number of atoms (and photons) involved can be as small as a few units. In fact, the N=1 limit of these systems corresponds to the enhanced spontaneous emission regime discussed above. A general study of these transient micromasers can be found in other references [2,26]. The general scope of these lectures being focussed on single atom effects, we will not consider these systems here. (ii)In a second version of micromasers [27], the atoms are continuously introduced one at a time in the cavity, couple to its mode and exit from the cavity before the next atom is introduced. If the cavity damping time is longer than the reciprocal of the atomic injection rate, each atom basically interacts with the field left by the previous one. Each atom-cavity coupling process is a coherent atom-field Rabi precession. Eventually a steady state builds up as a result of the competition between the atom field coupling and the cavity damping. Of course, a micromaser can also operate in a regime intermediate between the two limiting ones considered above: when the injection rate is increased, a few atoms at a time are present in the cavity and interact collectively with its mode. Rydberg atom micromasers have been operated on one photon [27] and two photon transitions [28]. The latter case is considered more specifically in Section 4.4.

4.2 Micromasers as systems to test the theory of measurement in Quantum Mechanics.

A remarkable feature of Rydberg atom masers is that the atoms play a double role in these sytems. They are the source of the field, and are also used to detect it. After each atom leaves the cavity, it is passed between metallic plates across which a static electric field is

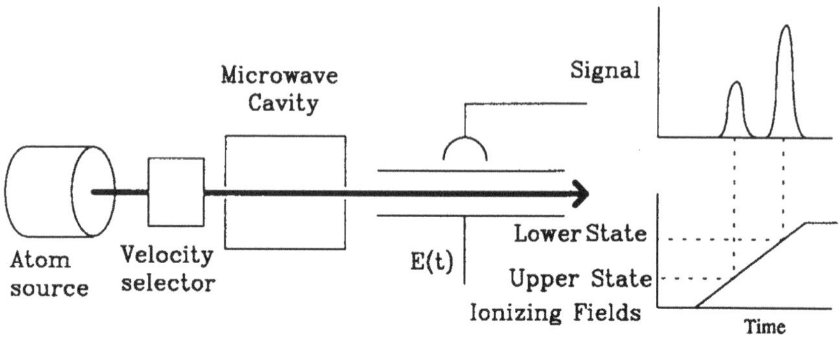

Figure 5: Set-up for an ideal micromaser experiment. The atoms are velocity selected and spend all the same time in the cavity. They are detected by the state selective field ionization method.

applied . This field ionizes the atom and the resulting electron is detected. Since the threshold ionizing field is slightly different for the two levels e and f involved in the transition, one can determine in this way the final atomic state. Two ionizing field zones can be used, the first being tuned to detect level e, the second level f [29]. Another method consists in applying a repetitive electric field ramp, the field reaching at different times in each ramp the threshold of ionization for levels e and f [28]. One then gets a time resolved electron signal from which the state of each atom exiting the cavity can be determined. Fig 5 presents the ideal set up used in this version of the experiment. When required, velocity selection of the incoming atoms can be achieved with the use of a Fizeau wheel [14,29] or by laser Doppler tuning. It is important to realize that the atomic detection is an intrinsically quantum mechanical measurement. By finding the atom in one level, one reduces the system wave function and influences the state of the field inside the cavity and thus the outcome of measurements made on subsequent atoms [30,31]. In this respect, the micromaser constitutes a simple prototype on which to test the postulates of the quantum theory of measurement.

A very simple illustration of these ideas is provided by the following simple system: consider a micromaser operating with a perfect lossless cavity and a 100% detection quantum efficiency (all atoms detected). After the first atom has crossed the initially empty cavity, the atom field system is in a linear superposition of levels $|e0\rangle$ and $|f1\rangle$, whose amplitudes depend upon the atom cavity coupling Ω and the atom cavity interaction time t_{int} (function of the atom velocity). If the atom is detected in level e, then the cavity is empty after the measure; if it is detected in level f, we know that the cavity contains exactly one photon. The second atom will then interact either with an empty cavity again, or with a cavity already excited. We can then resume the same analyzis. After the second atom has been detected, the field will be again in state 0 or 1 (if the first atom was detected in state e) or in state 1 or 2 (if the first atom had been detected in state f). Clearly, the photon number can only stay unchanged or increase by one unit at each atom crossing. It thus undergoes a random climbing of the energy ladder, the field always remaining in a pure number state (Fock state). In such an ideal experiment, the atoms crossing the cavity can be considered as simple measuring devices allowing us to perform a repetitive quantum measurement of the field oscillator. The fact that the same atoms are also source of the field sometimes complicates the experiment analyzis as discussed in next subsection.

4.3 Micromasers and non classical field statistics.

Fock states have no dispersion in energy and are thus quite different from ordinary laser field states which exhibit Poisson statistics and shot noise dispersion in photon number. In the simple model discussed in the last subsection, the field build up process ceases only if for some photon number n the Rabi angle $2\Omega\sqrt{(n+1)}t_{int}$ is an integer multiple of 2π. In this case, all subsequent atoms will undergo a complete oscillation and end up in level e without imparting any photon to the field. Photon number states such that this condition is fulfilled are called trapping states [32]. Obviously the above discussion oversimplifies the reality. Cavity losses cannot be neglected, as well as finite detection efficiency and velocity dispersion. Nevertheless, when all these effects are taken into account, the steady state solution of the system evolution equation may exhibit a subpoissonian character reminiscent of the Fock states [33].

It is important to understand how the statistical characteristics of the field (Poissonian or subpoissonian, Fock state etc) can be observed in a micromaser experiment. One interesting effect is the influence of the field statistics on the shape of the Rabi oscillation of a test atom crossing the cavity. The Rabi oscillation between levels e and f is a pure sinusoid with frequency $2\Omega\sqrt{(n+1)}$ for an n Fock state. On the other hand, if the field is in a superposition of Fock states, the Rabi oscillation is a sum of sinusoids of different frequencies (effect already discussed in the case of a thermal field in section 2 above). This sum exhibits collapses and revivals as the test atom field interaction time is increased [34]. This experiment requires to select and tune the velocity of the test atom across the cavity. The difficulty for a clear observation of this effect is that it is hard to change the velocity of the test atom without modifying the one of those that generate the field in the first place . When the atom velocity and hence the atom field interaction time is changed, the satistics of the field produced by the previous atoms is also modified, since it depends also on the atom field interaction time. This precludes a simple interpretation of the data . Nevertheless, the observed Rabi oscillation shows a good agreement with the result of a complete calculation taking into account all these complications and is a clear demonstration of the micromaser theory [14]. Another way of detecting a sub poisson field is to study in details the statistics of the atom population distribution at the exit of the cavity. This statistics reflects in an indirect way the one of the field. In this way, it has been recently possible to demonstrate the subpoissonnian character of the micromaser field [35].

These experiments lead to interesting potential applications . Fock states can be used to perform subtle tests of complementarity in quantum physics [36]. The basic idea is that a π pulse produced by a Fock state leaves a trace in the system (by changing n into n+1 or n-1), which is not the case of a Poisson field having a dispersion in n values. This trace can be used as a marker to know whether the atom has undergone or not such a pulse, thus precluding the observation of interference effects based on the indistinguishability of two channels. The same interference should on the other hand be conserved if a coherent poisson field is used instead of the Fock one. Although these experiments might be difficult to perform, they are not unrealistic with the present state of the art in micromaser technology. They make it very desirable to generate Fock states of the field. We discussed here only one possible method. Other ones are presently investigated. One very promising technique involves fast adiabatic passage of the atoms in the cavity: if the cavity frequency is swept across the atomic resonance while one (ore several atoms) are crossing it, one can show that these atoms must release an exact number of photons in it [37]. An experiment along these lines is in progress.

4.4 Two-Photon Micromasers.

The very strong coupling of Rydberg atoms to millimeter vawe transitions makes it possible to realize a very special kind of micromaser: the two-photon one. Such a micromaser operates on a two-photon transition between two levels having the same parity in a level configuration such as the one presented in Figure 6 . (Ordinary masers operate instead on one-photon transitions between levels with opposite parities). Such a quantum oscillator presents interesting features, due to its microscopic nature, and to the fundamental non linearity of the emission.

Two-photon lasers of masers have been suggested theoretically a long time ago [38]. Many interesting properties of this new quantum device have been pointed out since

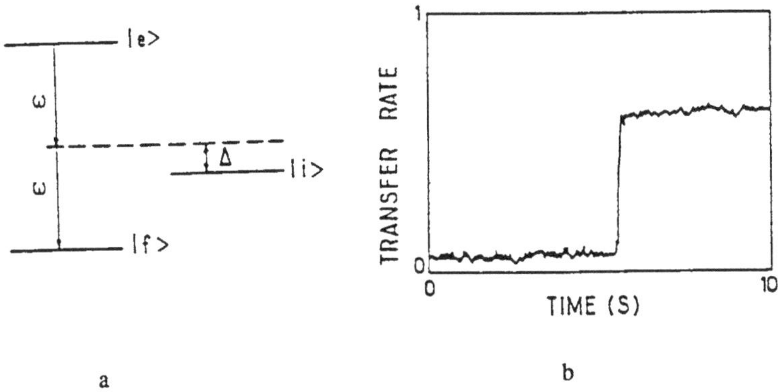

Figure 6: a) Energy configuration of the two-photon Rydberg atom maser. The upper (e) and lower level (f) of the maser are the 40S and 39S Rydberg states of Rb respectively. The intermediate level 39P is very close to the mid-energy point (detuning Δ)
b) Start up of the two-photon maser: transfer rate from level 40S to 39S versus time, for a single realization of the experiment. Time t=0 corresponds to the introduction of the inverted atoms in the cavity.

then [39]. These theoretical investigations prompted an important experimental effort aiming to the realization of such a laser. On "ordinary" atomic transitions between low lying levels, the weakness of the two-photon gain makes it difficult to reach the laser threshold. Even if the corresponding inversions densities are reached, laser emission is easily overcome by competing processes such as the one-photon superradiance to the intermediate level (|i> on Figure 6). This and other difficulties explain that only transient two-photon laser operation has been reported on such transitions [40].

The situation is radically different in the case of Rydberg atoms in superconducting cavities. The high Q cavity actually plays a double role in the experiment. If tuned to resonance on the two-photon transition, it enhances the atom-field coupling and sustains the maser field . The oscillation threshold (proportionnal to the reciprocal of the cavity quality factor) is therefore considerably reduced. On the other hand, with a low order mode cavity, there is no mode of the field resonant with the competing processes. For instance, the one-photon superradiance to the intermediate level |i> is suppressed, as discussed in section 2.1. In other words, one takes advantage at the same time of the two effects discussed in section 2: spontaneous emission enhancement in a tuned cavity, and spontaneous emission inhibition in a cavity out of resonance.

On a carefully selected two photon transition, with an intermediate level very close to the middle of the two photon transition, the two photon maser threshold is reduced down to a very low atomic flux value. There is, on the average, only a few atoms at a time in the cavity and a few tens of photons. In the experimental realization of the system [28], the maser action is detected indirectly by the above-described field ionization technique. The observed threshold is in good agreement with the theoretical predictions, and corresponds effectively to a quasi micromaser regime (no more than a few atom at a time in the cavity).

This new quantum device combines the interests of the micromasers, where the quantum nature of the cavity field plays a very important role, and the ones of a laser operating on an intrinsically non-linear emission process. Interesting effects, not found in "ordinary" lasers or micromasers, can be predicted theoretically [41].

In particular, the start-up dynamics of a two-photon micromaser is radically different from the one of an ordinary maser. In a semi classical analysis, in terms of gain and losses, one can show that the gain for small signals, proportionnal to the square of the field intensity I, is always smaller than the cavity losses (proportionnal to I). In other words, in this approach, the zero intensity point is a stable operating point for the two-photon maser, even well above threshold: a "semi classical" two photon laser starts only if a large enough field is initially injected in the cavity. In fact, the actual two photon micromaser starts spontaneously, without any injected field. The semi-classical approach failure is due to the microscopic nature of the system: the quantum fluctuations of the fied in the cavity are not negligible compared to the field necessary to start maser oscillation. A detailed quantum analyzis of the start-up process [41] shows that the dynamics of the cavity field mean intensity is the same as the one of a classical particle driven above a potential energy barrier by a random, fluctuating force. The mechanism of maser start-up is therefore reminiscent of the dynamics of an exoenergetic chemical reaction with an activation energy (Kramers process). Such processes can be very slow compared to the characteristic times of evolution of the system. One might thus expect long starting times for the two photon micromaser, even well above threshold. This is radically different from what is predicted for one-photon masers or micromasers, where start-up is basically a linear amplification of an initial field fluctuation. Such long starting times are actually observed in the experiment as shown on Figure 6b. The transfer rate between the two levels of the two-photon transition switches suddenly from zero (maser off) to a steady state value (maser on) a few seconds after the atomic beam has been switched on. This starting time is much greater than all the characteristic times of the experiment, which are all in the millisecond domain.

The sudden start up of the maser observed on Fig 6b can be attributed to the quantum measurement made each time an atom is detected, as discussed in section 4.2. As the starting time is much greater than the average time between the detection of successive atoms, one actually performs a repeated measurement process on the cavity field. This repeated measurement affects the field dynamics. A numerical simulation of the maser evolution, taking into account these "wave packets reductions" predicts a sudden switching of the maser field intensity, and therefore of the transfer rate, as actually observed. An exponential approach to the equilibrium is retrieved as an ensemble average over a large number of realizations of this experiment.

5. Conclusion: Perspectives of Cavity QED.

The simple system considered in this article -a quasi two level atom coupled to a single mode of the radiation field- will certainly be used for further fundamental studies and tests of basic Quantum Electrodynamics. Situations which have long been considered as belonging to purely "gedanken experiments" are now feasible. We already described in Section 4 repeated measurements on a single quantum oscillator. This experiment is so far limited by the weak efficiency of the atomic detector which misses a large fraction of the outgoing atoms. Techniques to increase this efficiency or to overcome its limitation are now being developed. Ideal repeated quantum measurement of an oscillator seem to be achievable in the near future, opening the way to various tests of basic laws. Production of non classical fields and particularly of squeezed fields using the concepts of Cavity QED is also a promising domain [42].

Most experiments so far have been restricted to microwaves and Rydberg atoms. Extension to the optical domain may become possible by developing new kinds of microscopic cavities (microscopic Fabry-Perot, Morphology Dependant Resonances or Mie resonances [43] of microscopic dielectric spheres) coupled to strong atomic transitions. This would open the way to an actual detection of the emitted field and to a direct determination of its statistical properties.

REFERENCES:

[1] S. HAROCHE In "New Trends in Atomic Physics" (Les Houches Summer School Lecture Notes Session 38). R. STORA and G. GRYNBERG editors, North Holland, Amsterdam (1984)

[2] S. HAROCHE and J.M. RAIMOND in "Advances in Atomic and Molecular Physics 20", B. BEDERSON and D.R. BATES editors, Academic Press, New York (1985)

[3] F. DIEDRICH, J. KRAUSE,G. REMPE, M.O. SCULLY and H. WALTHER I.E.E.E. Journal of Quantum Electronics 24,2147 (1988)

[4] M. BRUNE, P. GOY, L. DAVIDOVICH, S. HAROCHE I.E.E.E. Journal of Quantum Electronics 24,1323 (1988)

[5] H. WALTHER Physica Scripta T23, 165 (1988)

[6] S. HAROCHE and D. KLEPPNER Physics Today 42, 24 (1989)

[7] D. KLEPPNER Phys Rev Lett 47, 233 (1981)

[8] R.G. HULET, E.S. HILFER and D. KLEPPNER Phys. Rev. Lett. 55,2137 (1985)

[9] W. JHE, A. ANDERSON, E. HINDS, D. MESCHEDE, L. MOI and S. HAROCHE Phys. Rev. Lett. 58, 666 (1987)

[10] F. DEMARTINI, G. INNOCENTI, G. JACOBOVITZ and P. MATALONI Phys. Rev. Lett 59, 2955 (1987)

[11] G. GABRIELSE and H. DEHMELT Phys. Rev. Lett. 55, 67 (1985)

[12] D.J. HEINZEN, J. CHILDS, J.F. THOMAS and M. FELD Phys. Rev. Lett 58, 1320 (1987)

[13] P. GOY, J.M. RAIMOND, M. GROSS and S. HAROCHE Phys. Rev. Lett. 50, 1903 (1983)

[14] G. REMPE, H. WALTHER and N. KLEIN Phys. Rev. Lett. 58, 353 (1987)

[15] M. LEWENSTEIN and T. MOSSBERG Phys. Rev. A37, 2048 (1988)

[16] Y. ZHU, A. LEZAMA, T. MOSSBERG and M. LEWENSTEIN Phys. Rev. Lett. 61, 1946 (1988)

[17] B. MOLLOW Phys. Rev 188, 1969 (1969)

[18] J. DALIBARD, J. DUPONT-ROC and C. COHEN-TANNOUDJI J. Phys. (Paris) 43,1617 (1982)

[19] D. MESCHEDE, W. JHE and E. HINDS Phys. Rev. A to be published (1990)

[20] J.J. CHILDS, J. HUTTON, J. HEINZEN and M. FELD in "Technical digests of the Conference on Quantum Electronics and Laser Science", Baltimore (1989)

[21] J. SANCHEZ-MONDRAGON, N. NAROZHNY and J. EBERLY Phys. Rev. Lett 51, 550 (1983)

[22] Y. KALUZNY, P. GOY, M. GROSS, J.M. RAIMOND and S. HAROCHE Phys. Rev. Lett 51, 1175 (1983)

[23] M. RAIZEN, R. THOMPSON, R. BRECHA, H. KIMBLE and H. CARMICHAEL Phys. Rev. Lett 63, 240 (1989)

[24] E. JAYNES and F. CUMMINGS Proc. IEEE 51, 89 (1963)

[25] J.M. RAIMOND, P. GOY, M. GROSS, C. FABRE and S. HAROCHE Phys. Rev. Lett 49, 1924 (1982)

[26] L.MOI, P. GOY, M. GROSS, J.M. RAIMOND, C. FABRE and S. HAROCHE Phys. Rev. A27, 2043 and 2065 (1983)

[27] D. MESCHEDE, H. WALTHER and G. MULLER Phys. Rev. Lett 54, 551 (1985)

[28] M. BRUNE, J.M. RAIMOND, P. GOY, L. DAVIDOVICH and S.HAROCHE Phys. Rev. Lett. 59,1899 (1987)

[29] J. KRAUSE, M. SCULLY, T. WALTHER and H. WALTHER Phys. Rev A39, 1915 (1989)

[30] J. KRAUSE, M. SCULLY and H. WALTHER Phys. Rev. A36, 4547 (1987)

[31] P. MEYSTRE Optics Letters 12, 669 (1987); P. MEYSTRE and E.M. WRIGHT Phys. Rev. A37, 2524 (1988)

[32] P. MEYSTRE, G. REMPE and H. WALTHER Optics Letters 13, 1078 (1988)

[33] P. FILIPOWICZ, J. JAVANAINEN and P. MEYSTRE Phys. Rev. A34,3077 (1986)

[34] J. EBERLY, N. NAROZHNY and J. SANCHEZ-MONDRAGON Phys. Rev. Lett 44, 1323 (1980)

[35] G. REMPE and H. WALTHER in "Laser Spectroscopy IX" M.S. FELD editor, Springer Verlag (1989)

[36] M. SCULLY and H. WALTHER Phys. Rev. A39, 5229 (1989)

[37] J.M. RAIMOND, M. BRUNE, J. LEPAPE and S. HAROCHE in "Laser Spectroscopy IX" M.S. FELD editor, Springer Verlag (1989)

[38] A.M. PROKHOROV Science 149,828 (1965); P. SOROKIN and N. BRASLAU IBM J. Res. and Dev. 8,177 (1964)

[39] Z.C. ZWANG and H. HAKEN Z. Phys. B. 55,361 (1984) 56,77,83 (1984); H.P. YUEN Phys. Lett. 51A, 1 (1975); Phys. Rev A13,226 (1976)

[40] B. NIKOLAUS, D. ZHANG and P. TOSCHEK Phys. Rev. Lett. 47,171 (1981)

[41] L. DAVIDOVICH, J.M. RAIMOND, M. BRUNE and S. HAROCHE Phys. Rev. A36,3771 (1987)

[42] H. CARMICHAEL Phys. Rev. Lett. 55, 2790 (1985); A. HEIDMANN, J.M. RAIMOND and S. REYNAUD Phys. Rev. Lett. 54,326, (1985); H. KIMBLE, M. RAIZEN, L. OROZCO, MIN XIAO and T. BOYD in "Fundamentals of Quantum Optics II" F. EHLOTZKY editor, Springer Verlag (1987); M. LEWENSTEIN and T. MOSSBERG Phys. Rev. A38,1075 (1988); J.M. COURTY and S. REYNAUD Optics. Comm 72, 93 (1989)

[43] A. ASHKIN and J. DZIEDZIC Phys. Rev. Lett. 38,1351 (1977)

www.ingramcontent.com/pod-product-compliance
Ingram Content Group UK Ltd.
Pitfield, Milton Keynes, MK11 3LW, UK
UKHW020402150926
8624IPUK00004B/292

9 789971 509521